Finite Analytic Method in Flows and Heat Transfer

CRC Press
Taylor & Francis Group
Boca Raton London New York

CRC Press is an imprint of the
Taylor & Francis Group, an **informa** business

Finite Analytic Method in Flows and Heat Transfer

Ching Jen Chen
Florida A&M University - Florida State University

Richard Bernatz
Luther College

Kent D. Carlson
University of Iowa

Wanlai Lin
Emerson Electric Corporation

Published in 2000 by

Taylor & Francis
29 West 35th Street
New York, NY 10001

Published in Great Britain by

Routledge
11 New Fetter Lane
London EC4P 4EE

Library of Congress Cataloging-in-Publication Data

Finite analytic method in flows and heat transfer / Ching Jen Chen ... [et
al.].
 p. cm.
 ISBN 1-56032-898-3 (alk. paper)
 1. Fluid dynamics 2. Heat–Transmission–Mathematical models. I. Chen,
C.J. (Ching Jen), 1936-

QA911.F4385 2000
532'.05–dc21

                                                    00-037721
                                                    CIP

Contents

List of Figures

List of Tables

Preface

This book is the outgrowth of our research interest in computational fluid dynamics and heat transfer. The monograph contains the fundamental development of the *finite analytic method*, a relatively new numerical technique for solving various differential equations. The unique feature of the finite analytic method is the use of an analytic solution, on small subdomains, in the development of the numerical solution for the given differential equation. The entire problem domain is partitioned into a number of small subdomains, or elements. A "local" analytic solution is obtained on each small element in which the governing equation, if non-linear, is linearized. An algebraic form of the analytic solution is constructed, wherein the solution at a nodal value in the interior of the element is expressed as the sum of neighboring nodal values weighted by finite analytic coefficients. A system of these finite analytic algebraic equations is then solved to provide a numerical solution for the dependent variable at prescribed discrete locations within the domain.

There are two distinct features of the finite analytic solution. First, using an analytic solution as the basis for the algebraic equations results in "automatic" up-winding, and skew up-winding, in convective transport equations such as the Navier-Stokes equations. Second, when compared to traditional numerical methods such as finite differences, the finite analytic method is stable and accurate over a much broader range of flow and computational parameters such as Reynolds number and grid spacing.

The finite analytic method was conceived in 1977 when Dr. C. J. Chen and Dr. Peter Li (then a student of Dr. Chen) had difficulty obtaining the solution for a system of finite difference algebraic equations for the Navier-Stokes equations. The finite analytic method began as a result of their efforts to overcome these failings. The method was first used to solve simple two-dimensional Laplace equations, heat diffusion problems, and non-linear ordinary differential equations. Since then, the method has been extended by numerous graduate students and researchers to a variety of fluid flow and heat transfer problems, including two- and three-dimensional, laminar and turbulent applications.

This book consists of five parts and is intended to give a systematic coverage of knowledge needed for numerical computation of fluid flows and heat transfer. It is intended for graduate students studying computational fluid dynamics and heat transfer, or practioners interested in applying the numerical methods presented in the book to problems of their interest. The reader is expected to have some knowledge of differential equations and analytic methods, such as the method of separation of variables, for solving simple equations.

Part I begins with a brief history of computational machinery and a presentation of the governing equations for fluid flows and heat transfer. Some mathematical background in partial differential equations and the concept of well-posed problems is offered next. Part I concludes with a general discussion of numerical methods and a chapter on the finite difference method. Part II of the book is devoted to the principles of the finite analytic method and its development for various types of equations, including one-, two- and three-dimensional convective diffusion equations. The topics of stability and convergence are addressed in this part as well. Part II concludes with a chapter on the finite analytic method for hyperbolic equations, and another on the explicit finite analytic formulation.

Part III of the book concentrates on methods of coordinate generation for applications on complex domains. Recent developments in "automatic" grid generation for complex domains is included in this part as well. Solving for pressure in incompressible fluid applications is especially challenging since there is no explicit equation for pressure. A presentation of various schemes for accomplishing this task given in Part IV. The methods are divided into two groups; those for staggered grids, and those for non-staggered grids. Part IV concludes with a discussion on boundary conditions, especially those for pressure. Part V provides examples of the wide variety problems for which the finite analytic method has been used successfully, including two- and three-dimensional, laminar and turbulent fluid flows and heat transfer. Specific applications are as varied as the modeling of sea breeze and artificial heart blood flows, as well as ship hydrodynamics and heat transfer in three-dimensional arrays of electronic components.

WEB SITE: Readers interested in obtaining FORTRAN finite analytic code written by the authors and other researchers are invited to visit the finite analytic web site at **www.finiteanalytic.com**. E-mail comments and questions may also be forwarded to the authors at the web site.

ACKNOWLEDGEMENTS

Dr. Chen would like to thank the many friends, colleagues, and graduate students having interest in computational fluid flow and heat transfer, and interacting with him on the development and application of the finite analytic method. In particular, Dr. Chen would like to acknowledge his colleagues, Professors V. C Patel, T. F. Smith, Allen Chwang, L. D. Chen, and K. Atkinson and visiting scholars Li Wei, Bong-Min Zhen, Shiziong Zhang, Xiang Jin Zeng, and Dr. Tsai Whey-Fone for their encouragement and constructive criticism of the method during its development. Dr. Chen would also like to thank his graduate students for their hard work which offered much to the development of the finite analytic method. They include Peter Li, Hamid Naseri-Neshat, Hamn-Ching Chen, Kanwerdip Singh, Zahed Mohammad, Sheikholeslami, Tzong-Shyan Wung, Wu Sun Cheng, Kemokolam Obasih, Bahram Khalighi, Vahid Talaie, Ramiro Humberto

Bravo, Yuenong Xu, Zhuan Xu, Seok Ki Choi, Yousef Sa'ad Haik, Sheng-Yuh Jaw, Luke J. Chen, You Gou Kim, Hui-Chun Tien, and Weizhong Dai.

Dr. Chen thanks the NASA Lewis Center for the initial support of the development of the finite analytic method, and later, the United States Department of Energy for Support of the application of the finite analytic method, the United States Naval Sea System Command and the Office of Naval Research for further application of the finite analytic method in ship hydrodynamics. Thanks are extended to SCRI (Supercomputer Computations Research Institute) and ACNS (Academic Computing & Network Services) at Florida State University for the use of their computer facilities in the application of finite analytic method.

Dr. Chen is indebted to Dr. Akira Wada of Japan Central Research Institute of Electric Power Industry (CRIEPI) for recognizing the importance of computational methods and turbulence modeling, and inviting Dr. Chen to join CRIEPI in the summer of 1983 to initiate the draft of the notes on which this book is based. Dr. Chen also thanks Dr. Whey-Fone Tsai of Taiwan National Center for High Performance Computer (NCHC) for the invitation to give a finite analytic numerical method workshop in the summer of 1995.

Richard Bernatz is indebted to Dodi, who offered support and wisdom without hesitation through this seemingly endless book project.

The authors express their appreciation for the assistance and guidance given by Mr. Kenneth McCombs, Ms. Catherine M. Caputo and Mr. Tom Wang of Taylor & Francis in bringing our book about the finite analytic method to reality.

The authors are happy to receive any discussion or comments on the finite analytic method. They make no claims of the originality of all contents. Indeed, much was contributed by devoted students of Dr. Chen and the authors' colleagues. The authors attempt to present the finite analytic method and its application in a systematic and logic fashion so that readers may apply the method in their work.

Ching Jen Chen
Richard Bernatz
Kent D. Carlson
Wanlai Lin

Part I

Introduction to Computational Fluid Dynamics

Part One of the text gives a brief introduction to computational methods in fluid flow and heat transfer. Chapter 1 begins with a brief discussion about the methods of prediction in fluid flows and heat transfer. Next, the governing equations for flows and heat transfer are developed in Chapter 2, first for laminar flow and then for turbulent flow. The chapter concludes with a discussion of first- and second-order turbulence closure models. Chapter 3 discusses the classification of partial differential equations. Chapter 4 follows with a general mathematical introduction to well-posed problems. It ends with a discussion about well-posed problems in fluid flows and heat transfer. Chapter 5 is devoted to a discussion about various numerical methods and difficulties that arise with numerical techniques. Finally, details of the finite difference numerical method are presented in Chapter 6.

Chapter 1

Introduction

Fluid flow and heat transfer are integral components in a wide variety of science and engineering applications. Heat exchangers and fluid motion are important considerations in almost every furnace design. The same is true for air conditioning systems. Internal combustion engines, power stations and many chemical production systems require knowledge of fluid flows and heat transfer. Environmental applications include weather prediction and pollution transfer. The design of electronic equipment involves aspects of component cooling, another application of heat transfer and fluid flow. Numerical modeling of air flow around body shapes is an important part of automobiles and aircraft design. A comprehensive list would include health science applications such as the design of artificial hearts and lungs.

The broad range of important problems involving fluid and heat transfer explains the aggressive pursuit of methods to predict such quantities as fluid velocity, dynamic pressure, temperature, heat transfer rates and others. The purpose of this book is to introduce the finite analytic (FA) numerical method for fluid flow and heat transfer. The comprehensive presentation includes the detailed derivation of the solution method for transport equations, such as Navier-Stokes and energy equations, for one-, two-, and three dimensions, the description of how it may be adapted to irregular or complex domains and a results of its successful application to a variety of problems.

1.1   Methods of Prediction

Prediction methods fall into two broad categories: experimental and theoretical. Experimental methods usually include the construction of a scaled model, followed by direct measurement of the desired quantities. One drawback to an experimental approach is the scaling process may exclude one or more important aspects of the problem such as boiling or combustion. Though direct measurements are often desirable, the process is not free from instrumentation error as well.

	Advantages	Disadvantages
1	Fast speed	Need mathematical model
2	Economical (time, cost)	need accurate numerical methods
3	Wide range (parameter, size)	Computer limitation
4	Flexible (parameter, geometry)	High cost
5	Clean	Unreliable and difficult to see physics

Table 1.1: Advantages and disadvantages of numerical methods.

Theoretical methods begin with the development of a mathematical model involving differential equations for the dependent variables. These equations may include parameters that characterize the nature of the problem. In some cases, such as in turbulence models, separate experiments are conducted to determine the values for such parameters. The most challenging part of the theoretical method is solving the mathematical equations subject to initial and boundary conditions. Realistic equations for typical flow and transfer problems result in complicated non-linear terms making closed-form solutions extremely difficult, if not impossible.

Numerical methods typically use computer algorithms to approximate the solution to model equations at a finite number of prescribed locations in the problem domain. The ever improving speed and power of computing machinery make numerical methods more feasible for even the most complicated application. Table 1.1 summarizes some the advantages and disadvantages of numerical methods.

1.2 Numerical Methods

The objective of numerical, or computational, methods is to determine values of the dependent variables, such as velocity components, pressure, temperature, etc., at a finite number of prescribed locations within the problem domain. The initial step in any numerical procedure is to partition, or discretize, the problem domain into small, polygonal subdomains, or elements. The computational locations, called "nodes" or "grid points," are usually located at the vertices of the subdomains. The governing equations for the flow or heat transfer problem are then transformed, or discretized, into algebraic counterparts.

There are numerous numerical methods for deriving the algebraic forms of flow and heat transfer equations. They include, but are not limited to, finite difference, finite element, finite volume, and spectral methods. The finite analytic method is a relatively new procedure for solving flow and heat transfer governing equations. The unique feature of the finite analytic method is the incorporation of an analytic solution on the small sub-domains in the development of an algebraic counterpart for the given differential equation. There are two distinct advantages of the finite analytic solution. First, using an analytic solution as the basis for the algebraic equation results in "automatic up-winding," and skew up-winding, in convective transport equations such as Navier-Stokes. Second, the finite analytic method is stable and accurate over a much broader range of flow and computational parameters, such as Reynolds number and node spacing, than the finite difference method.

1.3   Purpose and Outline

A comprehensive presentation of the finite analytic method is the primary purpose of this book. This text combines and organizes important components of finite analytic flow modeling so scientists and engineers interested in computational fluid dynamics may understand and apply the method to problems of their interest. After the introductory topics in part I are presented, a careful introduction to the fundamentals of the finite analytic method for a variety of equations is presented in part II. Part III provides a background of grid generation techniques making application of the finite analytic method to irregular domains possible. An additional unique topic is the diagonal Cartesian method for flow and heat transfer on complex domains. This innovative technique incorporates diagonal line segments with the usual vertical and horizontal grid lines of Cartesian coordinates to approximate irregular boundaries. The boundary node specification may even be automated so as to reduce the amount of labor required by other common grid generation techniques.

Computational considerations, including methods for solving the coupled continuity and momentum equations as well as successful implementation of boundary conditions, are discussed in part IV. Part V concludes the book with selected applications based on the FA method including turbulent flow, turbulent heat transfer and flows on complex domains.

Chapter 2

Governing Equations

The mathematical modeling of fluid motion was first was first applied to inviscid flows when the Euler equation was derived in 1755. It evolved into a model for viscous flows through the work of Navier in 1823, and Stokes in 1845. Fourier developed a mathematical model for the conduction of heat in 1822, thus completing the foundation for a mathematical model for viscous-conducting fluids. From 1775, it took nearly a century of effort by many scientists, mathematicians, physicians and engineers to fully develop the viscous-conducting fluid model.

2.1 Stokes-Fourier Postulates

The postulates made by Navier, Stokes and Fourier for the derivation of the viscous-conducting fluid model have a strong bearing on the modeling of a wide range of fluid flows, including turbulent flows. The *Stokes-Fourier Postulates* are summarized and rephrased as follows:

1. A fluid is a continuum in local equilibrium. Molecular motions are averaged and thus detailed information on the dynamics of molecular collisions is lost. A model is required to recover the lost information (*modeling requirement*).

2. The diffusion of momentum and thermal energy by viscous fluid motion is proportional to the rate of deformation and gradient of temperature, respectively (*diffusion gradient model*).

3. A fluid is assumed to be isotropic (*isotropic molecular collision model*).

4. A fluid is assumed to be homogeneous. That is, the viscous stresses τ_{ij} and heat fluxes q_{ij} are not explicit functions of space or time.

5. When a fluid is at rest, the viscous stress is the hydrostatic pressure (*consistency and realizability requirement*).

6. When the flow is of pure dilatation , the average viscous stress $\bar{\tau}$ is equal to the pressure P (*Stokes hypothesis*). Specifically,

$$\bar{\tau} = \frac{1}{3}(\tau_{XX} + \tau_{YY} + \tau_{ZZ}) = -P \qquad (2.1)$$

7. The model moduli (density, viscosity, specific heat, thermal conductivity, etc.) require experimental calibration and determination (*uniqueness of moduli*).

The assumption of a fluid continuum in the first postulate eliminates the need for the description of the intermolecular forces and collisions. In order to recover the information lost by this postulate, a viscous-conducting fluid model must be introduced and the model moduli, such as viscosity and thermal conductivity, must be calibrated by experimentation. The same is true when averaging is imposed on the Navier-Stokes equations for modeling turbulent motions since the details of turbulent motions are eliminated. In order to recover the information lost during the averaging process, a turbulence model must be introduced and turbulence model moduli must be determined from experiments. It should be remarked that the Navier-Stokes equations are not restricted to laminar fluid motion and are capable of describing turbulent fluid motion. The turbulent flow motion predicted by solving the Navier-Stokes equations directly is known as a *direct simulation*.

2.2 The Navier-Stokes and Energy Equations

2.2.1 Compressible Flows

Using the Stokes-Fourier Postulates, the following governing equations are derived for compressible flows in three dimensions. They are based on the principles of conservation of mass, momentum and energy.

Continuity equation:

$$\frac{\partial \rho}{\partial t} + \frac{\partial \rho U_i}{\partial X_i} = 0 \qquad (2.2)$$

Momentum equations:

$$\rho \frac{DU_i}{Dt} = -\frac{\partial P}{\partial X_i} + \frac{\partial[\mu(\frac{\partial U_i}{\partial X_j} + \frac{\partial U_j}{\partial X_i}) - \frac{2}{3}\mu\delta_{ij}\frac{\partial U_l}{\partial X_l}]}{\partial X_j} + \rho g_i \qquad (2.3)$$

Energy equation:

$$\rho C_P \frac{DT}{Dt} = \frac{\partial(K\frac{\partial T}{\partial X_j})}{\partial X_j} + \frac{Dp}{Dt} + \mu\left[\left(\frac{\partial U_i}{\partial X_j} + \frac{\partial U_j}{\partial X_i}\right) - \frac{2}{3}\delta_{ij}\frac{\partial U_l}{\partial X_l}\right]\frac{\partial U_i}{\partial X_j} \qquad (2.4)$$

Equation of state:

$$\rho = f(P, T) \qquad (2.5)$$

The dependent variables are U_i (i^{th} component of the velocity vector), T (temperature), P (pressure), and ρ (density). The independent variables are X_i (the ith spatial variable) and t (time). Fluid properties are represented by μ (molecular dynamic viscosity), C_P (specific heat at constant pressure) and K (thermal conductivity). A gravitational force acting in the ith direction is represented by g_i. The Kronecker delta is represented by δ_{ij}, and the operator $\frac{D}{Dt}$ is the total, or substantial, derivative defined as

$$\frac{D}{Dt} = \frac{\partial}{\partial t} + U_i \frac{\partial}{\partial X_i} \tag{2.6}$$

Formal details of the derivation of the Navier-Stokes and energy equations are given by Schlichting [158], White [175], and Currie [59].

2.2.2 Incompressible Flows

For incompressible flows ($\frac{D\rho}{Dt} = 0$), Equations (2.2) - (2.4) simplify to

$$\frac{\partial U_i}{\partial X_i} = 0, \tag{2.7}$$

$$\frac{DU_i}{Dt} = -\frac{1}{\rho}\frac{\partial P}{\partial X_i} + \nu \frac{\partial^2 U_i}{\partial X_j \partial X_j} \quad \text{and} \tag{2.8}$$

$$\frac{DT}{Dt} = \alpha \frac{\partial^2 T}{\partial X_j \partial X_j} + \frac{\nu}{C_V}\left(\frac{\partial U_i}{\partial X_j} + \frac{\partial U_j}{\partial X_i}\right)\frac{\partial U_i}{\partial X_j}. \tag{2.9}$$

Note that the coupling of the momentum and temperature equations is lost for incompressible flows because the equation of state is no longer available.

In addition to the incompressible fluid assumption, the expressions $\nu = \frac{\mu}{\rho}$ (kinematic viscosity), C_V and $\alpha = \frac{K}{\rho C_P}$ (thermal diffusivity) in Equations (2.7) - (2.9) are constant. Note that $C_P = C_V$ for an incompressible fluid.

2.2.3 Vorticity and Stream Function

The momentum equations given in Equation 2.8 are for the three components of the velocity vector $\vec{V}$. The vector form of Equation (2.8) is

$$\frac{\partial \vec{V}}{\partial t} + (\vec{V} \cdot \nabla)\vec{V} = -\nabla\left(\frac{P}{\rho}\right) + \nu \nabla^2 \vec{V}. \tag{2.10}$$

Replacing the nonlinear term in Equation (2.10) by an equivalent form using vector identities, the vector equation becomes

$$\frac{\partial \vec{V}}{\partial t} + \nabla(\frac{1}{2}\vec{V} \cdot \vec{V}) - \vec{V} \times (\nabla \times \vec{V}) = -\nabla\left(\frac{P}{\rho}\right) + \nu \nabla^2 \vec{V}. \tag{2.11}$$

The *vorticity* $\vec{\omega}$ is defined as the curl of the velocity vector,

$$\vec{\omega} = \nabla \times \vec{V}. \tag{2.12}$$

Taking the curl of Equation (2.11) results in

$$\frac{\partial \vec{\omega}}{\partial t} - \nabla \times (\vec{V} \times \vec{\omega}) = \nu \nabla^2 \vec{\omega} , \tag{2.13}$$

where it should be noted that the curl of the gradient of any scalar is zero. Using vector identities, the second term on the left-hand side may be expanded to give

$$\nabla \times (\vec{V} \times \vec{\omega}) = \vec{V}(\nabla \cdot \vec{\omega} - \vec{\omega}(\nabla \cdot \vec{V}) - (\vec{V} \cdot \nabla)\vec{\omega} + (\vec{\omega} \cdot \nabla)\vec{V} . \tag{2.14}$$

Using the fact that $\nabla \cdot \vec{\omega} = 0$ (the divergence of the curl of any vector is zero) and that $\nabla \cdot \vec{V} = 0$ from the continuity equation, Equation (2.14) simplifies to

$$\frac{\partial \vec{\omega}}{\partial t} + (\vec{V} \cdot \nabla)\vec{\omega} = (\vec{\omega} \cdot \nabla)\vec{V} + \nu \nabla^2 \vec{\omega} . \tag{2.15}$$

For two-dimensional flows the vorticity vector $\vec{\omega}$ is perpendicular to the plane of the flow. Therefore $(\vec{\omega} \cdot \nabla)\vec{V} = 0$, and Equation (2.15) simplifies to

$$\frac{\partial \vec{\omega}}{\partial t} + (\vec{V} \cdot \nabla)\vec{\omega} = \nu \nabla^2 \vec{\omega} . \tag{2.16}$$

The *streamfunction* ψ is defined for the case of two-dimensional flows. The relationship between the velocity components U and V and ψ is $U = \psi_Y$ and $V = -\psi_X$, respectively.

2.2.4   Non-dimensional Governing Equations

The variables of space, time, velocity, pressure and temperature in Equations (2.7) - (2.9) can be made non-dimensional by the characteristic length L, characteristic time L/U_W, characteristic velocity U_W, dynamic pressure ρU_W^2 and temperature difference ΔT, respectively. More specifically, the non-dimensional variables are defined as

$$x = \frac{X}{L} , \qquad y = \frac{Y}{L} , \qquad \tau = \frac{tU_W}{L}$$

$$u = \frac{U}{U_W} , \qquad v = \frac{V}{U_W} , \qquad p = \frac{P}{\rho U_W^2}$$

and

$$\theta = \frac{T - T_{ref}}{\Delta T} .$$

The following non-dimensional governing equations are obtained using the non-dimensional variables above.

$$\frac{\partial u_i}{\partial x_i} = 0 , \tag{2.17}$$

$$\frac{Du_i}{D\tau} = -\frac{\partial p}{\partial x_i} + \frac{1}{Re} \frac{\partial^2 u_i}{\partial x_j \partial x_j} \qquad \text{and} \tag{2.18}$$

$$\frac{D\theta}{D\tau} = \frac{1}{Pe} \frac{\partial^2 \theta}{\partial x_j \partial x_j} + s^* . \tag{2.19}$$

The Reynolds number Re is defined as

$$Re = \frac{U_W L}{\nu} .$$

The Peclet number Pe is defined as $Pe = RePr$, where

$$Pr = \frac{\nu}{\alpha}$$

is the Prandtl number of the fluid. The term s^* in the energy equation is the source term.

2.3 Turbulent Navier-Stokes and Energy Equations

This section outlines the derivation of the Reynolds averaged governing equations for turbulent flows using the ensemble averaging method. To begin, various options available for calculating an averaged quantity are considered.

2.3.1 Averaging Processes

Reynolds Averaging

For a given velocity component u_i^*, let $u_i^* = U_i + u_i$, where u_i^* is the total or instantaneous value, U_i is the mean value and u_i is the fluctuating value. Calculating the mean value U_i may be done in terms of a short- or long-term average.

- Long-term averaging $(T \to \infty)$

$$\overline{u_i^*} = U_i = \frac{1}{T} \int_0^T u_i^* dt .$$

  In this case, U_i is not a function of time.

- Short-term averaging $(\Delta T$ short$)$

$$\overline{u_i^*} = U_i(t) = \frac{1}{\Delta T} \int_{-\frac{\Delta T}{2}}^{+\frac{\Delta T}{2}} u_i^*(t + t') dt' .$$

Here t' is the variable of integration over the ΔT interval centered at t. In this case, U_i is a function of time t and ΔT. If ΔT becomes large, the short-term average approaches the long-term average. The result of a short-term averaging is somewhat similar to a signal that has been smoothed or has had a high frequency component filtered out. It should also be noted that, $\overline{u_i} = 0$ and $\overline{u_i u_j} \neq 0$ in general. See Figure 2.1 for further information.

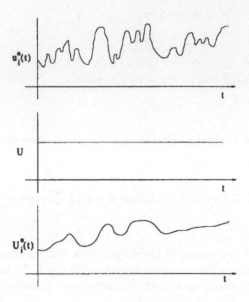

Figure 2.1: Long-term (U) and short-term $(U_i^*(t))$ Reynolds averaging.

Ensemble Averaging (Phase Averaging)

In a short-term averaging process, the calculation is performed once. However, the averaging depends on size of the filter time interval, ΔT, which is not known in advance. An alternative way of constructing an average for a time-dependent flow is to obtain an ensemble average, which requires one to repeat the calculation for N experiments, and then average the results of the N experiments, as indicated in Equation (2.20).

$$\overline{u_i^*} = U_i(t) = \lim_{N \to \infty} \frac{1}{N} \sum_{n=1}^{N} u_i^*(t, n) \,. \tag{2.20}$$

Refer to Figure 2.2 for a graphical interpretation of the ensemble averaging process. Equation 2.20 includes the short-term Reynolds average and is potentially more general. However, it is more difficult to produce since it requires N experiments. Further, N must be large enough so that the ensemble average is not dependent on N.

Using the definition of the ensemble average given by Equation (2.20), the following properties result.

1. $\overline{A u_i^*} = A \overline{u_i^*} = A U_i$

2. $\overline{u_i^* + u_j^*} = U_i + U_j$

3. $\overline{\partial u_i^* / \partial x_j} = \partial U_i^* / \partial x_j$

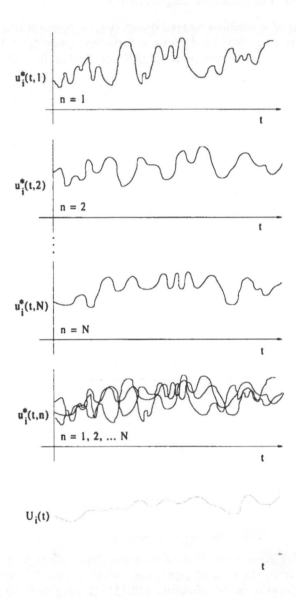

Figure 2.2: Ensemble averaging.

4. $\overline{u_i^* u_j^*} = U_i U_j + \overline{u_i u_j}$

2.3.2 Averaged Turbulence Equations

Turbulence equations of ensemble-averaged quantities for incompressible flows are derived in this section. Start by defining a turbulent quantity to be equal to the sum of its mean value (ensemble averaged) and fluctuating value. That is,

$$P^* = P + p, \quad T^* = T + \theta, \quad u_i^* = U_i + u_i \quad \text{and} \quad \tau_{ij}^* = \tau_{ij} + \tau_{ij}'.$$

Taking an ensemble average of Equations 2.7, 2.8 and 2.9, we have

$$\frac{\partial U_i}{\partial X_i} = 0, \tag{2.21}$$

$$\rho \frac{DU_i}{Dt} = \rho G_i - \frac{\partial P}{\partial X_i} + \frac{\partial \tau_{ij}}{\partial X_j} + \frac{\partial \tau_{ij}^t}{\partial X_j}, \tag{2.22}$$

wherein

$$\tau_{ij} = \mu \left(\frac{\partial U_i}{\partial X_j} + \frac{\partial U_j}{\partial X_i} \right)$$

and

$$\tau_{ij}^t = -\rho \overline{u_i u_j},$$

Where, τ_{ij} is modeled from the viscous fluid model while τ_{ij}^t is determined from a turbulence model.

$$\rho C_p \frac{DT}{Dt} = \tau_{ij} \frac{\partial U_i}{\partial X_j} - \frac{\partial q_i}{\partial X_i} + \frac{\partial q_i^t}{\partial X_i} + \Phi^t, \tag{2.23}$$

where

$$
\begin{aligned}
q_i &= -\kappa \partial T / \partial X_i, \\
q_i^t &= -C_p \rho \overline{u_i \theta}, \\
\Phi^t &= \overline{\tau_{ij}' \frac{\partial u_i}{\partial X_j}} \quad \text{and} \\
\tau_{ij}' &= \mu \left(\frac{\partial u_i}{\partial X_j} + \frac{\partial u_j}{\partial X_i} \right).
\end{aligned}
$$

Here, q_i is modeled from the Fourier postulates, while q_i^t and Φ^t need to be modeled from the turbulence model. Equations (2.21) - (2.23) give five equations for the five variables P, U, V, W and T. These equations include the variables $\overline{u_i u_j}$, $\overline{u_i \theta}$ and Φ^t. Therefore, the set of equations is not closed. Closing the set of turbulence equations is referred to as the *turbulence closure problem*.

Before the closure problem can be addressed, the ensemble average quantities $\overline{u_i u_j}$, $\overline{u_i \theta}$ and Φ^t must be derived. These additional ten unknowns (six from $\overline{u_i u_j}$, three from $\overline{u_i \theta}$ and one from Φ^t) can be derived from the fluctuation equations for u_i and θ. The fluctuation equations for u_i and θ are derived subtracting the time-averaged equation from the original equation. That is,

$$[\text{N-S}] - \overline{[\text{N-S}]} \rightarrow \text{Equation for } u_i$$

or

$$\rho\left(\frac{\partial u_i}{\partial t} + U_l\frac{\partial u_i}{\partial X_l} + u_l\frac{\partial U_i}{\partial X_l} + u_l\frac{\partial u_i}{\partial X_l}\right) = -\frac{\partial P}{\partial X_i} + \mu\left(\frac{\partial^2 u_i}{\partial X_l\partial X_l}\right) + \frac{\partial\rho\overline{u_i,u_l}}{\partial X_l}$$

$$(2.24)$$

and

$$[\text{Energy}] - \overline{[\text{Energy}]} \rightarrow \text{Equation for } \theta$$

or

$$\rho C_p\left(\frac{\partial\theta}{\partial t} + U_l\frac{\partial\theta}{\partial X_l} + u_l\frac{\partial T}{\partial X_l} + u_l\frac{\partial\theta}{\partial X_l}\right) = k\frac{\partial^2\theta}{\partial X_l\partial X_l} + \frac{\partial C_p\rho\overline{u_l\theta}}{\partial X_l} - \Phi^t \,. \quad (2.25)$$

2.3.3   Turbulence Transport Equations

In order to close the set of averaged turbulence equations, expressions will be derived for $\overline{u_iu_j}$, $\overline{u_i\theta}$, etc. from Equations 2.24 and 2.25. These are the second-order, one point correlation-based transport equations.

Reynolds Stress Transport Equations

The equation for $\overline{u_iu_j}$ is obtained by the following method:

- Multiply Equation 2.24, for the ith variable, by u_j.

- Multiply Equation 2.24, for the jth variable, by u_i.

- Add the results of the above two operations.

- Take the average of the total addition.

Following the above procedure, the equations obtained for $\overline{u_iu_j}$ are

$$\frac{D\overline{u_iu_j}}{Dt} = \frac{\partial}{\partial X_l}\left(-\overline{u_iu_ju_l} - \frac{\overline{p}}{\rho}(\delta_{jl}u_i + \delta_{il}u_j) + \nu\frac{\partial\overline{u_iu_j}}{\partial X_l}\right)$$

$$- \left(\overline{u_iu_l}\frac{\partial U_j}{\partial X_l} + \overline{u_ju_l}\frac{\partial U_i}{\partial X_l}\right) - 2\nu\overline{\frac{\partial u_i}{\partial X_l}\frac{\partial u_j}{\partial X_l}}$$

$$+ \frac{\overline{p}}{\rho}\left(\frac{\partial u_i}{\partial X_j} + \frac{\partial u_j}{\partial X_i}\right) \,.$$

$$(2.26)$$

In order to understand the physics of this equation, consider what the seven terms on the right hand side of this equation represent:

- The first two terms on the right represent the turbulent diffusion of the momentum.

- The third term represents the molecular diffusion of the momentum, and is usually negligible compared to the first two terms. However, it is the term with the highest derivative. Hence, it must be retained.

- The fourth and fifth terms represent the production of Reynolds stresses $\overline{u_i u_j}$ by the interaction between the Reynolds stresses and the gradients of the mean velocities.

- The sixth term represents the viscous dissipation of the Reynolds stresses.

- The seventh term represents the Pressure-Straining (PS) of the flow, which tends to restore the isotropic nature of the flow.

Turbulent Kinetic Energy Equation

Using Equation 2.26 for $\overline{u_i u_j}$, one can easily obtain the turbulent kinetic energy equation by setting $i = j$. The mean turbulent kinetic energy is denoted by $k = \overline{u_i u_i}/2$, while the fluctuating turbulent kinetic energy is denoted by $k' = u_i u_i / 2$. Then, the turbulent kinetic energy k equation is

$$\frac{Dk}{Dt} = \frac{\partial}{\partial X_l}\left(-\overline{k' u_l} - \frac{\overline{p u_l}}{\rho} + \nu\frac{\partial k}{\partial X_l}\right) - \overline{u_i u_l}\frac{\partial U_i}{\partial X_l} - \epsilon + 0 \, . \tag{2.27}$$

Note that the pressure-strain term goes to zero because of the continuity equation. The physics hidden in the equation is revealed when each of the terms on the right-hand side is studied carefully.

- The first two terms on the right represent the diffusion of the kinetic energy from high intensity to low intensity due to turbulent fluctuating motions.

- The third term represents the molecular diffusion of the turbulent kinetic energy.

- The fourth term represents the production of the turbulent kinetic energy due to the interaction of turbulent stress and the gradient of the mean flow velocity.

- The fifth term, ϵ, represents the rate of dissipation of the turbulent kinetic energy which tends to occur in the small eddy scale. ϵ is always positive, since

$$\epsilon = \mu\overline{\frac{\partial u_i}{\partial X_l}\frac{\partial u_i}{\partial X_l}} \, .$$

- The pressure-strain term of the Reynolds stress transport equation in the last term vanishes, and hence, its net contribution to the turbulent kinetic energy transfer rate is zero.

Rate of Dissipation Equation

The dissipation rate of turbulent kinetic energy,

$$\epsilon = \nu \overline{\frac{\partial u_i}{\partial X_l} \frac{\partial u_i}{\partial X_l}}$$

directly affects the growth rate of k. Thus, it is one of the most important turbulent quantities. It appears naturally in the derivation of the k equation. The equation for the rate of dissipation of the turbulent kinetic energy, ϵ, is obtained by the following approach.

- Perform the operation $\partial/\partial X_l$ on Equation 2.24 for the ith variable.

- Multiply the resulting equation by $\partial u_i/\partial X_l$.

- Take the ensemble average of the product and multiply by 2.

Following these steps, the ϵ equation obtained is

$$
\begin{aligned}
\frac{D\epsilon}{Dt} =\ & \frac{\partial}{\partial X_l}\left(-\overline{\epsilon' u_l} - \frac{2\nu}{\rho}\overline{\frac{\partial u_l}{\partial X_j}\frac{\partial p}{\partial X_j}} + \nu\frac{\partial \epsilon}{\partial X_l}\right) \\
& - 2\nu\overline{u_l \frac{\partial u_i}{\partial X_j}\frac{\partial^2 U_i}{\partial X_l \partial X_j}} - 2\nu\frac{\partial U_i}{\partial X_j}\left(\overline{\frac{\partial u_l}{\partial X_i}\frac{\partial u_l}{\partial X_j}} + \overline{\frac{\partial u_i}{\partial X_l}\frac{\partial u_j}{\partial X_l}}\right) \\
& - 2\nu\overline{\frac{\partial u_i}{\partial X_j}\frac{\partial u_i}{\partial X_l}\frac{\partial u_j}{\partial X_l}} - 2\overline{\left(\nu\frac{\partial^2 u_i}{\partial X_l \partial X_l}\right)^2}.
\end{aligned}
\tag{2.28}
$$

The fluctuation, or instantaneous, quantity of ϵ is given by

$$\epsilon' = \nu\frac{\partial u_i}{\partial X_l}\frac{\partial u_i}{\partial X_l}.$$

In order to understand the ϵ equation, let us examine the various terms on the right-hand side of the equation.

- The first two terms again represent the turbulent diffusion of ϵ.

- The third term represents the molecular diffusion of ϵ.

- The fourth and fifth terms represent the production of ϵ.

- The last two terms represent the destruction (source or sink) of the dissipation rate of the turbulent kinetic energy.

Reynolds Turbulent Heat Flux Equation

Defining the fluctuating viscous dissipation term as

$$\Phi' = \tau'_{ij}\frac{\partial u_i}{\partial X_j} = \tau'_{nm}\frac{\partial u_n}{\partial X_m},$$

the equation for the Reynolds turbulent heat flux, $\overline{u_i\theta}$, can be obtained using the following approach:

- Multiply Equation 2.24 for the ith variable, by θ.

- Multiply Equation 2.25 by u_i.

- Add the two products.

- Take the ensemble average of the resultant sum.

Thus, the $\overline{u_i \theta}$ equation can be written as

$$\frac{D\overline{u_i \theta}}{Dt} = \frac{\partial}{\partial X_l}\left(-\overline{u_i u_l \theta} - \delta_{il}\frac{\overline{P\theta}}{\rho} + \alpha \overline{u_i \frac{\partial \theta}{\partial X_l}} + \nu \overline{\theta \frac{\partial u_i}{\partial X_l}}\right)$$
$$- \left(\overline{u_i u_l}\frac{\partial \theta}{\partial X_l} + \overline{u_l \theta}\frac{\partial U_i}{\partial X_l}\right) - (\alpha + \nu)\overline{\frac{\partial u_i}{\partial X_l}\frac{\partial \theta}{\partial X_l}}$$
$$+ \overline{\frac{p\partial\theta}{\rho\partial X_i}} + \overline{\Phi' u_i} . \tag{2.29}$$

Here, α and ν are the thermal diffusivity and the kinematic viscosity, respectively.

2.4   Turbulence Closure

Reynolds [154] introduced the time-averaged Navier-Stokes equations in 1895. The averaging process produces a set of turbulent stresses τ_{ij}^t , known as *Reynolds stresses*. Since 1895, researchers and scientists have attempted to create a turbulence model for the Reynolds stresses to recover turbulent motion information lost in the averaging process. The problem of specifying unknowns $\overline{u_i u_j}$ is referred to as the *turbulence closure problem*.

Turbulence models were relatively simple and intuitive due to the lack of computing machinery in the early 20th century. Historically, the first attempt to model the Reynolds stresses was to express them as a function of the mean flow quantities. Since the $\overline{u_i u_j}$ terms are modeled directly without any additional differential equations, this model is also called the *zero equation model* or the *first-order closure model*.

2.4.1   First-Order Closure (Zero Equation Model)

The following are examples of first-order models.

- $-\overline{uv} = \nu_t \partial U / \partial Y$. (Boussinesq eddy viscosity model)

- $-\overline{uv} = l^2 \left|dU/dY\right| \partial U / \partial Y$. (Prandtl mixing length model)

- $-\overline{uv} = \kappa X \partial U / \partial Y$. (Prandtl wake mixing length model)

- $-\overline{uv} = \kappa^2 \left|(dU/dY)/(d^2U/dY^2)\right|^2 / \partial U / \partial Y$. (von Karman mixing length model)

First-order closure models were very popular in the early 20th century because of their simplicity. Unfortunately, these models are very limited by the geometry of the problem. That is, a particular model is only appropriate for a narrow set of problems with similar geometry.

2.4.2 Second-Order Closure

Turbulence modeling advanced quickly with the advent of high-speed computing after the World War II. From the late 1950s to the present, turbulence modeling in engineering progressed to the second-order turbulence closure model. The second-order turbulence transport quantities, namely the Reynolds stresses $\overline{u_i u_j}$, the Reynolds heat fluxes $\overline{u_i \theta}$, the turbulent kinetic energy k, and the rate of dissipation of the turbulent kinetic energy ϵ , are modeled by differential equations. These variables, in general, are related to the turbulent scale quantities such as the turbulent length l, time t and velocity u_i. Turbulence models with various combinations of the variables have been proposed. Examples include, $k - k^{1/2}/l$ (Kolmogorov, 1942) [107], $k - k/l^2$ (Spalding, 1982) [162], $k - kl$ (Ng and Spalding, 1972) [140], and $k - \epsilon$ (Launder and Spalding, 1974) [111]. The $k - \epsilon$ model has emerged as the most popular turbulence model, mainly because ϵ (the rate of dissipation of k) appears naturally in the k equation and can be derived from the averaged Navier-Stokes equations without introducing additional variables. Due to its popularity, the following discussion is restricted to the $k - \epsilon$ type turbulence models.

Second-order turbulence closure modeling greatly improved prediction capability over that of the simple first-order turbulence models. The basic postulates adopted by most turbulence models are analogous to the Stokes-Fourier postulates of the viscous-conducting fluid model. They are summarized below.

1. Ensemble averaged Navier-Stokes equations and the Fourier energy equation can properly describe the turbulent mean motion and turbulent transport properties. Instantaneous fluid motions are averaged and detailed information about fluid motions is lost. A model is required to recover the information lost (*modeling requirement*).

2. The turbulent diffusion of turbulent transport properties is proportional to the gradient of transport properties (*diffusion gradient model*).

3. Small turbulent eddies are isotropic (*isotropic dissipation model*).

4. All turbulent transport quantities are local functions of Reynolds stress, turbulent kinetic energy, rate of dissipation of turbulent kinetic energy, mean flow variables, and thermodynamics variables, ($\overline{u_i u_j}$, k, ϵ, $\overline{u_i \theta}$, U_i, ρ, P, T) (*one point correlation closure statement*).

5. All modeled turbulent phenomena must be consistent in symmetry, invariance, permutation, and physical observations (*consistency and realizability requirement*).

6. Turbulent phenomenon can be characterized by one turbulent scale, (k, ϵ), based on turbulent kinetic energy and its rate of dissipation, i.e., $u = \sqrt{k}$, $l = k^{3/2}/\epsilon$ and $t = k/\epsilon$ (*turbulence scale hypothesis*).

7. All turbulence model moduli, C_ϵ, $C_{\epsilon 1}$, $C_{\epsilon 2}$, C_k, C_1, C_2, C_T, C_{T1} and C_{T2}, require experimental calibration and determination (*uniqueness of moduli*).

If a turbulence model is complete, then the model moduli are unique and valid for any turbulent flow geometry and conditions. Presently, no such turbulence model is available. Consequently, various turbulence models, and hence model moduli, exist under the above postulates. Examples include a high Reynolds number model, a low Reynolds number model, a near wall model, and a two-scale model.

Some models make *ad hoc* adjustments to relax the aforementioned postulates. For example, one approach which modifies postulate 3 considers small turbulent eddies to be *anisotropic*. Also, postulate 6 may be modified by assuming multiple scales are required to characterize the turbulent phenomenon. For example, introduce as a second scale the *Kolmogorov scale* (ν, ϵ) for the smaller eddies, which gives $u = (\nu\epsilon)^{1/4}$, $l = (\nu^3/\epsilon)^{1/4}$ and $t = \sqrt{\nu/\epsilon}$. Thus, the $k - \epsilon$ scale is used to relate the mean flow kinetic energy to the dissipation of kinetic energy in the larger eddies, and the Kolmogorov scale (ν, ϵ) is used to relate the kinetic energy dissipation in the larger eddies to the viscous dissipation in the smaller eddies.

This section concludes with a listing of the differential equations of the second-order turbulence transport quantities. Each equation includes one or more model moduli. These are constants that are determined by experiment. For a more detailed development of the second-order turbulence equations, and turbulence modeling in general, the reader is referred to the book by Chen [51].

- The Reynolds stress, $\overline{u_i u_j}$, equation is given by

$$
\begin{aligned}
\frac{D\overline{u_i u_j}}{Dt} &= \frac{\partial}{\partial X_l}\left[\left(C_k \frac{k^2}{\epsilon} + \nu\right)\frac{\partial \overline{u_i u_j}}{\partial X_l}\right] - \left(\overline{u_i u_l}\frac{\partial U_j}{\partial X_l} + \overline{u_j u_l}\frac{\partial U_i}{\partial X_l}\right) \\
&\quad - \frac{2}{3}\delta_{ij}\epsilon - C_1 \frac{\epsilon}{k}\left(\overline{u_i u_j} - \frac{2}{3}\delta_{ij}k\right) \\
&\quad + C_2\left(\overline{u_i u_l}\frac{\partial U_j}{\partial X_l} + \overline{u_j u_l}\frac{\partial U_i}{\partial X_l} - \frac{2}{3}\delta_{ij}\overline{u_n u_m}\frac{\partial U_n}{\partial X_m}\right) \\
&= D_{ij} + P_{ij} - \frac{2}{3}\delta_{ij}\epsilon + PS ,
\end{aligned}
\tag{2.30}
$$

where $C_k = 0.09$, $C_1 = 2.3$ and $C_2 = 0.4$.

- The kinetic energy, k, equation is given by,

$$
\begin{aligned}
\frac{Dk}{Dt} &= \frac{\partial}{\partial X_l}\left[\left(C_k \frac{k^2}{\epsilon} + \nu\right)\frac{\partial k}{\partial X_l}\right] - \overline{u_i u_l}\frac{\partial U_i}{\partial X_l} - \epsilon \\
&= D_k + P_k - \epsilon ,
\end{aligned}
\tag{2.31}
$$

where $C_k = 0.09 \sim 0.11$.

- The rate of dissipation, ϵ, equation is given by

$$\frac{D\epsilon}{Dt} = \frac{\partial}{\partial X_l}\left[\left(C_\epsilon \frac{k^2}{\epsilon} + \nu\right)\frac{\partial \epsilon}{\partial X_l}\right] - C_{\epsilon 1}\frac{\epsilon}{k}\overline{u_i u_l}\frac{\partial U_i}{\partial X_l} - C_{\epsilon 2}\frac{\epsilon^2}{k}$$
$$= D_\epsilon + P_\epsilon - \Phi_\epsilon , \qquad (2.32)$$

where $C_\epsilon = 0.07$, $C_{\epsilon 1} = 1.45$ and $C_{\epsilon 2} = 1.92$.

- The Reynolds heat flux, $\overline{u_i\theta}$, equation is given by

$$\frac{D\overline{u_i\theta}}{Dt} = \frac{\partial}{\partial X_l}\left[\left(C_T\frac{k^2}{\epsilon} + \alpha\right)\frac{\partial \overline{u_i\theta}}{\partial X_l}\right] - \left(\overline{u_i u_l}\frac{\partial T}{\partial X_l} + \overline{u_l\theta}\frac{\partial U_i}{\partial X_l}\right)$$
$$\quad - C_{T1}\frac{\epsilon}{k}\overline{u_i\theta} + C_{T2}\overline{u_m\theta}\frac{\partial U_i}{\partial X_m}$$
$$= D_{u\theta} + P_{u\theta} - PT , \qquad (2.33)$$

where $C_T = 0.07$, $C_{T1} = 3.2$ and $C_{T2} = 0.5$.

2.5 Progress in Turbulence Modeling

This section begins by presenting several second-order turbulence closure models. This is followed by a discussion of some possible variations of these models. Then the discussion will shift to numerical modeling of turbulence in near-wall regions. Finally, a brief overview of the performance of current turbulence models will be presented, and the areas of turbulence modeling that need the most improvement will be mentioned.

2.5.1 Second-Order Turbulence Closure Models

Equations (2.21) - (2.23) and (2.30) - (2.33) are known as the *Reynolds Stress Model* (RSM) [80], or *differential model*. The RSM consists of 16 partial differential equations in 16 unknowns for three-dimensional problems, so it is very costly in terms of computational power and time to use this model to simulate turbulent flow.

The other second-order closure models that will be presented here are based on the RSM, but make further assumptions to reduce the number of PDEs that must be solved. One such model is the $k - \epsilon$-Algebraic ($k - \epsilon - A$) model, or *Algebraic Stress Model* (ASM). In the $k - \epsilon - A$ model, the quantities k and ϵ are still solved with Equations (2.31) and (2.32), but the Reynolds stresses $\overline{u_i u_j}$ and Reynolds heat fluxes $\overline{u_i\theta}$ are solved using approximated algebraic equations. The Reynolds stress equation is derived by assuming $\overline{u_i u_j}$ is proportional to k [157, 156]. Similarly, the Reynolds heat flux equation is derived by assuming $\overline{u_i\theta}$ is proportional to k. With these assumptions, the $\overline{u_i u_j}$ and $\overline{u_i\theta}$ equations become

$$\overline{u_i u_j} = k\left[\frac{\frac{2}{3}(C_1 - 1)\epsilon\,\delta_{ij} + (1 - C_2)P_{ij} + \frac{2}{3}C_2 P_k\delta_{ij}}{C_1\epsilon + (P_k - \epsilon)}\right] \qquad (2.34)$$

and

$$\overline{u_i\theta} = k \left[\frac{C_{T2}\overline{u_m\theta}\frac{\partial U_i}{\partial X_m} - \left(\overline{u_i u_l}\frac{\partial T}{\partial X_l} + \overline{u_l\theta}\frac{\partial U_i}{\partial X_l}\right)}{C_{T1}\epsilon + (P_k - \epsilon)} \right]. \tag{2.35}$$

For three-dimensional problems, Equations (2.34) and (2.35) constitute nine algebraic equations (six Reynolds stress equations and three Reynolds heat flux equations), which replace the corresponding nine partial differential equations in the RSM. This greatly reduces the computational effort required to solve turbulence problems.

An even simpler form of the $k - \epsilon - A$ model is the *equilibrium* $k - \epsilon - A$ model, which uses Equations (2.34) and (2.35) with the term $(P_k - \epsilon)$ in each equation set to zero. The equilibrium $k-\epsilon-A$ model is appropriate only for high shear-high temperature gradient flows and local equilibrium flows. In general, the $k - \epsilon - A$ and equilibrium $k - \epsilon - A$ models are suitable when the transport of $\overline{u_i u_j}$ and $\overline{u_i\theta}$ are not very important, since these terms have been either crudely modeled or neglected [97].

Another popular $k - \epsilon$ model is the $k - \epsilon$–Eddy Viscosity ($k - \epsilon - E$) model. As in the $k - \epsilon - A$ model, the $k - \epsilon - E$ model retains the PDEs for k and ϵ given in Equations (2.31) and (2.32). However, $\overline{u_i u_j}$ and $\overline{u_i\theta}$ are modeled using the Boussinesq eddy viscosity model, which yields

$$-\overline{u_i u_j} = \nu_t \left(\frac{\partial U_i}{\partial X_j} + \frac{\partial U_j}{\partial X_i} \right) - \frac{2}{3}\delta_{ij}k \tag{2.36}$$

and

$$-\overline{u_i\theta} = \frac{\nu_t}{Pr_t}\frac{\partial T}{\partial X_i}, \tag{2.37}$$

where

$$\nu_t = C_\mu \frac{k^2}{\epsilon}.$$

In Equations (2.36) and (2.37), $C_\mu = 0.09$ and the turbulent Prandtl number $Pr_t = 0.8 \sim 1.3$. In general, the performance of the $k - \epsilon - E$ model is not as good as the $k - \epsilon - A$ model in complex flows, but the two models perform about the same in boundary layer-like flows.

The $k - \epsilon - A$ and $k - \epsilon - E$ models are referred to as *two-equation models* because they both use PDEs for k and ϵ, and simpler equations for $\overline{u_i u_j}$ and $\overline{u_i\theta}$. The simplest second-order closure model is a one-equation model known as the *mixing length model*, or *k-equation model*. The only partial differential equation used for the turbulence quantities in this model is the equation for k given in Equation (2.31). In this equation, ϵ is given as

$$\epsilon = \frac{k^{3/2}}{l}, \tag{2.38}$$

where l is a mixing length, which must be provided. The Reynolds stresses and heat fluxes are obtained using Equations (2.36) and (2.37). Because the mixing length is problem dependent, the prediction capability of the k-equation model is lower than that of the other second-order closure models.

2.5.2 Modifications of Second-Order Closure Models

Each of the models discussed in the previous subsection predict some turbulent flows well, but do a poor job predicting others. Because of this, many researchers over the past few decades have attempted to modify the RSM, $k - \epsilon - A$ and $k - \epsilon - E$ models in order to improve their performance. Some of these efforts are briefly discussed here.

The first type of modification is the addition of cross-diffusion terms in the k and/or ϵ equations. For example, Jaw and Chen [96] proposed a $k - \epsilon - E$ model that included a term in the ϵ equation which accounted for the cross-diffusion of k. This term arises if one derives the turbulence equations assuming that small turbulent eddies can be anisotropic. This is more realistic than Postulate 3 in Section 2.4.2, which assumes such eddies are isotropic. The ϵ equation (Equation (2.32)) then becomes

$$\frac{D\epsilon}{Dt} = \frac{\partial}{\partial X_l}\left[\left(C_\epsilon \frac{k^2}{\epsilon} + \nu\right)\frac{\partial \epsilon}{\partial X_l}\right] - C_{\epsilon\,1}\frac{\epsilon}{k}\overline{u_i u_l}\frac{\partial U_i}{\partial X_l} - C_{\epsilon\,2}\frac{\epsilon^2}{k}$$
$$-C_{\epsilon\,3}\frac{\epsilon}{k}\frac{\partial}{\partial X_l}\left(C_k\frac{k^2}{\epsilon}\frac{\partial k}{\partial X_l}\right), \tag{2.39}$$

where $C_\epsilon = 0.11$, $C_{\epsilon 1} = 1.23$, $C_{\epsilon 2} = 1.92$ and $C_{\epsilon 3} = 1.67$.

This model uses the definition of ν_t given in Equation (2.36) with $C_\mu = 0.09$, and it uses the k equation (Equation (2.31)) with $C_k = 0.103$. Jaw and Chen found that the addition of the cross-diffusion term to the $k - \epsilon - E$ model does indeed improve its performance, especially in predicting flows where turbulent production and dissipation are not equal (i.e. non-equilibrium turbulent flows). A few other models which include cross-diffusion terms are mentioned in [97].

It should be noted that cross-diffusion terms only modify the k and/or ϵ equations. Since the standard $k - \epsilon - E$, $k - \epsilon - A$ and RSM models all use the same k and ϵ equations, cross-diffusion terms can be added to any of these models. For instance, Jaw and Chen [98] added a cross-diffusion term to the ϵ equation in an RSM model, and found that it improved the prediction of round jet flow.

The next type of modification considered here allows turbulence models to account for multiple time and length scales. The modification of model scales stems from the realization that turbulence is comprised of fluctuating motions with a large range of eddy sizes and time scales. It follows that models containing multiple scales will be able to better predict these turbulent motions. Chen and Singh [43, 44] proposed a two-scale model that adopted both the large eddy (energy-containing) scale (k,ϵ) and the small eddy (energy-dissipating) scale (ν, ϵ) in the modeling of the ϵ transport equation. The first scale is based on experimental evidence [69] that indicates large eddies contain most of the turbulent kinetic energy in the flow, and do not play a significant role in the dissipation of turbulent kinetic energy. The second scale is based on the Kolmogorov hypothesis that the dissipation of turbulent kinetic energy occurs primarily in small eddies. Chen and Singh reasoned that the dissipation or destruction term of the k and ϵ equations should be modeled with the (ν,ϵ) scale. Since the dissipation term in the k equation is ϵ, no alteration to the k equation is needed. The two-scale ϵ transport equation is

$$\frac{D\epsilon}{Dt} = \frac{\partial}{\partial X_l}\left[\left(C_\epsilon \frac{k^2}{\epsilon} + \nu\right)\frac{\partial \epsilon}{\partial X_l}\right]$$
$$-C_{\epsilon 1}\left(\frac{\epsilon}{\nu}\right)^{1/2}\overline{u_i u_j}\frac{\partial U_i}{\partial X_j} - C_{\epsilon 2}\left(\frac{\epsilon}{\nu}\right)^{1/2}\epsilon\ , \tag{2.40}$$

where $C_\epsilon = 2.0$, $C_{\epsilon 1} = 17.5 Re^{-1/2}$ and $C_{\epsilon 2} = 18.9 Re^{-1/2}$. Note that the coefficients $C_{\epsilon 1}$ and $C_{\epsilon 2}$ in this equation are functions of the Reynolds number, Re, based on the large characteristic length scale and the mean velocity. A similar two-scale model based on fractal dynamics and including a cross-diffusion term was developed by Jaw and Chen [97, 95, 94].

The two-scale turbulence model given in Equation (2.40) and the model developed by Jaw and Chen [95, 94] were applied to predict turbulent free shear flows, and improved results were obtained over comparable one-scale turbulence models. However, it was found that the predictions were sensitive to the applied boundary conditions. Further study of these models is needed. Several multi-scale turbulence models, including the two mentioned above, are discussed in [97].

In addition to the modifications listed above, many researchers have put considerable effort into developing different ways to model various components of the turbulence equations given in Equations (2.26) - (2.29). Jaw and Chen [97] give a thorough review of different models for Equation (2.26), the Reynolds stress equation. This review gives the reader some perspective on the extent of this type of research. Even with this extensive effort, a universal turbulence model is still not available.

2.5.3 Numerical Considerations in the Near-Wall Region

The different turbulence models discussed in this section clearly illustrate that accurate numerical simulation of turbulence is a very challenging problem. When rigid boundary surfaces exist in the solution domain, modeling becomes even more complicated. Consider a turbulent flow running parallel to a wall. Away from the wall, the flow has the turbulent characteristics considered so far in this chapter. As one approaches the wall, however, the fluid velocity must decrease so that it satisfies the no-slip condition, which states that the velocity of the fluid in contact with the wall must be zero. Therefore, in the near-wall region, the turbulent flow is reduced to laminar flow, and molecular viscosity begins to dominate the flow behavior. This presents a problem in turbulence modeling, since some of the postulates used to form the turbulence models now become questionable (e.g. isotropic diffusion, isotropic dissipation and the modeling of the pressure-strain term, PS, in Equation (2.30)). The rapid change in flow characteristics close to the wall also brings about an additional numerical difficulty. The sharp gradients in the transport properties created by this rapid change require that a large number of grid nodes must be placed in the near-wall region in order to accurately model this region. This increases the computational time and power required to simulate the flow. Three approaches are commonly used to alleviate these problems: wall functions, near-wall (or low Reynolds number) turbulence models, and two-layer models.

The *wall function* approach is the simplest of the three. Instead of solving the governing equations in the near-wall region, a semi-analytic solution is applied in this region. In doing so, it is possible to relate the surface boundary conditions at a given point on the wall to a point in the fluid a distance Y_p away from the wall, where Y_p is large enough to avoid the troublesome near-wall region. This effectively allows computation to start not at the boundary, but rather in the fluid at the first set of nodal points away from the boundary, beyond the near-wall region. The wall function approach begins by assuming that flow in the near-wall region is parallel to the wall and two-dimensional. If U_p is the fluid velocity at the first nodal point away from the wall, which is at a distance Y_p from the wall, then the wall function boundary conditions are specified as

$$u_p^+ = \frac{U_p}{U_\tau} = \frac{1}{\kappa}ln(Ey_p^+)$$

$$\theta_p^+ = \frac{T_w - T_p}{T_\tau} = 2.195ln(y_p^+) + 13.2Pr - 5.66 \quad \text{(from[101])}$$

$$-\overline{uv} = U_\tau^2, \quad k = \frac{U_\tau^2}{C_\mu^{1/2}}, \quad \epsilon = \frac{U_\tau^3}{\kappa Y_p}, \tag{2.41}$$

where

$$y_p^+ = \frac{U_\tau Y_p}{\nu}, \quad U_\tau = \left(\frac{\tau_w}{\rho}\right)^{1/2} \text{ and } T_\tau = \frac{\alpha q_w}{\kappa}.$$

In Equations (2.41), $\kappa = 0.41$, $E = 9.0$ and $C_\mu = 0.09$. Pr is the Prandtl number, and ν and ρ are the kinematic viscosity and density of the fluid, respectively. T_w, τ_w and q_w are the temperature, shear stress and heat flux at the wall. U_τ and T_τ are called the *friction velocity* and *friction temperature*, respectively. Equations (2.41) are valid for $y_p^+ > 10$.

The wall function approach is frequently used because it provides researchers a relatively simple way to avoid (1) using a large number of nodes, and (2) accounting for the invalidity of turbulence models in the near-wall region. However, this method does have a drawback. Because it was derived assuming that the flow is moving parallel to the wall, it is not valid for flows which contain regions of separation or stagnation points. Still, since the number of nodes for which the wall functions are not appropriate is generally far smaller than the total number of wall nodes, this approach is still commonly used for engineering applications.

Near-wall or low Reynolds number turbulence models constitute the opposite end of the modeling spectrum from wall functions. Low Reynolds number turbulence models account for the inaccuracy of turbulence models in the near-wall region by modifying the turbulence equations to account for the differing behaviors of diffusion, dissipation, and pressure redistribution near walls. In addition, these models solve the governing equations all the way to the wall. An extensive review of several low Reynolds number models, of both RSM and $k - \epsilon - E$ type, is given by Jaw and Chen [97]. They note that these models generally require rather high numerical resolution near the wall, typically 60 to 100 grid points across the boundary layer alone. Also, they point out that these models perform poorly in boundary layer flows with an adverse pressure gradient.

The final approach to the near-wall turbulence modeling problem is somewhere between the first two in complexity. In the *two-layer model*, a standard $k - \epsilon$ model is used away from walls, and a one-equation (k-equation) turbulence model similar to the one mentioned at the end of Section 2.5.1 is used in the near-wall region. As in the k-equation model, the turbulent kinetic energy k is modeled with its transport equation, Equation (2.31). The Reynolds stresses are given by Equation (2.36), except that the eddy viscosity is given as

$$\nu_t = C_\mu k^{1/2} l_\mu .\tag{2.42}$$

The dissipation rate ϵ is given by

$$\epsilon = \frac{k^{3/2}}{l_\epsilon} .\tag{2.43}$$

In Equations (2.42) and (2.43), the length scales l_μ and l_ϵ are considered to be functions of viscosity in terms of the turbulence Reynolds number $R_y = k^{1/2} y / \nu$, where y is the normal distance from the wall. Wolfshtein [177] proposed

$$l_\mu = C_1 y \left[1 - exp \left(-\frac{R_y}{A_\mu} \right) \right]\tag{2.44}$$

and

$$l_\epsilon = C_1 y \left[1 - exp \left(-\frac{R_y}{A_\epsilon} \right) \right] ,\tag{2.45}$$

with $C_1 = \kappa C_\mu^{-3/4}$, where $\kappa = 0.42$ is the von Karman constant and $C_\mu = 0.09$. Based on numerical tests, Chen and Patel [53] assigned the model coefficients values of $A_\mu = 70$ and $A_\epsilon = 5.08$.

In two-layer modeling, the two models have to be matched at some location. This is generally done at the edge of the viscous sublayer, far enough from the wall that viscous effects have become negligible. This corresponds roughly to $80 < y^+ < 120$, where y^+ is defined as in Equation (2.41). In general, two-layer models have been reported to predict more promising results than either low-Reynolds number models or the wall function approach [97].

2.5.4 Overview and Areas for Future Work

Compared to first-order turbulence models, second-order turbulence models have greatly improved the accuracy and predictive capability of numerical simulations of turbulent flows. As this section implies, there are currently a large number of turbulence models being developed in the research community and in use in the industrial sector. The choice of which model to use is based on a trade-off between accuracy and simplicity, and also on what type of flows are going to be modeled. Each of the models discussed in this section applies successfully to some turbulent flows while giving unsatisfactory predictions of other flows, especially for flows that are very different from those for which the model was calibrated. Modifications to the $k - \epsilon$ models, such as the inclusion of cross-diffusion terms and use of two-scale modeling concepts, have improved their performance. However, it still must be concluded that at present, no unique turbulence model exists that can provide satisfactory predictions of all turbulent flows.

It is apparent, then, that fundamental improvements must still be made in second-order turbulence closure models. The weakest point in the modeling of $\overline{u_i u_j}$, k and ϵ is the ϵ equation. The modeling of the destruction terms (the last line of Equation (2.28)) is questionable. After modeling, these terms become the last two terms on the right-hand side of Equation (2.32). The production terms (the middle line of Equation (2.28)) are omitted, which is also questionable. The modeling of the pressure-strain term in the Reynolds stress equation (the last line of Equation (2.26), modeled by PS in Equation (2.30)) can also be improved.

Finally, it should be mentioned that another option exists for modeling turbulent flows besides the ensemble averaged Navier-Stokes and energy equations. The derivations of the instantaneous Navier-Stokes and energy equations given in Equations (2.7) - (2.9) do not rule out turbulent flow. Two methods that use these equations are *large eddy simulation* (LES) [134] and direct numerical simulation (DNS) [126]. LES, which assumes a filter model to represent the behavior of small eddies, calculates the three-dimensional, time-dependent large eddy structure in turbulent flows using Equations (2.7) - (2.9). DNS, on the other hand, just directly solves Equations (2.7) - (2.9). At present, both approaches consume too much computer memory and time for most practical applications. For example, to capture fine scale turbulent eddies, DNS requires the grid size to be smaller than the finest turbulent eddies. In general, the grid number required by DNS is roughly equal to the turbulent Reynolds number of the flow raised to the $\frac{9}{4}$ power. For a flow with turbulent Reynolds number $R_T = 10^5$, which is common in engineering applications, the grid number required would be larger than 10^{11}. This is beyond current practical computing capabilities. However, as computers become faster and more powerful, and LES and DNS techniques become more refined, these approaches may rise in popularity.

Mathematical modeling for fluid flow and heat transfer, either laminar or turbulent, usually results in one or more partial differential equations. Solving these equations, either analytically or numerically, is often a formidable task. A brief, albeit useful, introduction to partial differential equations is given in the next chapter.

Chapter 3

Classification of PDEs

This chapter introduces some general topics concerning partial differential equations (PDEs). The governing equations for fluid flow and heat transfer presented in Chapter 2 are specific examples of PDEs. Both analytical and numerical solution methods for PDEs usually depend on the classification of the PDE. Therefore, this chapter begins with some terminology used in classifying PDEs. Then, attention is turned toward first-order PDEs in Section 3.2, and second-order PDEs in Section 3.3, and their classification based on characteristics.

3.1 Terminology

A partial differential equation (PDE) is an equation involving one or more partial derivatives of an unknown function of several independent variables. Suppose the dependent variable u is a function of the spatial variables x, y and z, and time t so that $u = u(x, y, z, t)$. The general form of a PDE for u is shown in Equation (3.1).

$$F(x, y, z, t; u, u_x, u_y, u_z, u_t, u_{xx}, u_{yy}, ...) = 0 \qquad (3.1)$$

The *order* of a PDE is the highest order derivative in Equation (3.1). The PDE is said to be *linear* if the function F is algebraically linear in each variable u, u_x, u_y,..., and if the coefficients of u, u_x, u_y,... are functions of the independent variables only. An equation that is not linear is said to be *nonlinear*. A nonlinear equation is *quasilinear* if it is linear in the highest order derivatives. Examples of PDEs with identification of their order and linearity are given in Table 3.1.

3.2 First-Order Equations and Characteristics

A mathematical model for an engineering problem may consist of a single PDE, or a set of PDEs that must be solved simultaneously. In fluid flows, for example, the set of equations

$$u_x + v_y = 0 , \qquad (3.2)$$

PDE	Order	Linearity
$u_y u_x + v u_y = xy$	1st	nonlinear
$u u_x + v u_y = -p_x$	1st	quasilinear
$x u_x + y u_y = u^2 + 1$	1st	quasilinear
$x u_x + y u_y = -px$	1st	linear
$\nabla^2 p = 2(u_x v_y - u_y v_x)$	2nd	linear
$u u_x^2 + v u_y^2 = -p_x + u_{yy} + u_{xx}$	2nd	quasilinear
$u u_x + v u_y = -p_x + (u_{yy})^2 + u_{xx}$	2nd	nonlinear
$[u - \frac{\partial(k^2/\varepsilon)}{x}]k_x + [v - \frac{\partial(k^2/\varepsilon)}{y}]k_y$ $= \frac{k^2}{\varepsilon}(k_{xx} + k_{yy}) + p_k$	2nd	quasilinear

Table 3.1: PDEs: order and linearity.

$$uu_x + vu_y = -x^2 \tag{3.3}$$

has the general form

$$a_1 u_x + b_1 u_y + c_1 v_x + d_1 v_y = q_1 , \tag{3.4}$$

$$a_2 u_x + b_2 u_y + c_2 v_x + d_2 v_y = q_2 , \tag{3.5}$$

where $a_1 = d_1 = 1$, $a_2 = u$, $b_2 = v$, $q_2 = -x^2$, and all other coefficients are zero.

For the general case, if a_1, a_2, b_1, b_2, c_1, and c_2 are functions of x and y only, then Equations (3.4) and (3.5) are linear. If a_1, a_2, b_1, b_2, c_1, c_2 are functions of x, y, u and v only, then Equations (3.4) and (3.5) are quasilinear. If any of a_1, a_2, b_1, b_2, c_1, c_2 is a function of u_x, u_y, v_x, or v_y, then Equations (3.4) and (3.5) are nonlinear.

In order to solve for u and v from Equations (3.4) and (3.5), the following question arises: *Given u and v on a curve* $\Gamma(x, y) = 0$, *as shown in Figure 3.1, are* u_x, u_y, v_x *and* v_y *uniquely determined from Equations (3.4) and (3.5)?* If so, solutions for u and v at a location P "near" the curve can be found using Taylor series expansion. That is,

$$u_p = u|_\Gamma + u_x|_\Gamma \Delta x + u_y|_\Gamma \Delta y + \cdots \tag{3.6}$$

and

$$v_p = v|_\Gamma + v_x|_\Gamma \Delta x + v_y|_\Gamma \Delta y + \cdots , \tag{3.7}$$

with

$$\Delta \hat{n} = \Delta x \hat{i} + \Delta y \hat{j} . \tag{3.8}$$

Suppose u and v are differentiable on Γ. Then, along $\Gamma(x, y) = 0$ we have

$$a_1 u_x + b_1 u_y + c_1 v_x + d_1 v_y = q_1 , \tag{3.9}$$

$$a_2 u_x + b_2 u_y + c_2 v_x + d_2 v_y = q_2 , \tag{3.10}$$

$$dx u_x + dy u_y + 0 + 0 = du \tag{3.11}$$

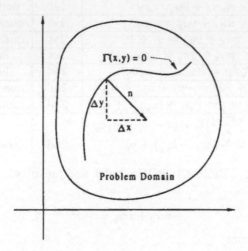

Figure 3.1: Characteristic curve for a PDE.

and

$$0 + 0 + dx v_x + dy v_y = dv .$$ (3.12)

This is a system of four equations in the four unknowns. It is possible to solve for u_x, u_y, v_x and v_y uniquely, if

$$\Delta = \begin{vmatrix} a_1 & b_1 & c_1 & d_1 \\ a_2 & b_2 & c_2 & d_2 \\ dx & dy & 0 & 0 \\ 0 & 0 & dx & dy \end{vmatrix} \neq 0 .$$ (3.13)

If $\Delta = 0$, there is no unique solution for u_x, v_x, u_y and v_y on $\Gamma(x,y) = 0$. In this case

$$(a_1 c_2 - a_2 c_1) \left(\frac{dy}{dx}\right)^2 + (-a_1 d_2 + a_2 d_1 -$$

$$b_1 c_2 + b_2 c_2) \left(\frac{dy}{dx}\right) + (b_1 a_2 - b_2 d_1) = 0$$ (3.14)

or

$$A \left(\frac{dy}{dx}\right)^2 + B \left(\frac{dy}{dx}\right) + C = 0 ,$$ (3.15)

where

$$A = a_1 c_2 - a_2 c_1 ,$$ (3.16)

$$B = -a_1 d_2 + a_2 d_1 - b_1 c_2 + b_2 c_2$$ (3.17)

Discriminant	Type	Char. Families	Comput. Ease
$B^2 - 4AC > 0$	Hyperbolic	two	difficult
$B^2 - 4AC = 0$	Parabolic	one	less difficult
$B^2 - 4AC < 0$	Elliptic	zero	even less difficult

Table 3.2: Types of first-order PDEs.

and

$$C = b_1 a_2 - b_2 d_1. \tag{3.18}$$

The quadratic Equation (3.15) for $\frac{dy}{dx}$ yields

$$\frac{dy}{dx} = \frac{-B \pm \sqrt{B^2 - 4AC}}{2A}. \tag{3.19}$$

Equation (3.19) can be integrated to find $\Gamma(x, y) = 0$ along which $\Delta = 0$. It is along such curves that we are unable to determine u_x, u_y, v_x and v_y uniquely. Therefore, we are unable to use the Taylor series expansion to determine u at a location near the Γ curve. Note that the set of curves in the xy plane satisfying $\Gamma(x, y) = constant$ and Equation (3.19) defines a family of curves known as *characteristics* of the original set of PDEs.

First order PDEs are characterized by the nature of the discriminant in Equation (3.19) and, consequently, by the number of families of characteristics associated with the PDE. If $B^2 - 4AC > 0$, there are two distinct families of characteristics on which u_x, v_x, u_y and v_y are not uniquely determined. In this case, the system represented by Equations (3.4) and (3.5) is called *hyperbolic* . If $B^2 - 4AC = 0$, there is a single family of characteristics on which u_x, v_x, u_y and v_y are not uniquely determined, and the system is called *parabolic* . When $B^2 - 4AC < 0$, there are no families of characteristics on which u_x, v_x, u_y and v_y are not uniquely defined in the real domain. The system of Equations (3.4) and (3.5) is called *elliptic* in this case.

Table 3.2 summarizes the classification of first-order PDEs based on the discriminant appearing in Equation (3.19). PDEs associated with the majority of engineering problems are too complex to solve analytically (that is, where the solution process ends with a formula for the dependent variable as a function of the independent variables). In these cases, numerical methods are used to find approximate values for the dependent variable at a finite number of prescribed (discrete) locations within the problem domain. The last column in Table 3.2 indicates the ease in computing a solution to each type of equation. Note that the difficulty decreases as the number of families of characteristics decreases. Issues such as stability and convergence are important in numerical solution techniques. Often they depend on the order, linearity, and type of PDE being solved.

3.3 Second-Order Equations and Characteristics

Examples of typical second order PDEs encountered in fluid flow and heat transfer problems are

$$uT_x + vT_y = \frac{\partial[(C_T \frac{k^2}{\epsilon} + \alpha)\frac{\partial T}{\partial x}]}{\partial x} + \frac{\partial[(C_T \frac{k^2}{\epsilon} + \alpha)\frac{\partial T}{\partial y}]}{\partial y} + S \qquad (3.20)$$

and

$$uu_x + vu - y = \frac{1}{Re}(U_{xx} + u_{yy}) - P_x + \frac{1}{Fr}. \qquad (3.21)$$

These equations are specific examples of the general second-order equation

$$Au_{xx} + Bu_{xy} + Cu_{yy} = f(x, y, u, u_x, u_y) . \qquad (3.22)$$

The linearity of Equation (3.22) is determined by the following guidelines.

- If A, B, C are functions of x and y only, and f is linear in u, u_x and u_y, then Equation (3.22) is linear.

- Equation (3.22) is quasilinear if either of the following is true: (a) Any one of A, B or C is a function of u, u_x, or u_y and not a function of u_{xx}, u_{yy}, or u_{xy}. (b) f is nonlinear

- If either A, B or C is a function of u_{xx}, u_{yy}, or u_{xy} , then Equation (3.22) is nonlinear.

In order to solve Equation (3.22) the following question is posed: *Given u, u_x and u_y on a curve $\Gamma(x, y) = 0$, under what conditions are u_{xx}, u_{yy} and u_{xy} uniquely determined from Equation (3.22)?* The functions u, u_x and u_y on the curve Γ supply what is referred to as *Cauchy data* for the problem.

If the higher order derivatives can be determined from the Cauchy data on Γ, we can find a solution for u using a Taylor series expansion. That is, if P is an arbitrary point "near" the curve Γ, then

$$u_p = u|_\Gamma + u_x|_\Gamma \Delta x + u_y|_\Gamma \Delta y + u_{xx}|_\Gamma \frac{\Delta x^2}{2} + u_{xy}|_\Gamma \Delta x \Delta y + u_{yy}|_\Gamma \frac{\Delta y^2}{2} + ... \quad (3.23)$$

To examine when it is possible to solve for u_{xx}, u_{yy} and u_{xy} from Equation (3.22) with the Cauchy data given on $\Gamma(x, y) = 0$, assume that u is continuous and twice differentiable on Γ. If so, along $\Gamma(x, y) = 0$ it follows that

$$Au_{xx} + Bu_{xy} + Cu_{yy} = f , \qquad (3.24)$$

$$dxu_{xx} + dyu_{xy} + 0 = du_x \qquad (3.25)$$

and

$$0 + dxu_{xy} + dyu_{yy} = du_y. \qquad (3.26)$$

Discriminant	Type
$B^2 - 4AC > 0$	Hyperbolic
$B^2 - 4AC = 0$	Parabolic
$B^2 - 4AC < 0$	Elliptic

Table 3.3: Types of second-order PDEs.

Here, A, B, C, f, dx, dy, du_x, du_y are known along $\Gamma(x, y) = 0$. It is possible to solve for unique u_{xx}, u_{xy}, u_{yy} from the above three equations if

$$\Delta = \begin{vmatrix} A & B & C \\ dx & dy & 0 \\ 0 & dx & dy \end{vmatrix} \neq 0. \tag{3.27}$$

Unique solutions for u_{xx}, u_{xy} and u_{yy} on $\Gamma(x, y) = 0$ are not possible if $\Delta = 0$. Consequently, u_P cannot be determined using the Taylor series expansion. If $\Delta = 0$, then

$$A(\frac{dy}{dx})^2 - B(\frac{dy}{dx}) + C = 0 \tag{3.28}$$

and

$$\frac{dy}{dx} = \frac{-B \pm \sqrt{B^2 - 4AC}}{2A}. \tag{3.29}$$

It is possible to integrate Equation 3.29 to find $\Gamma(x, y) = 0$ along which $\Delta = 0$. As in the case of first-order PDEs, the family of curves satisfying $\Gamma(x, y) = constant$ and Equation (3.29) define a set of characteristics for the original PDE given in Equation (3.22).

The classification of Equation (3.22) is based on the value of the discriminant $B^2 - 4AC$ and, therefore, the number of families of characteristics associated with the equation. These classifications are summarized in Table 3.3.

If the discriminant in Equation (3.29) is positive, the PDE has two distinct families of characteristics, and the equation is called *hyperbolic*. If the Cauchy data is given along any of the characteristics, it is insufficient in defining a unique solution to the problem. Further, any characteristic from one family is nowhere tangent to any characteristic from the other family. The families may be used to define a new coordinate system on which the original PDE reduces to a simple *canonical* form (see Garabedian [71]).

If the discriminant in Equation (3.29) is zero, the PDE has a single family of characteristics on which the Cauchy data does not guarantee a unique solution. The PDE is classified as *parabolic* in this case. Finally, if the discriminant is negative, the PDE has no family of characteristics. Therefore, Cauchy data may be given on any curve in the xy plane and the Taylor series expansion will yield a solution for an arbitrary point near the curve. The PDE is called *elliptic* in this case.

The names hyperbolic, parabolic and elliptic come from the similarity of the PDE to the corresponding second order algebraic equation. Examples are shown in Table 3.3.

Coefficients	Type	PDE Example	Algebraic
$B = 1, A = 1, C = 2$	Elliptic	$u_{xx} + 2u_{yy} = f$	$x^2 + 2y^2 = constant$
$B = 0, A = 1, C = 0$	Parabolic	$u_{xx} = f$	$x^2 = y$
$B = 0, A = 1, C = -1$	Hyperbolic	$u_{xx} - u_{yy} = f$	$x^2 - y^2 = constant$

Table 3.4: Examples of second-order PDEs.

Classification of PDEs with more than two independent variables can be done by analyzing two variables at a time. For example, the equation

$$u_t + uu_x + vu_y = \nu(u_{xx} + u_{yy}) - \frac{1}{\rho}p_x \qquad (3.30)$$

is parabolic in (x,t) and (y,t), and elliptic in (x,y).

Chapter 4

Well-Posed Problems

This chapter discusses what it means for a problem involving a partial differential equation to be well-posed. After defining the three criteria for a well-posed problem in the first section, examples are given in the next three sections to illustrate these concepts. The rest of the chapter is devoted to discussion and examples of well-posed, and not well-posed (ill-posed), problems in fluid flows and heat transfer.

4.1 Well-Posed Problems

For the sake of discussion, suppose a problem consists of a PDE for u as a function of one or more spatial variables and time, with some information (data) about u at spatial and temporal boundaries. The problem is said to be *well-posed* if each the following is true:

1. a solution for u exists (*existence*),

2. a solution for u is unique (*uniqueness*),

3. a solution for u is stable (*stability*).

4.1.1 Example of Existence

The non-dimensional Navier-Stokes and energy equations for incompressible flows derived in Section 2.2.2 are listed below. Note the source term in the energy equation has been dropped, and T, not θ, represents the nondimensional temperature. Parameters Re and Pe are the Reynolds number and Peclet numbers, respectively.

$$\frac{\partial u_i}{\partial x_i} = 0 \tag{4.1}$$

$$\frac{Du_i}{Dt} = \frac{\partial p}{\partial x_i} + \frac{1}{Re}\frac{\partial^2 u_i}{\partial x_j \partial x_j} \tag{4.2}$$

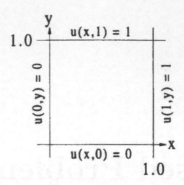

Figure 4.1: Boundary conditions ensure uniqueness.

$$\frac{DT}{Dt} = \frac{1}{Pe} \frac{\partial^2 T}{\partial x_j \partial x_j}$$ (4.3)

The existence of a solution for each of the five unknowns (u, v, w, p and T) is assured because there are five independent equations derived from independent conservation laws. However, proper initial and boundary conditions are required for a unique solution.

4.1.2   Example of Uniqueness

For purposes of illustration, consider the two-dimensional Laplace equation

$$u_{xx} + u_{yy} = 0,$$ (4.4)

on the domain defined by $0 \leq x \leq 1$ and $0 \leq y \leq 1$. Any linear function $u(x,y) = Ax + By + C$ (where A, B and C are constants) solves Equation (4.4). Without specifying sufficient boundary conditions, there is no *unique* solution to this problem.

Now, consider the boundary conditions shown in Figure 4.1 for Equation (4.4). Note that there are two conditions each for x and y. Consequently, a unique solution to Equation (4.4) can be determined to be $u(x,y) = x + y$.

A necessary condition for a solution to be uniquely determined for a given problem is the number of conditions specified for the dependent variable with respect to an independent variable most equal the highest order derivative of the dependent variable with respect to that independent variable. For example, the PDE

$$u_{xxx} + u_{yy} = 0$$

requires conditions for u, or derivatives of u, at three x locations and two y locations for any hope of determining a unique solution. The PDE

$$u_{xx} - u_y = 0$$

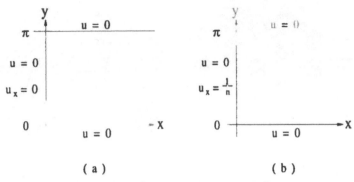

Figure 4.2: An example of instability.

requires specification of u, or derivatives of u, at two x locations and one y location. The condition on y in this example is like an "initial condition" even though y is a spatial variable. In most engineering applications, an initial condition is associated with the temporal variable t, and u is given for $t = 0$.

Existence and uniqueness of a solution are not sufficient to make a problem well-posed, as illustrated by the next example.

4.1.3 Example of Instability

Consider the following boundary value problem with the PDE

$$u_{xx} + u_{yy} = 0, \tag{4.5}$$

and boundary conditions as indicated in Figure 4.2 part (a). Note that $u = 0$ is a solution for this boundary value problem.

Next, consider the boundary value problem having the same PDE, Equation (4.5), and boundary conditions as indicated in Figure 4.2 part (b). The solution in the case is $u = \frac{1}{n^2} sinh(nx) sin(ny)$. Now, for large n, $u_x = \frac{1}{n} \to 0$, and the boundary value problem in Figure 4.2 part (b) will approach the boundary value problem in Figure 4.2 part (a). However, at some positive value of $x = \delta$, $\lim_{n \to \infty} u(\delta, y) = \infty$ inside the problem region. Thus, the problem is unstable in the sense that as the boundary condition at $x = 0$ in part (b) approaches the condition $u_x = 0$ as n becomes large, the corresponding solution to part (b) does *not* approach the solution to the problem in part (a).

The balance of this chapter focuses on the concept of well-posed problems in physics and engineering. Existence and uniqueness of solution for some specific problems are examined, and some ill-posed problems are discussed.

4.2 Existence and Physical Problems

If a physical or a engineering phenomenon is observed, then one can be sure that independent mathematical equations that properly model all dependent

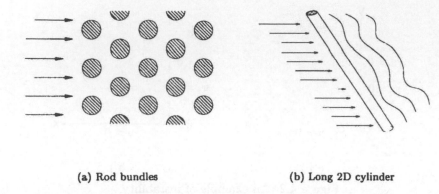

(a) Rod bundles   (b) Long 2D cylinder

Figure 4.3: Flow and existence

variables of the phenomenon should provide the existence of a realistic solution. If a phenomenon is unknown, the conservation laws (mass, momentum and energy) should provide the framework for equations for which a solution exists. However, there may be certain characteristics of the phenomenon one may not know *a priori*. In such cases, assumptions should be made with care.

As an example, for flows past rod bundles (see Figure 4.3(a)), one must use problem geometry along with the Reynolds and Prandtl numbers to decide if the flow is

- laminar or turbulent,

- steady or unsteady,

- two dimensional or three dimensional,

- periodic or irregular in time and spacing.

An investigator's understanding of the physics, along with any previous experience and experimental observation are often crucial in making proper assumptions.

Flow complexities arise primarily from the non-linear convective terms in the Navier-Stokes equations. Without the convective terms

- the flow will not be turbulent because the time averaged Navier-Stokes equations will not have Reynolds Stress terms $\overline{u_i u_j}$,

- there will be no vortex stretching and cascade phenomena ($\vec{w} \cdot \nabla \vec{v}$),

- there will be no Coriolis and centrifugal forces ($\frac{u^2}{r}, \frac{uv}{r}$).

Whether the flow is assumed to be laminar or turbulent is typically determined by experiment or stability analysis of laminar flow solutions. If the flow is not known to be steady it is safer to assume it is unsteady. The question of

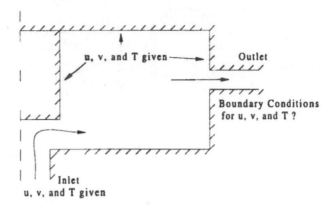

Figure 4.4: Heat exchanger.

spatial dependence (i.e. is the flow one-, two-, or three-dimensional) is usually determined from the problem geometry. However, this is not fool-proof as demonstrated by the flow past a long 2D cylinder. Figure 4.3(b) shows that this "two-dimensional" flow has, in fact, a three-dimensional feature.

4.3   Uniqueness and the Downstream Condition

As pointed out in Section 4.1.3, an elliptic PDE must be posed as a boundary value problem with two boundary conditions for each spatial variable. The same is true for the Navier-Stokes equations. However, one of the difficulties in solving the Navier-Stokes equations on some domains is the specification of a downstream boundary condition. The heat exchanger domain shown in Figure 4.4 is an example of such a problem.

There are certain principles one can use to overcome the difficulty in specifying downstream boundary conditions.

- **Avoid ambiguous locations.** Specify the downstream condition far removed from the main problem region. Coordinate transformations may make it possible to specify the downstream condition at infinity.

- **Use integral conditions.** The conservation laws, including mass and momentum, are used again. For example

$$m_{inlet} = m_{outlet} = \int_0^h \rho U_I \, dy, \tag{4.6}$$

$$q_{inlet} + q_{wall} + q_{outlet} = 0, \tag{4.7}$$

$$q_{outlet} = \int_0^h \rho C_P T U \, dy + \int_0^H , k \frac{\partial T}{\partial Y}\bigg|_0 \, dy \tag{4.8}$$

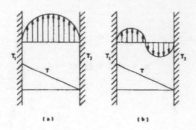

Figure 4.5: Convection between vertical plates.

$$q_{outlet} = -(q_{inlet} + q_{wall}).$$ (4.9)

• **Use fully developed conditions.** That is,

$$\frac{\partial u_i}{\partial x} = 0,$$ (4.10)

where u_i is any appropriate velocity component. This condition is frequently used if the downstream positions are sufficiently removed from the main problem region.

• **Use the no diffusion condition.** The second-order derivative terms ($\frac{\partial^2 u_i}{\partial x^2}$) in the Navier-Stokes equations make the equations elliptic. However, the convection terms are strong near the downstream region. Therefore, it is reasonable to assume $\frac{\partial^2 u_i}{\partial x^2} = 0$. That is, the problem becomes essentially parabolic near the outlet region.

4.4   Some Improperly Posed Problems

We conclude this chapter with two examples of ill-posed problems in engineering. When dealing with improperly posed problem, basic strategies include determining what part, if any, of the improperly posed solution is useful. In the case of multiple solutions, one may have to choose a "proper" solution using engineering and physical concepts. For more specifics about addressing ill-posed problems, please refer to the book by Larrentier [110].

Free convection between vertical plates , as shown in Figure 4.5, is an example of a physical phenomena that may result in an improperly posed problem. If the governing equations are properly given, and no boundary condition is give along the y direction, two different solutions are obtained. One solution has net flow (Figure 4.5 part (a)), another solution has zero net flow (Figure 4.5 part (b)).

Chapter 5

Numerical Methods

The mathematical equations for most fluid flow and heat transfer problems are too complex for analytical solution techniques. Even in simpler cases, numerical methods may be preferred. Numerical methods usually begin by dividing, or *discretizing*, the problem domain into a number of small subdomains, or *elements*, defined by grid lines. The solution for the governing equations is not found for every point in the problem domain. Instead, a solutions is determined for a finite number of discrete points called *nodes*. Nodes may be located at the intersection of grid lines (*cell-vertex nodes*), or inside the elements defined by the grid lines (*cell-centered nodes*).

A crucial step in any numerical method is the transformation of the PDE on an arbitrary subdomain into an algebraic equation . This transformation process is usually what distinguishes one numerical method from the another. The algebraic equation on each subdomain usually relates the value of the dependent variable at one interior node to values of the same variable at surrounding nodes. A system of equations is constructed when the equations for all, or a group, of subdomains are combined. This system is solved to provide the solution for the dependent variable at the discrete nodal locations.

The accuracy of the numerical solution depends, primarily, on each of these steps. That is,

- the number and arrangement of nodes (*grid generation*),

- the derivation of the algebraic equation (*numerical method*), and

- the solution to the linear system of algebraic equations (*algebraic solution method*).

The remainder of this chapter is devoted to brief introductions to various coordinate generation and numerical methods.

5.1   Grid Generation

There is no unified way of generating a computational grid. Often, it depends on the researcher's preference. The process usually begins by selecting a coordinate system (i.e., Cartesian, cylindrical, or spherical). For sake of simplicity, many researchers prefer a uniform Cartesian grid for any problem. This approach does not require *a priori* knowledge of the physical phenomena of the problem, which may be an advantage in some problems and a disadvantage in others. In some cases, domain geometry makes cylindrical or spherical coordinate systems appropriate. Techniques such as conformal mapping, boundary-fitting, unstructured grids, multigrids, or adaptive grids may be used for complex geometries as well.

There are guiding principles for choosing coordinate types and grid resolution. In terms of grid resolution, it is very important to have sufficient nodal density, or domain resolution, in regions where dependent variables have steep gradients. For example, the velocity, temperature and pressure may have large gradients in regions near walls, inlets and outlets, corners, and mixing or shear layers.

Proper choice of coordinate systems may alleviate some numerical simulation difficulties, especially those caused by *numerical diffusion* . It is advantageous to have coordinate lines running parallel to a particular boundary or flow direction.

5.2   Numerical Methods

Many numerical methods have some similarities in development and application. Depending upon the details of classification, there are three to eight different numerical methods one may use to solve a PDE. A brief introduction to three of these methods, the finite difference, the finite element, and the finite analytic, is given in this section. Each method has advantages and disadvantages. Often, the method used by a researcher depends strongly on the personal knowledge of that method.

Suppose $F(\phi) = g(x, y)$ represents a general PDE, where ϕ is a function of x and y, and F is either a linear or non-linear operator. As a means of introduction and comparison of the three methods, we will outline the steps involved in generating an counterpart algebraic equation to the PDE on a small subdomain.

5.2.1   Finite Difference Method

The *finite difference* (FD) numerical method is, historically, the first attempt at a numerical solution for a PDE. As the name implies, derivative terms in the PDE are approximated by differences in the respective variable over finite differences in the independent variable.

1. The problem domain is discretized into small subdomains as shown in Figure 5.1(a). The nodes marked by the solid circles are included in the

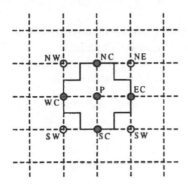

(a) FD subdomain

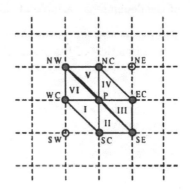

(b) FE subdomain

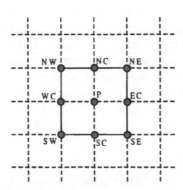

(c) FA subdomain

Figure 5.1: Subdomains for various methods.

element. The nodes with open circles are excluded.

2. Derivative terms in F are replaced with finite difference expressions.

3. Any nonlinear terms in F are linearized on the subdomain using the value of the dependent variable at node P.

4. The four nodes on the subdomain boundary are used to express u at node P as a function of the value of ϕ at those nodes. That is,

$$\phi_P = C_N\phi_N + C_E\phi_E + C_S\phi_S + C_W\phi_W.$$

The coefficients $C_N, C_E, ...$ in this equation are functions of the nodal spacing only.

The primary advantage of the FD method is the ease in constructing the algebraic equation to the PDE. Perhaps the major disadvantage of the FD method is that the resulting algebraic equation does not incorporate values of ϕ at the four corner nodes, NW, NE, SE, and SW. This may cause serious problems in convective flow situations when flow runs askew of the coordinate directions. Chapter 6 provides added detail on the FD method where the algebraic equation is derived for the unsteady, one-dimensional heat equation, and formulas for the coefficients are presented.

5.2.2   Finite Element Method

The *finite element* (FE) method has its origins in the fields of solid mechanics and structural analysis. Application of the FE method to fluid flow and heat transfer problems was enhanced by the first symposium [1] on the FE method devoted to fluid flow and heat transfer. A finite element formulation is the result of either the *calculus of variations* or *weighted residual methods*. The calculus of variations involves the equivalence of a differential equation and a related functional. That is, the function that minimizes the functional is exactly the function that solves the differential equation. One drawback of the variational approach is that not every differential equation has a related functional. The outline that follows is for the weighted residual method.

1. The problem domain is discretized into small subdomains as shown in Figure 5.1(b). Nodes marked with the solid circles are included in the element. Nodes marked by the open circles are excluded.

2. Any non-linear terms in the governing PDE are locally linearized using the value of the dependent variable at node P

3. Approximate ϕ using simple linearly independent *basis* or *shape* functions, ψ_i. $\phi \approx \sum_i \alpha_i\psi_i = Ax + By + C$. The coefficients A, B and C are determined by requiring that the approximate solution match ϕ at the nodes of the triangular region I in Figure 5.1(b). This means that the coefficients α_i are just the nodal values of ϕ.

4. The difference between the true and the approximate solution is called the *residual*, denoted by ϵ. That is, $\epsilon = F(Ax + By + C) - g(x, y)$.

5. Minimize the residual on region I by solving

$$\int_I w_i(x, y)\epsilon\, dx dy = 0$$

using "weight" functions $w_i(x, y)$. The form of the weight functions determines different procedures. The *Galerkin* procedure defines

$$w_i = \frac{\partial \phi}{\partial \alpha_i} = \psi_i.$$

6. Repeat the third through fifth steps for regions II-VI shown in Figure 5.1(b).

7. Form the sum of the integral expressions for the six regions of the FE subdomain and set it equal to zero. Solving this equation for ϕ_P gives an algebraic equation for this value in terms of the six nodal values of ϕ on the boundary of the subdomain. That is,

$$\phi_P = \sum_i C_i\phi_i, \qquad i = N, E, SE, ..., NW.$$

The number of neighboring nodes on the subdomain boundary depends on the shape functions used in the third step. The FE subdomain in Figure 5.1(b) has 6 boundary nodes included in the algebraic equation for u at node P.

The FE method varies in detail depending on the shape functions used in the third step. Advantages of the FE method include the natural incorporation of boundary conditions in the variational technique, and the ability of the triangular elements to approximate irregular boundaries. However, the FE method suffers from the same problem the FD method has when convective terms in a governing PDE are important. Note that the FE formulation outlined in this example results in an algebraic equation for ϕ at node P that does not include the value of u at node SW node. This means a strong flow from the SW would result in no influence of the u value at node SW on ϕ at node P, which is physically unrealistic. There are many excellent texts on the FE method. Among them are ones by Zienkiewicz [184], Finlayson [66], and Comini et al. [58].

The *control volume method* is another weighted residual method. The weight functions used in the third step are $w_i(x, y) = 1$. A very good source for details on the control volume method is the book by Patankar [145].

5.2.3 Finite Analytic Method

The *finite analytic* (FA) method was proposed by Chen et al., [40] in 1981. The method is based on finding an analytic solution to a linear, or linearized, PDE on a small FA subdomain using the method of separation of variables. The steps are outlined below.

1. The problem domain is discretized into small subdomains as shown in Figure 5.1(c). Nodes marked by the solid circles are included in the element. The open circles denote nodes that are excluded.

2. The governing PDE is linearized, if needed, by using a representative quantity on the small subdomain.

3. Proper boundary conditions are constructed using interpolation and three nodal quantities on each boundary.

4. The analytic solution to the linear PDE, with boundary conditions constructed in step three, is found using the method of separation of variables and the principle of superposition.

5. An algebraic equation relating ϕ at node P to ϕ at the eight surrounding nodes is derived by evaluating the analytic solution from the previous step at the center node P. The algebraic equation has the form

$$\phi_P = \sum_i C_i \phi_i, \qquad i = N, NE, EC, SE, ..., WC, NW.$$

The FA method is sometimes referred to as the *exact finite element method* because the analytic solution obtained in the FA method is not based on a prescribed shape function as in the FE method. One of the advantages of the FA method is that its algebraic equation for ϕ at P involves all eight surrounding nodes. Further, the "weight" given each node is a function of distance from P *and* the magnitude of any convective term. One drawback to the FA method is that the analytic solution involves calculating partial sums that include exponential terms. However, today's more powerful computers make coefficient calculation less of an issue, and the increased accuracy the FA method provides for highly convective flows outweighs computation cost concerns. A detailed development of the finite analytic numerical method is presented in Part II of the text.

Chapter 6

The Finite Difference Method

Finite difference methods are examined in detail in this chapter. The connection of finite difference methods to the definition of the derivative and Taylor's theorem is developed. Formulas for backward, centered, and forward differences are introduced, and a discussion of the accuracy of the methods is presented. Formulas for higher-order derivatives are given as well. The chapter concludes with the development of the finite difference formula for the unsteady, one-dimensional heat equation. The formulas for the finite difference coefficients will be the basis of a comparison with corresponding finite analytic coefficients presented in Chapter 8.

6.1 Discretization

The first step in a finite difference procedure for solving a PDE is to replace the continuous problem domain by a mesh, or grid, of discrete locations. As an example, suppose the objective is to solve a PDE on the domain Ω on which u is a function of x and y. The first step is to establish a grid of nodal locations on Ω by replacing $u(x, y)$ by $u(i\Delta x, j\Delta y)$. Points are located according to values of i and j, and difference equations are usually written in terms of the general point $P(i, j)$ and its neighbors.

Taking $u_{i,j}$ to be located at node P in Figure 5.1(a), the four "near neighbor" values for $u(x, y)$ can be expressed in terms of $u(x_o, y_o)$ as

$$u_{EC} = u_{i+1,j} = u(x_o + \Delta x, y_o),$$
$$u_{WC} = u_{i-1,j} = u(x_o - \Delta x, y_o),$$
$$u_{NC} = u_{i,j+1} = u(x_o, y_o + \Delta y) \quad \text{and}$$
$$u_{SC} = u_{i,j-1} = u(x_o, y_o - \Delta y).$$

Many different finite difference representations are possible for any given PDE.

47

It is usually impossible to establish a "best" form on an absolute basis. First of all, the accuracy of a difference scheme may depend on the exact form of the equation and the problem being solved. Secondly, the "best" scheme may be determined by the objective. That is, is the objective to optimize accuracy, economy or programming simplicity?

The idea of a finite difference representation for a derivative can be introduced by the definition of the partial derivative of u with respect to x at $x = x_o$ and $y = y_o$,

$$\frac{\partial u}{\partial x} = \lim_{\Delta x \to 0} \left[\frac{u(x_o + \Delta x, y_o) - u(x_o, y_o)}{\Delta x} \right]$$

$$= \lim_{\Delta x \to 0} \left[\frac{u_E - u_P}{\Delta x} \right].$$

If u is continuous and differentiable in a neighborhood of (x_o, y_o), it is expected that

$$\frac{u(x_o + \Delta x, y_o) - u(x_o, y_o)}{\Delta x}$$

will be a reasonable approximation to $\frac{\partial u}{\partial x}$ for a sufficiently small Δx.

The finite difference approximation can be put on a more formal basis through the use of Taylor's formula with a remainder. Developing a Taylor series expansion for $u(x_o + \Delta x, y_o)$ about $P_o = (x_o, y_o)$ gives

$$u(x_o + \Delta x, y_o) = u(x_o, y_o) + \left.\frac{\partial u}{\partial x}\right|_{P_o} \Delta x + \left.\frac{\partial^2 u}{\partial x^2}\right|_{P_o} \frac{\Delta x^2}{2} + \cdots$$

$$+ \left.\frac{\partial^{n-1} u}{\partial x^{n-1}}\right|_{P_o} \frac{\Delta x^{n-1}}{(n-1)!} + \left.\frac{\partial^n u}{\partial x^n}\right|_{\xi} \frac{\Delta x^n}{(n)!} \tag{6.1}$$

for $x_o \le \xi \le x_o + \Delta x$. The last term in Equation (6.1) can be identified as the remainder. The forward difference form for the derivative is found by rearranging Equation (6.1):

$$\left.\frac{\partial u}{\partial x}\right|_{P_o} = \frac{u(x_o + \Delta x, y_o) - u(x_o, y_o)}{\Delta x} - \left.\frac{\partial^2 u}{\partial x^2}\right|_{P_o} \frac{\Delta x}{2} - \cdots . \tag{6.2}$$

Switching now to the i, j notation for brevity, we consider

$$\left.\frac{\partial u}{\partial x}\right|_{i,j} = \frac{u_{i+1,j} - u_{i,j}}{\Delta x} + truncation\ error, \tag{6.3}$$

where $\frac{u_{i+1,j} - u_{i,j}}{\Delta x}$ is obviously the finite-difference representation for $\frac{\partial u}{\partial x}|_{i,j}$. The *truncation error* (TE) is the difference between the partial derivative and its finite difference representation. The limiting behavior of the truncation error is characterized by the *order of* (0) notation, whereby

$$\frac{\partial u}{\partial x}|_{i,j} = \frac{u_{i+1,j} - u_{i,j}}{\Delta x} + 0(\Delta x) = \frac{u_E - u_P}{\Delta x} + O(\Delta x). \tag{6.4}$$

Writing the truncation error as $0(\Delta x)$ means $|TE| \leq K\|\Delta x|$ for $\Delta x \to 0$ (Sufficiently small Δx), and some positive real constant K. As a practical matter, the *order* of the truncation error in this case is the largest power of Δx common to all terms in the truncation error.

A more general definition of the "O" notation is that $f(x) = O[\phi(x)]$ implies there exists a constant K, independent of x, such that $|f(x)| \leq K|\phi(x)|$ for all x in domain S, where f and ϕ are real or complex functions defined in S. Frequently, S is restricted by $x \to \infty$ (sufficiently large x), or as is most common in finite difference applications, $x \to 0$ (sufficiently small x). More details on the "O" notation can be found in the classic book by Whittaker and Watson [176].

Note that $O(\Delta x)$ gives little indication about the *exact* size of the truncation error, but rather how it behaves as Δx tends toward zero. If another difference expression is such that $TE = O(\Delta x^2)$, the truncation error of the latter representation would be smaller that the former for "small" Δx. It must be stressed that we are assured this is true only if the grid is "sufficiently" refined.

6.2 Central, Backward and Forward Differences

Various representations can be found for $\frac{\partial u}{\partial x}|_{i,j}$. For example we could expand "backwards"

$$u(x_o - \Delta x, y_o) =$$
$$u(x_o, y_o) - \frac{\partial u}{\partial x}\Big|_{P_o} \Delta x + \frac{\partial^2 u}{\partial x \partial x}\Big|_{P_o} \frac{\Delta x^2}{2} - \frac{\partial^3 u}{\partial x^3}\Big|_{P_o} \frac{\Delta x^3}{6} + \cdots, \quad (6.5)$$

and obtain the *backwards difference* representation

$$\frac{\partial u}{\partial x}\Big|_{i,j} = \frac{u_{i,j} - u_{i-1,j}}{\Delta x} + 0(\Delta x). \quad (6.6)$$

Further, we can subtract Equation 6.5 from Equation 6.1, rearrange, and obtain the *central difference*

$$\frac{\partial u}{\partial x}\Big|_{i,j} = \frac{u_{i+1,j} - u_{i-1,j}}{2\Delta x} + 0((\Delta x)^2). \quad (6.7)$$

An approximation to the second derivative of u with respect to x is obtained by adding Equations 6.1 and 6.5, then rearranging to give

$$\frac{\partial^2 u}{\partial x^2}\Big|_{i,j} = \frac{u_{i+1,j} - 2u_{i,j} + u_{i-1,j}}{(\Delta x)^2} + 0((\Delta x)^2). \quad (6.8)$$

It should be emphasized these are only a few examples of the many ways first and second derivatives may be approximated.

It is convenient to utilize difference operators to represent specific finite difference expressions when particular forms are used repeatedly. Define the first forward difference of $u_{i,j}$ with respect to x at the point (i,j) as

$$\Delta_x u_{i,j} = u_{i+1,j} - u_{i,j}. \tag{6.9}$$

Thus, we can express the forward difference approximation for the first partial derivative as

$$\left.\frac{\partial u}{\partial x}\right|_{i,j} = \frac{u_{i+1,j} - u_{i,j}}{\Delta x} + 0(\Delta x) = \frac{\Delta_x u_{i,j}}{\Delta x} + 0(\Delta x). \tag{6.10}$$

Similarly, derivative with respect to another variable, such as y, can be represented by

$$\frac{\Delta_y u_{i,j}}{\Delta y} = \frac{u_{i+1,j} - u_{i,j}}{\Delta y}. \tag{6.11}$$

The first backward difference of $u_{i,j}$ with respect to x at (i,j) is denoted by

$$\nabla_x u_{i,j} = u_{i,j} - u_{i-1,j}. \tag{6.12}$$

It follows that the first backward difference approximation to the first derivative can be written as

$$\left.\frac{\partial u}{\partial x}\right|_{i,j} = \frac{u_{i,j} - u_{i-1,j}}{\Delta x} + 0(\Delta x) = \frac{\nabla_x u_{i,j}}{\Delta x} + 0(\Delta x). \tag{6.13}$$

The central difference operators $\bar{\delta}$, δ, and δ^2 are defined as

$$\bar{\delta}_x u_{i,j} = u_{i+1,j} - u_{i-1,j} , \tag{6.14}$$

$$\delta_x u_{i,j} = u_{i+\frac{1}{2},j} - u_{i-\frac{1}{2},j} \tag{6.15}$$

and

$$\delta_x^2 u_{i,j} = \delta_x(\delta_x u_{i,j}) = u_{i+1,j} - 2u_{i,j} + u_{i-1,j} , \tag{6.16}$$

respectively. An averaging operator, μ, is defined as

$$\mu_x u_{i,j} = \frac{u_{i+\frac{1}{2},j} + u_{i-\frac{1}{2},j}}{2} . \tag{6.17}$$

It is convenient to have specific operators for common central differences although two of them can be easily expressed in terms of first-difference operators

$$\bar{\delta}_x u_{i,j} = \Delta_x u_{i,j} + \nabla_x u_{i,j} \tag{6.18}$$

and

$$\bar{\delta}_x^2 u_{i,j} = \Delta_x u_{i,j} - \nabla_x u_{i,j} = \Delta_x \nabla_x u_{i,j}. \tag{6.19}$$

Using the newly defined operators, the central difference representation for the first partial derivative can be written as

$$\left.\frac{\partial u}{\partial x}\right|_{i,j} = \frac{u_{i+1,j} - u_{i-1,j}}{2\Delta x} + 0((\Delta x)^2) = \frac{\bar{\delta}_x u_{i,j}}{2\Delta x} + 0((\Delta x)^2), \tag{6.20}$$

and the central difference representation of the second derivative as

$$\frac{\partial^2 u}{\partial x^2}\bigg|_{i,j} = \frac{u_{i+1,j} - 2u(i,j) + u_{i-1,j}}{(\Delta x)^2} + 0((\Delta x)^2) = \frac{\delta_x^2 u_{i,j}}{(\Delta x)^2} + 0((\Delta x)^2). \quad (6.21)$$

Higher-order forward and backward difference operators are defined as

$$\delta_x^n u_{i,j} = \Delta_x(\Delta_x^{n-1} u_{i,j}) \quad (6.22)$$

and

$$\nabla_x^n u_{i,j} = \nabla_x(\nabla_x^{n-1} u_{i,j}). \quad (6.23)$$

As an example, a forward second derivative approximation is given by

$$\begin{aligned}
\frac{\delta_x^2 u_{i,j}}{(\Delta x)^2} &= \frac{\Delta_x(u_{i+1,j} - u_{i,j})}{(\Delta x)^2} \\
&= \frac{u_{i+2,j} - 2u(i+1,j) + u_{i,j}}{(\Delta x)^2} \\
&= \frac{\partial^2 u}{\partial x^2}\bigg|_{i,j} + 0(\Delta x).
\end{aligned}$$

Forward and backward difference approximations to derivatives of any order can be obtained from

$$\frac{\partial^n u}{\partial x^n}_{i,j} = \frac{\delta_x^n u_{i,j}}{\Delta x}^n + 0(\Delta x) \quad (6.24)$$

and

$$\frac{\partial^n u}{\partial x^n}_{i,j} = \frac{\nabla_x^n u_{i,j}}{\Delta x}^n + 0(\Delta x). \quad (6.25)$$

Central difference representations of derivatives of order three and greater may be expressed in terms of Δ, ∇ or δ. A more complete development on the use of difference operators can be found in many textbooks on numerical analysis such as Hildebrand [82].

Most of the PDEs arising in fluid mechanics and heat transfer involve first and second partial derivatives only. In the majority of cases these derivatives are approximated using values of the dependent variable at only two or three grid points. With these restrictions, the most frequently used first-derivative approximations on a *uniform grid* ($\Delta x = h = constant$) are

$$\frac{\partial u}{\partial x}\bigg|_{i,j} = \frac{u_{i+1,j} - u_{i,j}}{h} + 0(h),$$

$$\frac{\partial u}{\partial x}\bigg|_{i,j} = \frac{u_{i,j} - u_{i-1,j}}{h} + 0(h),$$

$$\frac{\partial u}{\partial x}\bigg|_{i,j} = \frac{u_{i+1,j} - u_{i-1,j}}{2h} + 0(h^2),$$

$$\frac{\partial u}{\partial x}\bigg|_{i,j} = \frac{-3u_{i,j} + 4u_{i+1,j} - u_{i+2,j}}{2h} + 0(h^2),$$

Derivative	Finite Difference Representation
$\dfrac{\partial^3 u}{\partial x^3}_{i,j}$	$\dfrac{u_{i+2,j}-2u_{i+1,j}+2u_{i-1,j}-u_{i-2,j}}{2h^3}+0(h^2)$
$\dfrac{\partial^4 u}{\partial x^4}_{i,j}$	$\dfrac{u_{i+2,j}-4u_{i+1,j}+6u_{i,j}-4u_{i-1,j}+u_{i-2,j}}{h^4}+0(h^2)$
$\dfrac{\partial^2 u}{\partial x^2}_{i,j}$	$\dfrac{-u_{i+3,j}+4u_{i+2,j}-5u_{i+1,j}+2u_{i,j}}{h^2}+0(h^2)$
$\dfrac{\partial^3 u}{\partial x^3}_{i,j}$	$\dfrac{-3u_{i+4,j}+14u_{i+3,j}-24u_{i+2,j}+18u_{i+1,j}-5u_{i,j}}{2h^3}+0(h^2)$
$\dfrac{\partial^2 u}{\partial x^2}_{i,j}$	$\dfrac{2u_{i,j}-5u_{i-1,j}+4u_{i-2,j}-u_{i-3,j}}{h^2}+0(h^2)$
$\dfrac{\partial^3 u}{\partial x^3}_{i,j}$	$\dfrac{5u_{i,j}-18u_{i-1,j}+24u_{i-2,j}-14u_{i-3,j}+3u_{i-4,j}}{2h^3}+0(h^2)$
$\dfrac{\partial u}{\partial x}_{i,j}$	$\dfrac{-u_{i+2,j}+8u_{i+1,j}-8u_{i-1,j}+u_{i-2,j}}{12h}+0(h^4)$
$\dfrac{\partial^2 u}{\partial x^2}_{i,j}$	$\dfrac{-u_{i+2,j}+16u_{i+1,j}-30u_{i,j}+16u_{i-1,j}-u_{i-2,j}}{12h^2}+0(h^4)$

Table 6.1: Difference approximations using more than three points.

$$\left.\frac{\partial u}{\partial x}\right|_{i,j} = \frac{3u_{i,j}-4u_{i-1,j}+u_{i+2,j}}{2h}+0(h^2) \quad \text{and}$$

$$\left.\frac{\partial u}{\partial x}\right|_{i,j} = \frac{1}{2h}\left(\frac{\delta_x \bar{u}_{i,j}}{1+\delta_x^2/6}\right)+0(h^4).$$

The most common three-point second derivative approximations for a uniform grid, $\Delta x = h = constant$ are

$$\left.\frac{\partial^2 u}{\partial x^2}\right|_{i,j} = \frac{u_{i,j}-2u_{i+1,j}+u_{i+2,j}}{h^2}+0(h),$$

$$\left.\frac{\partial^2 u}{\partial x^2}\right|_{i,j} = \frac{u_{i,j}-2u_{i-1,j}+u_{i-2,j}}{h^2}+0(h),$$

$$\left.\frac{\partial^2 u}{\partial x^2}\right|_{i,j} = \frac{u_{i+1,j}-2u_{i,j}+u_{i-1,j}}{h^2}+0(h^2) \quad \text{and}$$

$$\left.\frac{\partial^2 u}{\partial x^2}\right|_{i,j} = \frac{\delta_x^2 u_{i,j}}{h^2(1+\delta_x^2/12)}+0(h^4).$$

Table 6.1 gives finite difference approximations when more than three points are used, and Table 6.2 lists the finite difference approximations for mixed partial derivatives.

6.3   Unsteady, One-Dimensional Heat Equation

In this section, we derive the finite difference algebraic equation for the unsteady, one-dimensional heat equation. Not only will this show details of the process, but the formulas derived for the FD coefficients will be compared with the finite analytic coefficients for the same equation in Section 8.

The dimensionless heat equation, derived in Section 2.2.4, is shown in Equation (6.26) for the case of one spatial variable x. The symbol ϕ represents the

Derivative	Finite Difference Representation
$\frac{\partial^2 u}{\partial x \partial y}\Big|_{i,j}$	$\frac{1}{\Delta x}\left[\frac{u_{i+1,j}-u_{i+1,j}}{\Delta y} - \frac{u_{i,j}-u_{i,j-1}}{\Delta y}\right] + 0(\Delta x, \Delta y)$
$\frac{\partial^2 u}{\partial x \partial y}\Big|_{i,j}$	$\frac{1}{\Delta x}\left[\frac{u_{i,j+1}-u_{i,j}}{\Delta y} - \frac{u_{i-1,j+1}-u_{i-1,j}}{\Delta y}\right] + 0(\Delta x, \Delta y)$
$\frac{\partial^2 u}{\partial x \partial y}\Big|_{i,j}$	$\frac{1}{\Delta x}\left[\frac{u_{i,j}-u_{i,j-1}}{\Delta y} - \frac{u_{i-1,j}-u_{i-1,j-1}}{\Delta y}\right] + 0(\Delta x, \Delta y)$
$\frac{\partial^2 u}{\partial x \partial y}\Big|_{i,j}$	$\frac{1}{\Delta x}\left[\frac{u_{i+1,j+1}-u_{i+1,j}}{\Delta y} - \frac{u_{i,j+1}-u_{i,j}}{\Delta y}\right] + 0(\Delta x, \Delta y)$
$\frac{\partial^2 u}{\partial x \partial y}\Big|_{i,j}$	$\frac{1}{\Delta x}\left[\frac{u_{i+1,j+1}-u_{i+1,j-1}}{2\Delta y} - \frac{u_{i,j+1}-u_{i,j-1}}{2\Delta y}\right] + 0(\Delta x, \Delta y^2)$
$\frac{\partial^2 u}{\partial x \partial y}\Big|_{i,j}$	$\frac{1}{\Delta x}\left[\frac{u_{i,j+1}-u_{i,j-1}}{2\Delta y} - \frac{u_{i-1,j+1}-u_{i-1,j-1}}{2\Delta y}\right] + 0(\Delta x, \Delta y^2)$
$\frac{\partial^2 u}{\partial x \partial y}\Big|_{i,j}$	$\frac{1}{2\Delta x}\left[\frac{u_{i+1,j+1}-u_{i+1,j-1}}{2\Delta y} - \frac{u_{i-1,j+1}-u_{i-1,j-1}}{2\Delta y}\right] + 0(\Delta x^2, \Delta y^2)$
$\frac{\partial^2 u}{\partial x \partial y}\Big|_{i,j}$	$\frac{1}{2\Delta x}\left[\frac{u_{i+1,j+1}-u_{i+1,j}}{\Delta y} - \frac{u_{i-1,j+1}-u_{i-1,j}}{\Delta y}\right] + 0(\Delta x^2, \Delta y)$
$\frac{\partial^2 u}{\partial x \partial y}\Big|_{i,j}$	$\frac{1}{2\Delta x}\left[\frac{u_{i+1,j}-u_{i+1,j-1}}{\Delta y} - \frac{u_{i-1,j}-u_{i-1,j-1}}{\Delta y}\right] + 0(\Delta x^2, \Delta y)$

Table 6.2: Difference approximations for mixed partial derivatives.

dimensionless temperature, and Pe is the Peclet number.

$$\frac{\partial \phi}{\partial t} + u\frac{\partial \phi}{\partial x} = \frac{1}{Pe}\frac{\partial^2 \phi}{\partial x^2} \tag{6.26}$$

If $u = 0$ and $Pe = 1$, Equation (6.26) simplifies to Equation (6.27).

$$\frac{\partial \phi}{\partial t} = \frac{\partial^2 \phi}{\partial x^2} \tag{6.27}$$

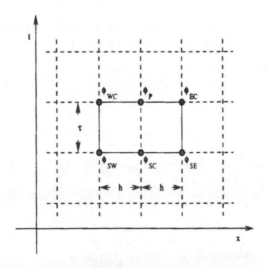

Figure 6.1: FD subdomain for one-dimensional heat equation.

The FD subdomain for this equation is shown in Figure 6.1. The spacing is uniform in x, given by h, and t, represented by τ. The FD algebraic version of

Equation (6.27) on this representative subdomain is constructed by replacing the two partial derivatives in the equation with FD approximations. For the unsteady term,

$$\frac{\partial \phi}{\partial t} \approx \frac{\phi_P - \phi_{SC}}{\tau}. \tag{6.28}$$

The second partial is approximated as

$$\frac{\partial^2 \phi}{\partial x^2} \approx \frac{\phi_x^+ - \phi_x^-}{h}, \tag{6.29}$$

where

$$\phi_x^+ = \frac{(1-\theta)\phi_{SE} + \theta\phi_{EC} - [(1-\theta)\phi_{SC} + \theta\phi_P]}{h}, \tag{6.30}$$

and

$$\phi_x^- = \frac{(1-\theta)\phi_{SC} + \theta\phi_P - [(1-\theta)\phi_{SW} + \theta\phi_{WC}]}{h}. \tag{6.31}$$

The expression θ in Equations (6.30) and (6.31) ranges in value from zero to one. If $\theta = 0$, the resulting equation is fully *explicit*. That is, the values used in the FD quotients are taken as "known" quantities from the previous time step. If $\theta = 1$, the resulting formula is fully *implicit*, and the quantities are considered as "unknowns" from the current time step. Intermediate values of θ result in a formula that is *semi-implicit*.

If Equations (6.30) and (6.31) are substituted into Equation (6.29), the following FD formula for the second partial results.

$$\frac{\partial^2 \phi}{\partial x^2} \approx \frac{1}{h^2}[(1-\theta)\phi_{SE} + \theta\phi_{EC} - [(1-\theta)\phi_{SC} + \theta\phi_P]$$
$$-[(1-\theta)\phi_{SC} + \theta\phi_P - [(1-\theta)\phi_{SW} + \theta\phi_{WC}]]]. \tag{6.32}$$

When the right-hand sides of Equation (6.28) and (6.32) are substituted into Equation (6.27), the following FD equation results.

$$\frac{\phi_P - \phi_{SC}}{\tau} = \frac{1}{h^2}\left[(1-\theta)\phi_{SE} + \theta\phi_{EC} \right. \tag{6.33}$$
$$-2(1-\theta)\phi_{SC}$$
$$\left. -2\theta\phi_P + (1-\theta)\phi_{SW} + \theta\phi_{WC}\right].$$

Equation (6.34) may be simplified by collecting like terms of ϕ, letting $\lambda = \frac{\tau}{h^2}$ and solving for ϕ_P. The result is

$$\phi_P = C_{SW}\phi_{SW} + C_{SC}\phi_{SC} + C_{SE}\phi_{SE} + C_{WC}\phi_{WC} + C_{EC}\phi_{EC}, \tag{6.34}$$

where

$$C_{SW} = C_{SE} = \frac{\lambda(1-\theta)}{(1+2\lambda\theta)},$$

$$C_{SC} = \frac{[1 - 2\lambda(1-\theta)]}{1 + 2\lambda\theta} \quad \text{and} \tag{6.35}$$

$$C_{WC} = C_{EC} = \frac{\lambda\theta}{(1+2\lambda\theta)}.$$

Note that Equation (6.34) gives the value of ϕ_P as a function of the five surrounding nodes in the FD subdomain. The coefficients, or *weights*, of the locations are functions of $\lambda = \frac{\tau}{h^2}$ and θ, the later is a measure of how implicit the scheme. When $\theta = \frac{1}{2}$, the resulting coefficients are those for the *Crank-Nicolson* formula. The *Forsythe and Wasow* formula for the coefficients results when $\theta = \frac{(6\lambda-1)}{12\lambda}$. The coefficient formulas shown in Equation (6.34) are plotted as a function of λ for specific θ values and compared with finite analytic values for the same problem in Section 8.

6.4   Error and Stability

6.4.1   Error Types

A numerical technique such as the finite difference method calculates an approximation ϕ^* to the true solution ϕ in almost all cases. The *error* is the difference between the approximate and true solutions, $\phi^* - \phi$. There are at least two sources for the error. The truncation error is that which is created by approximating a continuous or infinite operation with a finite approximation. As outlined in section 6.1, the truncation error in the finite difference method is found by considering the terms 'truncated" from the Taylor series expansion used to express a given derivative. As shown in section 6.1, the truncation error is typically proportional to the temporal or spatial step size.

Round-off error results when the available number of decimal places used to represent a number is less than the places required (representing $\frac{1}{3}$ by 0.3333333, for example). The storage capacity for number representation in modern computing machinery usually means round-off errors are relatively small compared to truncation error. However, it is possible that the effect of round-off error becomes very significant as the number of calculations increase. Consequently, care must be used in decreasing step size (either spatially or temporally) in an effort to reduce truncation error.

Numerical algorithms usually involve many steps or iterations in which truncation error, round-off error and other inaccuracies may accumulate. Numerical analysts refer to this as *accumulation error*. Accumulation error is difficult to isolate and measure. It is conceivable that different sources of error may offset each other instead of simply combining to increase the error.

6.4.2   Stability

The *stability* of a numerical method concerns how any form of error, or perturbation, behaves as the numerical procedure continues. A numerical method is said to be *stable* if the maximum value of $\phi^* - \phi$ over the computation domain goes to zero (and does not increase exponentially) with increasing calculations.

The stability of a finite difference representation of a given PDE with prescribed boundary conditions may be investigated in in at least two different ways. One method is to consider the finite difference representation of both the

PDE and boundary condition in a matrix form for which eigenvalue analysis is used to study stability. A second method uses Fourier series representation of possible errors to track its behavior with respect to the finite difference representation of the PDE only. This analysis technique is often referred to as the *von Neumann* method.

von Neumann stability analysis for the finite difference representation of one-dimensional heat equation reveals the explicit form is conditionally stable; the required condition being $\frac{\tau}{h^2} \leq \frac{1}{2}$. The implicit form is *unconditionally* stable. Semi-implicit forms (see Equations 6.36) with $\frac{1}{2} \leq \theta < 1$ are unconditionally stable. Note this includes the Crank-Nicolson method, with $\theta = \frac{1}{2}$. Conditional stability results for $0 \leq \theta < \frac{1}{2}$, with the requirement

$$\frac{\tau}{h^2} \leq \frac{1}{2(1 - 2\theta)} \ .$$

Additional detail concerning von Neumann stability analysis may be found in the book by Özişik [143].

6.5   Two-Dimensional Heat Equation

The two-dimensional heat equation is considered in this section as a way of introducing an important solution technique called the *alternating-direction implicit* method. It is an effort to blend the stability inherent in fully implicit methods with the minimal resource needs of an explicit method. This is especially beneficial in the case of multidimensional applications.

The two-dimensional heat equation, in the absence of motion, reduces to

$$\frac{\partial \phi}{\partial t} = \frac{1}{Pe} \left(\frac{\partial^2 \phi}{\partial x^2} + \frac{\partial^2 \phi}{\partial y^2} \right), \tag{6.36}$$

where ϕ represents temperature, and Pe is the Peclet number. Notation is simplified using the following convention:

$$\phi(x, y, t) = \phi(i\Delta x, j\Delta y, n\Delta t) = \phi_{i,j}^n.$$

The fully implicit finite difference representation of Equation 6.36 is (letting $Pe = 1$ for sake of argument)

$$\frac{\phi_{i,j}^{n+1} - \phi_{i,j}^n}{\Delta t} = \frac{\phi_{i-1,j}^{n+1} - 2\phi_{i,j}^{n+1} + \phi_{i+1,j}^{n+1}}{(\Delta x)^2} + \frac{\phi_{i,j-1}^{n+1} - 2\phi_{i,j}^{n+1} + \phi_{i,j+1}^{n+1}}{(\Delta y)^2}. \tag{6.37}$$

This formulation treats ϕ at *every* nodal location as an unknown, denoted by superscripts of "$n + 1$". Consequently, a system of N^2 linear equations must be solved simultaneously for a grid of size $N \times N$. The resulting $N^2 \times N^2$ matrix is sparse, but not banded, so that it must be solved in either a direct or indirect way. This requires significant computer resources for applications requiring many nodal locations.

In an effort to reduce the size and complexity of the resulting linear system without reverting to the conditionally stable explicit formulation, Peaceman and Rachford [149] introduced the alternating-direction implicit (ADI) method. With their approach, a single row or column of nodes from the computational grid is considered to be unknown. It is on this isolated row or column where calculation of the unknown temperature for time step $n + 1$ is focused. The resulting finite difference representation in the ADI approach is

$$\frac{\phi_{i,j}^{n+1} - \phi_{i,j}^{n}}{\Delta t} = \frac{\phi_{i-1,j}^{n+1} - 2\phi_{i,j}^{n+1} + \phi_{i+1,j}^{n+1}}{(\Delta x)^2} + \frac{\phi_{i,j-1}^{n} - 2\phi_{i,j}^{n} + \phi_{i,j+1}^{n}}{(\Delta y)^2}, \qquad (6.38)$$

where updated values for ϕ are determined for row j. When all such equations for row j are collected, a system of N equations results with the only non-zero entries filling the three central diagonals. Consequently, the majority of the entries in the in the $N \times N$ coefficient matrix need not be stored, significantly reducing computer resource needs. Additionally, the banded form of the matrix allows for a relatively simple solution procedure.

Because not all the nodal locations are updated simultaneously, and the value at one nodal location depends on the value at its four near neighbors (not just those on the row or column under consideration), an iterative sweep must be made of the rows or columns until subsequent iterates change less than some prescribed parameter. Once the convergence parameter is satisfied for all nodes, the solution process for ϕ^{n+1} at all nodes is complete. The process is repeated for the next time step, with one exception. If rows are swept in one time step, the columns of nodes are swept in the next, hence the term "alternating" in the method title.

Peaceman and Rachford [149] found the ADI method for 2D heat transfer flows is stable for all time steps Δt. However, the method appears to be unstable for certain values of Δt in three-dimensional applications [91]

The ADI method will be incorporated in the solution algorithm based on the finite analytic method presented in part II of the book.

6.6   Exercises for Part I

1. Classify the following partial differential equation:

$$F_{zz} + y^3 F_{yy} = F_z + FF_y + x^2 - y^3.$$

   Are there any real characteristics? If so, find the equation of the characteristics and sketch several of them.

2. Verify that

$$\frac{\partial^2 f}{\partial x^2} \doteq \frac{1}{h^2}\left(-f_{i+3} + 4f_{i+2} - 5f_{i+1} + 2f_i\right),$$

   and that the expression has an error of order h^2.

3. Derive the vorticity-streamfunction equations for two-dimensional, incompressible fluid flow from the Navier-Stokes equations:

$$u_x + v_y = 0,$$

$$u_t + u \cdot u_x + v \cdot u_y = -p_x + \frac{1}{Re}\left(u_{xx} + u_{yy}\right),$$

$$v_t + u \cdot v_x + v \cdot v_y = -p_y + \frac{1}{Re}\left(v_{xx} + v_{yy}\right).$$

4. (a) Show that a 9 point FD approximation for the elliptic PDE $u_{xx} = u_{yy} = 0$ on a uniform grid is

$$u_E + u_W + u_N + u_S + 4(u_{NE} + u_{NW} + u_{SE} + u_{SW}) - 20u_p = 0.$$

   (b) Show that the truncation error for the above equation is of order $O(h^6)$.

5. Establish the truncation error of the following finite difference approximation to u_y at $p_{i,j}$ for a uniform mesh:

$$u_y \doteq \frac{-3u_{i,j} + 4u_{i,j+1} - u_{i,j+2}}{2\Delta y}.$$

   What is the order of the truncation error?

6. Show that the explicit finite difference equation for

$$\frac{\partial f}{\partial t} - \frac{\partial^2 f}{\partial x^2} \doteq \frac{1}{k}\left(f_{i,t+1} - f_{i,t}\right) - \frac{1}{h^2}\left(f_{i+1,t} - 2f_{i,t} + f_{i-1,t}\right)$$

   has an accuracy of $O(k^2) + O(h^4)$ provided the ratio k/h^2 is chosen to be 1/6.

7. Describe an *upstream* (i.e., *windward*) difference. When and why is it used?

8. Find and sketch characteristics of the PDE $u_t + t \cdot u_t = \sin x$.

9. (a) Derive a finite difference approximation using the Crank-Nicolson method for $\phi_t = \phi_{xx} + \phi_{yy}$.

   (b) What method would propose to use to solve the resulting system of finite difference algebraic equations? Explain why.

10. Explain what it means for a numerical method to be *strongly implicit*.

11. (a) Find a five point finite difference approximation for $T_{xx} + T_{yy} = 0$ in a nonuniform grid system shown in Figure 6.2(a).

    (b) What is the accuracy of the difference equation?

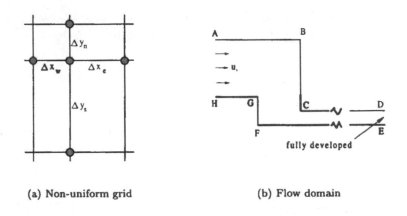

(a) Non-uniform grid   (b) Flow domain

Figure 6.2: Figures for exercises 11 and 12.

12. As an engineer you are asked to solve the two dimensional channel flow as depicted if Figure 6.2(b). The inlet velocity at $\overline{AH}$ is uniformly given as u_i. The flow is considered to be fully developed at the outlet $\overline{DE}$ The governing equations for incompressible laminar flow are

$$
\begin{aligned}
u_x + u_y &= 0, \\
\rho(u \cdot u_x + v \cdot u_y) &= -p_x + \mu(u_{xx} + u_{yy}) \quad \text{and} \\
\rho(u \cdot v_x + v \cdot v_y) &= -p_y + \mu(v_{xx} + v_{yy}).
\end{aligned}
$$

Formulate a well-posed problem for this channel flow.

13. Classify the following equations, and determine if there are any real characteristics. If so, find the equation for the characteristics and sketch several.

    (a) $F_{xx} + y^3 F_{yy} = F_x + FF_y + x^2 - y^3$.

(b) $F_{xx} - x^2 F_{yy} = (F_x)^2 + x^2 - y^2$.

14. (Computer Exercise) Consider the following governing PDE:

$$u_{xx} + u_{yy} - u_x - u_y = 0,$$

with boundary conditions specified in Figure 6.3(a). Write a computer program to solve the given problem using finite difference methods and line by line iteration on a uniform 11×11 grid. Do so for each of the following cases:

(a) Central differences for u_{xx}, u_{yy}, u_x and u_y.

(b) Central differences for u_{xx} and u_{yy}, and backward differences for u_x and u_y.

Print results for $u(x, y)$ in an 11×11 table for both cases. Provide a contour plot as well specifying levels for u of 0.0, 0.2, 0.4, 0.6, 0.8 and 1.0.

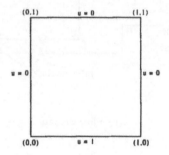

(a) Domain and boundary conditions

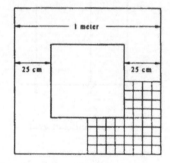

(b) Square chimney

Figure 6.3: Figures for exercises 14 and 15.

15. (Computer Exercise) A chimney manufacturer is interested in predicting the heat loss through it brick chimney walls. The cross-section geometry of the chimney under consideration is shown below in Figure 6.3(b).

The manufacturer claims that the steady-state heat loss will increase as the overall temperature difference increases and as the thermal conductivity, density, and specific heat of the wall increase. However, he has no quantitative data to substantiate his claim.

Suppose the temperature on the inside of the chimney is $325°$ C, the heat transfer coefficient on the outside is 100 $W/(m^2\,°K)$, and the ambient air temperature is $25°$ C. Additionally, suppose the following numbers are associated with the brick chimney walls: $K = 4W/(m^2\,°K)$, $\rho = 2000kg/m^3$, and $C = 480J/(kg°K)$. Complete the following.

(a) Calculate the heat loss per meter of chimney height.

(b) Plot the isotherms using a an interval of $\Delta T = 25^\circ C$.

(c) Calculate the change in heat loss if the conductivity of the material is decreased by 20%.

Describe your method of computation, and be sure to check grid independence with a tolerance of 10^{-3} degrees. Calculations need only to be made on the area designated by the grid in Figure 6.3(b). Why?

16. (Computer Exercise) Consider the following PDE for vorticity ω:

$$\frac{\partial \omega}{\partial t} + u \cdot \frac{\partial \omega}{\partial x} = \nu \frac{\partial \omega}{\partial x^2} \, ,$$

with initial and boundary conditions

$$\begin{array}{lll} \text{I.C.:} & t = 0, & \omega = \sin\frac{\pi x}{L}, \\ \text{B.C.:} & x = 0, & \omega = 0, \\ \text{B.C.:} & x = L, & \omega = 0. \end{array}$$

The kinematic viscosity ν is $10^{-2} \, ft^2 sec$, the reference velocity U of 10 $ft \cdot sec^{-1}$ and length L of 1 give a Reynolds number of $Re = \frac{LU}{\nu}$.

(a) Derive an explicit two-time level finite difference formula for the governing PDE.

(b) Compute the solution for ω with a uniform spatial increment of $\Delta x = 0.1$ and temporal increment $\Delta t \leq 0.5$.

(c) Plot results for $t = 0, 0.1, 0.2, 0.3, 0.4$ and 0.5.

Part II

The Finite Analytic Method

Part II of the book is devoted to the finite analytic (FA) method, a relatively new numerical method for solving the Navier-Stokes and energy equations for various flow and heat transfer problems. The method is developed for a variety of flow situations. The unifying concept in the various applications is the use of a local analytic solution as the basis of the algebraic formulation.

Chapter 7 begins by introducing the fundamental principles of the finite analytic method. The specifics of the method are detailed for one-dimensional problems in Chapter 8, two-dimensional problems in Chapter 9, and three-dimensional problems in Chapter 10. In each case, an example of how the method is used to solve a flow or heat transfer problem is presented.

The important analysis topics of consistency, stability, and convergence are addressed in Chapter 11 for the case of the unsteady, two-dimensional transport equation. Application of the FA method to hyperbolic PDEs, such as in supersonic flows, is presented in Chapter 12. Part II concludes with the explicit finite analytic method presented in Chapter 13. This numerical solution method is highly effective for problems involving convection dominated flows.

Chapter 7

Basic Principles

This chapter begins by casting the Navier-Stokes and energy equations presented in Chapter 2 in the form of a general three-dimensional transport equation. The FA method is presented for equations of this type in subsequent chapters. The second part of this chapter concerns the basic principles of the FA method. An overview of the primary steps in the FA method is given for the solution of a general nonlinear, second-order PDE.

7.1   The Transport Equation

The Navier-Stokes and energy equations used in numerical flow and heat transfer modeling may be expressed in the form of a general, three-dimensional transport equation. The non-dimensional form of the equation is

$$R\phi_t + \underbrace{R(u\phi_x + v\phi_y + w\phi_z)}_{convection\ terms} = \underbrace{\phi_{xx} + \phi_{yy} + \phi_{zz}}_{diffusion\ terms} + R \cdot f . \qquad (7.1)$$

The dependent variable in Equation (7.1) is ϕ, representing such quantities as a velocity component u, v or w, or temperature T. The source term is given by the product of f and R, where f may be a pressure gradient or buoyancy term, and R may be the dimensionless Reynolds or Peclet number. The last three terms on the left hand side of Equation (7.1) are *convection terms*. The first three on the right hand side are *diffusion terms*.

The nonlinear convection terms on the left-hand side of Equation (7.1) make analytical solutions of the transport equation difficult, if not impossible, in all but a few cases. Therefore, numerical methods are used in nearly all practical flow and heat transfer problems that include some form of Equation (7.1) in the governing equations. As mentioned in Chapter 5, a numerical solution of a PDE begins with a discretization of the given domain into smaller subdomains defined by grid lines and their points of intersection called nodes.

The discretization of the problem domain plays an essential role in numerical accuracy when the convection and diffusion terms in Equation (7.1) are consid-

ered.   The convection component of Equation (7.1) produces an asymmetric
phenomenon because the upstream conditions have a greater influence on ϕ at
node P than the downstream values.   If a numerical scheme does not capture
this property, physically unrealistic solutions may occur.   For example, the central difference scheme has been found to be stable only at low cell Reynolds
numbers (Reynolds number based on the representative cell velocity and cell
size), and unstable at higher (greater than 2) cell Reynolds numbers.

Upwind difference schemes, such as that used by Spalding [161], are stable
and do capture the asymmetric nature of the strongly convective flow.   However,
the numerical solution may exhibit significant *numerical* diffusion due to truncation error and an improper up-winding technique.   Schemes with high order
accuracy, such as the quadratic upstream (QUICK) scheme of Leonard [113],
can eliminate some false diffusion.   However, such methods may not accurately
simulate flow and heat transfer if the convection runs skew of the grid lines.

The finite analytic method is the result of efforts by Chen et al., [40], [37],
[35] to improve numerical solutions for applications with strong convection.   The
method uses analytic solutions to linearized versions of the governing PDEs
on small subdomains.   One very import result is that FA scheme provides an
automatic "analytic" upwinding effect, while minimizing the false diffusion by
eliminating the truncation error.

7.2   FA Fundamentals

Similar to other numerical domain techniques, the finite analytic method begins by discretizing the total domain of a problem into many small elements,
or subdomains, as depicted in Figure 7.1.   The fundamental principle of the finite analytic method is finding an analytic solution to the governing differential
equation on each of the nine-node subdomains, as shown in Figure 7.1.   The
necessary boundary conditions are constructed using the values of the dependent variable at the eight exterior nodes of the subdomain.   When the local
analytic solution is evaluated at the interior node P, it gives an algebraic equation relating the value of the dependent variable at node P to the values at the
subdomain boundary nodes.

The finite analytic method differs from the finite difference and finite element
methods in that the approximating algebraic form of the governing differential
equation is obtained from an analytic solution.   The solution for the dependent
variable at all of the discrete nodal locations on the entire domain is obtained
by assembling the local analytic solutions linked through the relationship of
adjacent nodes.   The finite analytic numerical solution is achieved by solving
the resulting system of algebraic equations.   In the case of nonlinear problems,
the governing differential equation is linearized on the subdomain before it is
solved analytically.   The "local linearization" does not remove the nonlinearity
of the equation on the larger scale as terms are allowed to vary from one element
to the next.

To illustrate the basic principles of the FA method, consider the two dimen-

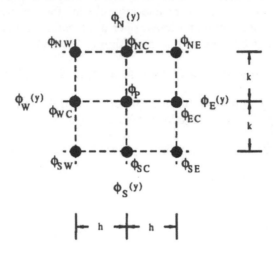

Figure 7.1: Finite analytic subdomain.

sional PDE $F(\phi) = g$, where F is a second order, linear or non-linear, elliptic partial differential operator and g is a non-homogeneous term that depends only on the independent variables x and y. The PDE is to be solved on a given domain D. Suppose the boundary conditions on D are properly specified so that the problem is well-posed.

The region D is divided into small rectangular subdomains defined by grid lines that run parallel to the coordinate axes. A typical FA subdomain is depicted in Figure 7.1, with interior node $P(i,j)$ and neighboring boundary nodes EC (east center), WC (west center), SC (south center), NC (north center), NE (northeast), NW (northwest), SE (southeast), and SW (southwest).

Once D has been divided into simple rectangular subdomains, an analytic solution on each subdomain may be obtained. Let the linear, or linearized, governing equation on a single subdomain of size $2h \times 2k$ be $L(\phi) = g$. The linear nature of the equation means an analytic solution may be obtained on the subdomain as a function of the boundary conditions. That is,

$$\phi = f(\phi_N(x), \phi_S(x), \phi_E(y), \phi_W(y), h, k, x, y, g) , \qquad (7.2)$$

where ϕ_N, ϕ_S, ϕ_E and ϕ_W are, respectively, the northern, southern, eastern and western boundary conditions of the subdomain, and $2h$ and $2k$ are, respectively, the x and y dimensions of the subdomain. For numerical purposes, the boundary functions ϕ_N, ϕ_S, ϕ_E and ϕ_W may be approximated in terms of the nodal values along the corresponding boundary. As an example, on the southern boundary of the FA element $\phi_S = \phi_S(\phi_{SE}, \phi_{SC}, \phi_{SW}, x)$. The analytic solution for ϕ on the subdomain, based on the southern boundary function, will be a function of x and y, and the southern nodal quantities of ϕ. Nodal values of ϕ on the other boundaries are incorporated in the analytic solution for ϕ in a similar way using the principle of superposition for linear operators. Substituting these conditions for each of the four boundaries into Equation (7.2) yields

$$\phi = f(\phi_{EC}, \phi_{WC}, \phi_{NC}, \phi_{SC}, \phi_{NE}, \phi_{NW}, \phi_{SW}, \phi_{SE}, h, k, x, y, g) . \qquad (7.3)$$

Equation (7.3) represents the analytic solution for ϕ on the entire FA element by incorporating the local boundary values, ϕ_{EC}, ϕ_{WC}, ..., ϕ_{SE}. Extracting this dependence explicitly, one can obtain the nine point FA formula for ϕ_P when Equation (7.3) is evaluated at point P. This can be written in the form

$$\begin{aligned}
\phi_P =\ & C_{EC}\phi_{EC} + C_{WC}\phi_{WC} + C_{NC}\phi_{NC} + C_{SC}\phi_{SC} \\
& + C_{NE}\phi_{NE} + C_{NW}\phi_{NW} + C_{SW}\phi_{SW} + C_{SE}\phi_{SE} + C_P g_P . \quad (7.4)
\end{aligned}$$

The terms labeled $C_{EC}, C_{WC}, \ldots, C_{SE}, C_P$ in Equation (7.4) multiplying the corresponding neighboring nodal values $\phi_{EC}, \phi_{WC}, \ldots, \phi_{SE}$, are the *finite analytic coefficients*. The details for determining these FA coefficients will be presented in subsequent chapters in this part of the text.

Similar formulas, like that given by Equation (7.4), may be derived for each nodal point in the whole domain D. These equations may be combined to form a system of linear algebraic equations relating the interior nodal value of ϕ to the neighboring nodal values of ϕ. The resulting system of algebraic equations is solved in conjunction with the boundary conditions of the original problem to provide the FA numerical solution for the entire problem domain.

Details of the FA method will be given for the one-, two-, and three-dimensional transport equation in the next three chapters. Specifically, formulas are developed for calculating the FA coefficients in Equation 7.4.

Chapter 8

The One-Dimensional Case

The FA numerical solution for the one-dimensional transport equation is presented in this chapter. The derivation of the FA coefficients is discussed in some detail. The chapter ends with a comparison of the FA and FD coefficients for the unsteady, one-dimensional heat equation.

8.1 The One-Dimensional Transport Equation

Equation (7.1) reduces to the unsteady, one-dimensional convective transport equation when the dependent variable ϕ is a function of only one spatial variable x.

$$R\phi_t + Ru\phi_x = \phi_{xx} + Rf(x, t) \tag{8.1}$$

In the nondimensional Equation (8.1), the velocity u in the convection term on the left-hand side may be a function of the independent variables x and t, and the dependent variable ϕ. For example, Burgers' equation is a simple model for the one-dimensional fluid flow. It is obtained from Equation (8.1) with $u = \phi$, $R = Re$ (Reynolds number), and $f = 0$. The governing equation for transient heat conduction is obtained from Equation (8.1) by setting $u = 0$, $R = Pe$ (Peclet number) and $f = 0$.

For numerical purposes, the problem domain D is divided into many small subdomains, or elements. The size of an arbitrary element is $2h \times \tau$, where $h = \Delta x$, and $\tau = \Delta t$. If need be, Equation (8.1) is linearized on each of the small elements. Then, an analytic solution is sought for the linear equation within each small element. A typical one time-step FA element is shown in Figure 6.1, including node P and its neighboring nodal points WC, EC, SW, SC and SE.

Complex initial and boundary conditions on the small element may be approximated by simple initial and boundary functions so that an analytic solution for ϕ can be derived on the element. However, even for such simple initial and boundary conditions, the analytic solution may still be difficult to obtain due

--

to the complicated dependence of u and f on independent or dependent variables. In this situation, the one-dimensional transport equation is linearized by approximating the convective velocity as a constant over the small element. Then, Equation (8.1) becomes

$$2A\phi_x + B\phi_t = \phi_{xx} + F\,, \tag{8.2}$$

where $A = \frac{1}{2}RU$, $B = R$, and U and F are representative constant values of u and Rf, respectively, over the FA element.

8.2   Finite Analytic Solution

Equation (8.2) is a linear partial differential equation with constant coefficients. The source term F may be absorbed by a change of variable. Thus, only the homogeneous convective transport equation

$$\phi_{xx} = 2A\phi_x + B\phi_t \tag{8.3}$$

is considered here. For the one time-step formulation, the initial condition ϕ_I and boundary conditions, represented by ϕ_W and ϕ_E, on the FA element shown in Figure 6.1, are

$$\phi(x,0) = \phi_I(x) = a_S(e^{2Ax} - 1) + b_S x + c_S\,, \tag{8.4}$$

$$\phi(-h,t) = \phi_W(t) = a_W + b_W t\,, \tag{8.5}$$

and

$$\phi(h,t) = \phi_E(t) = a_E + b_E t\,, \tag{8.6}$$

where the node P is taken as the origin in the FA element coordinate system. Note that the constant terms, linear terms, and exponential term of the initial and boundary conditions given in Equations (8.4) to (8.6) are solutions to Equation (8.3). These functions are chosen from the many that satisfy Equation (8.3) because each is monotonic and smooth between the nodes. If one expects a solution that is not monotonic on a given boundary, then further discretization is needed to assure this condition.

The coefficients in Equations (8.4) to (8.6) are solved in terms of the nodal values on the element boundaries. For example, the coefficients a_E and b_E in Equation (8.6) are determined by using the boundary conditions $\phi(h,0) = \phi_{SE}$ and $\phi(h,\tau) = \phi_{EC}$ in Equation (8.6). The two equations which result can then be solved for a_E and b_E. Note the southern boundary is a "time" boundary and actually is an initial condition. Formulas for the boundary function coefficients are

$$a_S = \frac{\phi_{SE} + \phi_{SW} - 2\phi_{SC}}{4\sinh^2 Ah}\,,$$

$$c_S = \phi_{SC},$$

$$b_S = \frac{\phi_{SE} - \phi_{SW} - \coth Ah(\phi_{SE} + \phi_{SW} - 2\phi_{SC})}{2h}\,,$$

$$a_W = \phi_{SW},$$

$$b_W = \frac{\phi_{WC} - \phi_{SW}}{\tau},$$

$$a_E = \phi_{SE} \quad \text{and}$$

$$b_E = \frac{\phi_{EC} - \phi_{SE}}{\tau}. \tag{8.7}$$

Equation (8.3), with Equations (8.4) to (8.6), is solved analytically by the method of separation of variables [24]. Details of the process are given in Appendix A. Evaluating the solution for point P will result in a 6-point FA algebraic equation giving the nodal value ϕ_P as a function of its neighboring nodal values, as shown in Equation (8.8).

$$\phi_P = C_{WC}\phi_{WC} + C_{EC}\phi_{EC} + C_{SW}\phi_{SW} + C_{SE}\phi_{SE} + C_{SC}\phi_{SC} \tag{8.8}$$

The coefficients in Equation (8.8), determined by evaluating the local analytic solution at point P, are given by

$$C_{WC} = e^{Ah}S_1, \quad C_{EC} = e^{-Ah}S_1, \quad C_{SW} = e^{Ah}S_2,$$

$$C_{SE} = e^{-Ah}S_2 \text{ and } C_{SC} = 4Ah\cosh(Ah)\coth(Ah)P_2,$$

where

$$S_1 = \frac{Bh^2}{\tau}(P_2 - Q_2) + Q_1,$$

$$S_2 = \frac{Bh^2}{\tau}(Q_2 - P_2) - 2Ah\coth(Ah)P_2,$$

$$P_2 = \sum_{m=1}^{\infty} \frac{(-1)^{m+1}\lambda_m h e^{-2F_m\tau}}{[(Ah)^2 + (\lambda_m h)^2]^2},$$

$$Q_i = \sum_{m=1}^{\infty} \frac{(-1)^{m+1}\lambda_m h}{[(Ah)^2 + (\lambda_m h)^2]^i} \quad (i = 1, 2),$$

$$F_m = \frac{A^2 + \lambda_m^2}{B} \quad \text{and}$$

$$\lambda_m = \frac{(2m-1)\pi)}{2h}.$$

The FA coefficient formulas involve three infinite sums, Q_1, Q_2 and P_2. However, it is possible to derive closed-form formulas for Q_1 and Q_2,

$$Q_1 = \frac{1}{e^{Ah} + e^{-Ah}}$$

and

$$Q_2 = \frac{e^{Ah} - e^{-Ah}}{2Ah(e^{Ah} + e^{-Ah})^2}.$$

Formulas similar to Equation (8.8) can be used to calculate new values for ϕ at each node of the domain. Ultimately, a system of algebraic equations is assembled to calculate ϕ values for all nodes at the new (or current) time step. Equation (8.8) is implicit because the nodal values ϕ_P, ϕ_{WC} and ϕ_{EC} are unknown quantities from the current time step. The nodal values ϕ_{SW}, ϕ_{SC} and ϕ_{SE} are known from the previous time-step. The tridiagonal matrix associated with the system of algebraic equations formed from Equation (8.8) is easily solved to give values of ϕ at the current time step [114].

8.3   Hybrid Finite Analytic Solution

A hybrid solution method using both finite analytic and finite difference techniques may be used to reduce the manipulation effort and computational time in solving the one-dimensional convective transport equation. The unsteady (time derivative) term in Equation (8.3) may be approximated by a simple finite difference formula

$$B\phi_t = B\frac{\phi_P - \phi_{SC}}{\tau} = g \, . \tag{8.9}$$

Equation (8.3) on the FA subdomain (Figure 6.1) is reduced to a steady-state convective transport equation with the unsteady term absorbed in a constant source term g,

$$\phi_{xx} = 2A\phi_x + g \tag{8.10}$$

as indicated in Equation (8.10).

The finite analytic algebraic equation can be derived for Equation (8.10) as

$$\phi_P = \frac{e^{Ah}\phi_{WC} + e^{-Ah}\phi_{EC}}{e^{Ah} + e^{-Ah}} - \frac{tanh(Ah)}{2Ah}gh^2 \, . \tag{8.11}$$

Substituting the expression for g from Equation (8.9) into Equation (8.11), a 4-point hybrid FA formula for the element shown in Figure 6.1 is obtained. It is

$$\phi_P = \frac{1}{1 + b_{SC}}(b_{WC}\phi_{WC} + b_{EC}\phi_{EC} + b_{SC}\phi_{SC}) \, , \tag{8.12}$$

where

$$b_{SC} = \frac{Bh^2}{2\tau}\frac{tanhAh}{Ah} \, ,$$

$$b_{WC} = \frac{e^{Ah}}{e^{Ah} + e^{-Ah}}$$

and

$$b_{EC} = \frac{e^{-Ah}}{e^{Ah} + e^{-Ah}} \, .$$

8.4   FA and FD Coefficient Comparison

A comparison between the FA and FD solutions for the one-dimensional heat equation is given in this section. The finite difference solution of the heat equation $\phi_{xx} = \phi_t$ ($A = 0, B = 1$ in Equation (8.3)) can be cast in the form of Equation (8.8). In this formulation, finite difference approximations of the derivatives are substituted into the heat equation, resulting in the following formulas for the coefficients C_{EC}, C_{WC}, C_{SC}, C_{SW} and C_{SE} (see Section 6.3):

$$C_{SC} = \frac{[1 - 2\lambda(1 - \theta)]}{1 + 2\lambda\theta}, \tag{8.13}$$

$$C_{SW} = C_{SE} = \frac{\lambda(1 - \theta)}{(1 + 2\lambda\theta)} \tag{8.14}$$

and

$$C_{WC} = C_{EC} = \frac{\lambda\theta}{(1 + 2\lambda\theta)}. \tag{8.15}$$

For these formulas, $\lambda = \tau/h^2$ and θ is an *implicit-explicit* parameter [42]. For example, when $\theta = 0$, Equation (8.8), with coefficients defined in Equations (8.13) to (8.15), gives an explicit finite difference formula. When $\theta = \frac{1}{2}$, the Crank-Nicolson (C-N) formula is obtained. An implicit formula results for $\theta = 1$. When $\theta = (6\lambda - 1)/12\lambda$, the Forsythe and Wasow (F-W) formula is obtained.

First, consider the explicit FD method ($\theta = 0$). The coefficients in Equations 8.13 to 8.15 for this case are obtained using a forward difference in the temporal derivative ϕ_t, and a central difference in the spatial derivative ϕ_{xx}. The coefficients for C_{SE} ($= C_{SW}$) and C_{SC} for the explicit FD method are shown in Figure 8.1(a). Note from Equation 8.15 that both C_{WC} and C_{EC} are zero in this case. The explicit finite difference technique is a suitable method for the one dimensional heat equation for λ less than 0.5. However, as λ increases to 0.5, the value of C_{SC} decreases to zero, and becomes negative as λ grows larger than 0.5. This explains the instability of the explicit FD method for $\lambda \geq 0.5$. Note also that both coefficients grow without bound as λ grows.

The finite difference coefficients for the implicit case ($\theta = 0$) are plotted as a function of λ in Figure 8.1(b). In this case, the coefficients C_{SW} and C_{SE} are zero, meaning the values of ϕ at these respective nodes at the current time step have no influence on the future value of ϕ at node P. The nonzero coefficients C_{EC} and C_{SC} in the implicit case are nonnegative and bounded for all values of λ, which mean the implicit scheme is stable for all λ as well. As λ tends to zero, which implies τ tends to zero, the value of C_{EC} decreases to a limit of 0.07, while the value of C_{SC} increases to around 0.82. These tendencies are consistent with the idea that the smaller the time-step, the greater the influence of ϕ at node P at the current time step, and the lesser the influence of ϕ at the neighboring nodes at the next time step.

The Crank-Nicolson formula for determining the FD coefficients uses difference expressions involving values of ϕ at both the current and future time-step. In fact, they are incorporated equally resulting in equal values for C_{SE} and C_{EC}, as indicated in Figure 8.1(c). One advantage of the Crank-Nicolson method is the incorporation of all five neighboring nodes in the calculation of the new value of ϕ_P. However, for $\lambda \geq 1.0$ C_{SC} becomes negative and the method loses its stability. But this "semi-implicit" method has a greater range of stability than the fully explicit case shown in Figure 8.1(a).

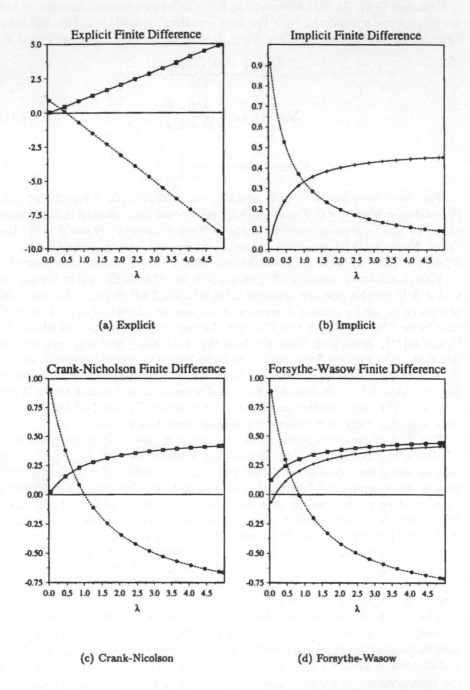

(a) Explicit

(b) Implicit

(c) Crank-Nicolson

(d) Forsythe-Wasow

Figure 8.1: FD coefficients C_{SE} ($\square$), C_{EC} (+) and C_{SC} ($\circ$) as a function of $\lambda = \frac{\tau}{h^2}$.

Coefficients for the Forsythe-Wasow formulas for the finite difference method are presented in Figure 8.1(d). Like the Crank-Nicolson approach, this semi-implicit method results in nonzero coefficients for nodes at both time steps. However, unlike the C-N case, the values for the current and future time step nodes are not identical. Coefficients for the southeast node (from the current time step) are always slightly great than those for the east-central node. Note that as λ increases, which means the spacing in time becomes larger than the distance between adjacent nodes, the values for the east-central approaches the value of the south east coefficient. Actually, one would expect the weight of the east central node to eventually surpass that of the southeast node as λ becomes sufficiently large. Perhaps the most troubling feature of the F-W coefficients, though, is C_{SC} becoming negative λ greater than 0.82, and negative values of C_{EC} for $\lambda \leq 0.08$.

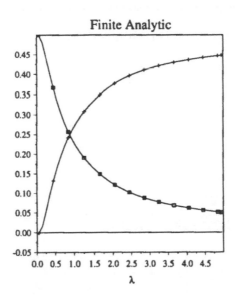

Figure 8.2: FA coefficients C_{SE} ($\square$) and C_{EC} (+) as a function of $\lambda = \frac{\tau}{h^2}$.

The last set of coefficients for the one dimensional heat equation are those derived using the finite analytic method, and are plotted as a function of λ in Figure 8.2. First note that both C_{SE} ($=C_{SW}$) and C_{EC} ($=C_{WC}$) are positive and bounded above by 0.5. The coefficients are equal for $\lambda = 1$, which is reasonable since the temporal and spatial spacing is equal. As λ approaches zero, the values of C_{EC} and C_{SE} approach zero and one, respectively. Again, this is reasonable because the problem could be considered independent in time, approaching a Laplace-type equation at the current time step, where the value of ϕ at node P is simply the average of its two closest neighboring nodes at SE and SW. Conversely, as λ becomes very large, the step in time is increasing large, in a relative way, to the spatial step. Therefore, the influence of the values of ϕ at SE and SW on the new value of ϕ at P should become less and less. For "large" λ, the problem is more Laplace-like case at the next time step, where ϕ at node P should be the average of the two nearest nodes, those at EC and WC.

Comparing Figures 8.1(c) and 8.2, the Crank-Nicolson and FA coefficients are closest in value when λ is approximately 0.7. It has been shown [114] that if λ is set to $1/\sqrt{20}$ the error in the Forsythe-Wasow formula is reduced to $O(\Delta x^6)$. It is at this value of λ that the FA and Forsythe-Wasow coefficients agree most closely as well.

8.5　Burgers' Equation

A comparison of the FA and exact solutions can be made for Burgers' equation

$$u_t + uu_x = \alpha u_{xx} \tag{8.16}$$

for

$$-\infty < x < \infty \quad \text{and} \quad t > 0$$

The initial conditions are

$$u(x, 0) = \begin{cases} 1 & \text{for} \quad x \le 0, \\ 0 & \text{for} \quad x > 0, \end{cases}$$

and boundary conditions

$$u(-\infty, t) = 1 \quad \text{and} \quad u(\infty, t) = 0.$$

The exact solution for the above problem is

$$u(x, t) = \left[1 + e^{\left[\frac{1}{2\alpha}(x - \frac{1}{2}t) \right]} \frac{erfc\left(\frac{-x}{2\sqrt{\alpha t}} \right)}{erfc\left(\frac{x-t}{2\sqrt{\alpha t}} \right)} \right]^{-1}. \tag{8.17}$$

Figure 8.3 gives a comparison of the FA and the exact solution for the case of $\alpha = 0.01$. The first graph is for the dimensionless time $t = 3.0$, and the second is at time $t = 12.0$. In the finite analytic solution process, the convective term u in each local element is approximated by a representative constant (area-averaged $A = \frac{\bar{u}}{2\alpha}$) known from previous time steps, so that a marching process can be used without iteration at each time step.

The solution obtained by the FA formulation agrees very well with the exact solution, including the wave shape and the propagation of the wave "front." Errors in either the shape of the wave or the speed of the frontal propagation may be reduced by employing a better estimated convective velocity term, $\bar{u}$, based on two or more time step interpolation so that the nonlinearity of the governing equation can be more accurately simulated.

It is important to point out that the FA solutions show neither oscillation nor overshooting, phenomena which are encountered in many FD solutions.

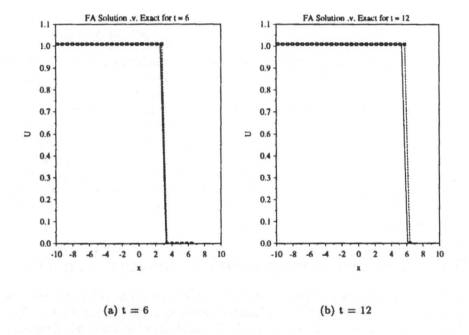

(a) t = 6   (b) t = 12

Figure 8.3: Comparison of the exact (solid line) and FA solutions (□) for Burgers' equation.

Chapter 9

The Two-Dimensional Case

The chapter is devoted to showing how the finite analytic method is used to create an algebraic solution for the two-dimensional transport equation. In order to incorporate an analytic solution in this process, the nonlinear transport equation must first be linearized on small FA subdomains. This process is described in the first section. Next, the derivation of the FA algebraic equation for an FA element with uniform grid spacing is outlined. Then it is shown how the results for the transport equation can be used to solve Poisson's and Laplace's equations. Because nonuniform grid methods are useful in many engineering applications, the adaptation of the FA coefficient formulas for nonuniform grids is given. The chapter concludes by applying the FA method to solve the problem of flow and heat transfer in a two-dimensional cavity.

9.1 The Two-Dimensional Transport Equation

In the two-dimensional transport equation shown in Equation (9.1), the dependent variable ϕ is a function of spatial variables x and y, and time t. The dimensionless parameter R represents the Reynolds number when ϕ is a velocity component, and the Peclet number when ϕ represents temperature.

$$R(\phi_t + u\phi_x + v\phi_y) = \phi_{xx} + \phi_{yy} + R \cdot f(x, y, t, \phi_j) \tag{9.1}$$

Note that the nonhomogeneous term f in Equation (9.1) may be a function of any dependent variable ϕ_j, including ϕ itself.

In order to derive an analytic solution for this case, Equation (9.1) must be linearized on the small FA element. The two planes in Figure 9.1 represent different time-steps in the solution process. The ϕ quantities on the $t = m - 1$ plane are known values from the previous time step. Our objective is to calculate ϕ values on the $t = m$ plane.

The process begins by letting $u = U + u'$ and $v = V + v'$ in Equation (9.1). Constants U and V are representative values for u and v, respectively, for the FA element. They may be the values of u and v at node P of the FA element,

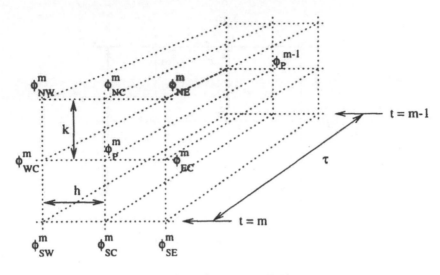

Figure 9.1: Two-dimensional FA element.

or may represent an element area average of the quantities. Substituting these expressions for u and v in Equation 9.1, and rearranging terms gives

$$R \cdot u\phi_x + R \cdot v\phi_y = \phi_{xx} + \phi_{yy} + F - R \cdot \phi_t , \qquad (9.2)$$

where

$$F = R \cdot f(x, y, t, \phi_j) - R \cdot u' \phi_x - R \cdot v' \phi_y .$$

The term $-R \cdot u' \phi_x - R \cdot v' \phi_y$ in F can be thought of as a higher order correction term for the velocity . If the FA element is sufficiently small, this correction term is negligible, and $F \approx R \cdot f$.

The ease in finding an analytic solution to the transport equation is increased greatly by replacing the time derivative in Equation (9.2) with a finite difference approximation. Also, the nonhomogeneous term is replaced by a representative constant

$$F_P = R \cdot f(x, y, t, \phi_j^{n-1})|_P ,$$

the value of F at node P on the $t = m - 1$ time plane. With these substitutions, Equation (9.2) becomes

$$2A\phi_x + 2B\phi_y = \phi_{xx} + \phi_{yy} + g , \qquad (9.3)$$

where A, B, and g are defined as

$$2A = R \cdot u_P, \qquad\qquad 2B = R \cdot v_P$$

and

$$g = F_P - R(\phi_P^m - \phi_P^{m-1})/\tau .$$

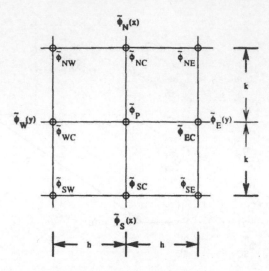

Figure 9.2: FA element for two-dimensional transport equation.

The last step in the transformation of the transport equation is to introduce a change of variable to make Equation (9.3) homogeneous. Defining

$$\tilde{\phi} = \phi + \frac{g}{2(A^2 + B^2)}(Ax + By) ,$$ (9.4)

and substituting for ϕ in Equation (9.3) gives

$$\tilde{\phi}_{xx} + \tilde{\phi}_{yy} = 2A\tilde{\phi}_x + 2B\tilde{\phi}_y$$ (9.5)

for the FA subdomain. This linear, homogeneous form of the two-dimensional transport equation will be solved analytically on the FA element. This will be done first for uniform grid spacing.

9.2   FA Solution on Uniform Grids

The elliptic nature in the spatial variables of Equation (9.5) requires that boundary conditions be specified on all four sides of the FA element. Although many appropriate functions may be used to approximate the boundary condition on a given side, it is best to choose a function that satisfies the governing equation. A constant function, a linear function (such as $Ay - Bx$), and an exponential function (such as $e^{2Ax+2By}$) all satisfy Equation (9.5) . These functions can be considered as "natural" or "basic" forms of solutions for Equation (9.5). These smooth, monotonic functions are appropriate for sufficiently refined discretizations of the problem domain because we would expect a smooth and monotonic solution between adjacent nodes.

As an example, the western boundary function $\tilde{\phi}_W$, on the FA element in

Figure 9.2 may be approximated by

$$\tilde{\phi}_W(y) = a_W(e^{2By} - 1) + b_W y + c_W \ . \tag{9.6}$$

The constants a_W, b_W and c_W are specified by using the boundary requirements $\tilde{\phi}_W(-k) = \tilde{\phi}_{SW}$, $\tilde{\phi}_W(0) = \tilde{\phi}_{WC}$ and $\tilde{\phi}_W(k) = \tilde{\phi}_{NW}$ in Equation (9.6). The three equations which result are then solved for a_W, b_W and c_W, giving

$$a_W = \frac{\tilde{\phi}_{NW} + \tilde{\phi}_{SW} - 2\tilde{\phi}_{WC}}{4\sinh^2 Bk} \ , \tag{9.7}$$

$$b_W = \frac{1}{2k}\left[\tilde{\phi}_{NW} - \tilde{\phi}_{SW} - \coth(Bk)(\tilde{\phi}_{NW} + \tilde{\phi}_{SW} - 2\tilde{\phi}_{WC}\right] \tag{9.8}$$

and

$$c_W = \tilde{\phi}_{WC} \ . \tag{9.9}$$

The boundary functions for the north, south and east sides ($\tilde{\phi}_N$, $\tilde{\phi}_S$ and $\tilde{\phi}_E$) can be similarly approximated.

Equation (9.5) is solved on the FA element by setting

$$\tilde{\phi} = \tilde{\phi}^W + \tilde{\phi}^N + \tilde{\phi}^E + \tilde{\phi}^S \ , \tag{9.10}$$

where each function on the right-hand side solves one of four *subproblems*. For example, $\tilde{\phi}^W$ solves the subproblem

$$\tilde{\phi}_{xx} + \tilde{\phi}_{yy} = 2A\tilde{\phi}_x + 2B\tilde{\phi}_y \ ,$$

with boundary conditions

$$\tilde{\phi}(-h, y) = \tilde{\phi}_W(y)$$

and

$$\tilde{\phi}(x, k) = \tilde{\phi}(h, y) = \tilde{\phi}(x, -k) = 0$$

on the FA element. An analytic solution for $\tilde{\phi}_W$ is found using the method of separation of variables (see Appendix B). The functions $\tilde{\phi}^N$, $\tilde{\phi}^E$ and $\tilde{\phi}^S$ solve similarly posed problems. The sum of the four solutions is a solution to Equation (9.5) on the FA element by the *principle of superposition* for linear, homogeneous differential equations. The result is an analytic solution for $\tilde{\phi}$ as a function of x, y and each of the nodal quantities of ϕ through the boundary function approximations. Evaluating the solution at the center node P ($x = y = 0$ on the FA element), results in an algebraic formula for $\tilde{\phi}$ at node P on the $t = m$ time plane. That is,

$$\tilde{\phi}_P^m = \tilde{\phi}_P^m(\tilde{\phi}_{NC}, \tilde{\phi}_{NE}, \cdots, \tilde{\phi}_{NW}) \ ,$$

where the values of $\tilde{\phi}$ at the boundary nodes may be from either time plane, depending on how implicit the formula is. The formula can be re-arranged by grouping terms involving the same nodal quantity ($\tilde{\phi}_{NC}$, $\tilde{\phi}_{NE}$, etc.). This gives the following equation, which relates the value of the dependent variable at node P to the corresponding values of its eight neighboring nodes:

$$\bar{\phi}_P^m = \sum_{j=NE}^{NC} C_j \bar{\phi}_j \tag{9.11}$$

Substituting for each $\bar{\phi}$ in Equation (9.11) using Equation (9.4) gives

$$\phi_P = C_{EC}\phi_{EC} + C_{WC}\phi_{WC} + \cdots + C_{SW}\phi_{SW} + C_{SE}\phi_{SE} + C_P g , \tag{9.12}$$

with the finite analytic coefficients $C_{EC}, C_{WC}, \cdots, C_{NE}$ given by

$$
\begin{aligned}
C_{EC} &= EBe^{-Ah} , & C_{NE} &= Ee^{-Ah-Bk} , \\
C_{WC} &= EBe^{Ah} , & C_{NW} &= Ee^{Ah-Bk} , \\
C_{SC} &= EAe^{Bk} , & C_{SE} &= Ee^{-Ah+Bk} , \\
C_{NC} &= EAe^{-Bk} , & C_{SW} &= Ee^{Ah+Bk}
\end{aligned} \tag{9.13}
$$

and

$$
\begin{aligned}
C_P = \; & \frac{Ah}{2(A^2+B^2)}[C_{NW} + C_{WC} + C_{SW} - C_{NE} - C_{EC} - C_{SE}] \\
& + \frac{Bk}{2(A^2+B^2)}[C_{SW} + C_{SC} + C_{SE} - C_{NW} - C_{NC} - C_{NE}] ,
\end{aligned} \tag{9.14}
$$

where

$$
\begin{aligned}
E &= \frac{1}{4\cosh(Ah)\cosh(Bk)} - AhE_2\coth(Ah) - BkE_2'\coth(Bk) , \\
EA &= 2Ah\frac{\cosh^2 Ah}{\sinh(Ah)}E_2 , \\
EB &= 2Bk\frac{\cosh^2 Bk}{\sinh(Bk)}E_2' , \\
E_2 &= \sum_{m=1}^{\infty} \frac{-(-1)^m(\lambda_m h)}{[(Ah)^2 + (\lambda_m h)^2]^2\cosh(\mu_m k)} , \\
E_2' &= \sum_{m=1}^{\infty} \frac{-(-1)^m(\lambda_m' k)}{[(Bk)^2 + (\lambda_m' k)^2]^2\cosh(\mu_m' h)} , \\
\mu_m &= \sqrt{A^2 + B^2 + \lambda_m^2} , \\
\lambda_m &= \frac{(2m-1)\pi}{2h} , \\
\mu_m' &= \sqrt{A^2 + B^2 + (\lambda_m')^2} \quad \text{and} \\
\lambda_m' &= \frac{(2m-1)\pi}{2k} .
\end{aligned}
$$

2×10^{-5}	1×10^{-5}	1×10^{-9}
0.23854	P	1×10^{-5}
0.52286	0.23854	2×10^{-5}

Table 9.1: FA coefficients for $2Ah = 10$ and $2Bk = 10$.

Notice that two series summations (E_2 and E_2') must be evaluated to find the FA coefficients. However, by substituting $\tilde{\phi} = Ay - Bx$ (a solution of Equation (9.5)) into Equation 9.11, the following relationship between E_2 and E_2' results (see Appendix B):

$$E_2' = \left(\frac{h}{k}\right)^2 E_2 + \frac{Ak \cdot tanh(Bk) - Bh \cdot tanh(Ah)}{4AkBk \cdot cosh(Ah)cosh(Bk)} . \qquad (9.15)$$

Hence, only E_2 needs to be evaluated numerically, and then E_2' can be found using Equation(9.15). Additionally, it has been found that for most applications, ten terms of the summation are enough to achieve an accuracy of 10^{-6}.

Finally, substituting the finite difference approximation of the unsteady term into Equation (9.12) yields

$$\phi_P^m = \left[\sum_{j=NE}^{NC} C_j \phi_j^m - C_P F_P + \frac{C_P R \phi_P^{m-1}}{\tau} \right] / (1 + \frac{C_P R}{\tau}) , \qquad (9.16)$$

where j denotes the boundary nodes $NE, EC, \ldots NC$.

In general the finite analytic coefficients, C_{nb} are functions of the local cell Reynolds numbers $2Ah$ and $2Bk$. The cell Reynolds number may be different from one element to another due to differences in center node velocities u_P and v_P, or grid sizes h and k, which may vary in the case of nonuniform grids.

Coefficients for two flow scenarios for the case $f = 0$ and $h = k$ are shown in Tables 9.2 and 9.2. If these FA coefficients are multiplied by 100, the resulting value may be interpreted as the percentage influence of a given boundary node value ϕ_{nb} on the interior node value ϕ_P, under the given convective vector with components $2Ah$ and $2Bk$. Table 9.2 shows that when there is strong convection from the southwest corner (cell Reynolds numbers of $2Ah = 2Bk = 10$), the FA solution correctly indicates the strong influence of the southwest boundary node SW, whose coefficient value is $C_{SW} = 0.52286$, on the interior node P. The downstream node NE has practically zero influence, as indicated by its coefficient value of $C_{NE} = 10^{-9}$.

Table 9.2 shows the values of the FA coefficients when the convection comes directly from the west side (cell Reynolds numbers of $2Ah = 100, 2Bh = 0$). The influence of node WC on node P is dominant ($C_{WC} = 0.98$), while the other two up-wind nodes, NW and SW, have very little influence ($C_{NW} = C_{SW} = 0.01$). The other boundary node coefficients are negligible at this high cell Reynolds number.

These two examples illustrate the inherent "up-winding" character of the FA solution coefficients. The FA coefficient formulas given in Equations (9.13)

0.01000	1×10^{-11}	1×10^{-48}
0.98000	P	1×10^{-44}
0.01000	1×10^{-11}	1×10^{-48}

Table 9.2: FA coefficients for $2Ah = 100$ and $2Bk = 0$.

- (9.14) show that the coefficients are determined, in part, by the direction and magnitude of the velocity through the A and B terms. The first example shows this to be true even when the convection is running skew of the coordinate grid lines. This response to strong convection in the FA coefficients is a distinct feature of the FA method.

FA coefficients derived using piecewise-exponential boundary functions of the form

$$\tilde{\phi}_W = a_W e^{2By} + c_W , \qquad (9.17)$$

where

$$a_W = \frac{\tilde{\phi}_{SW} - \tilde{\phi}_{WC}}{e^{-2Bk} - 1} , \quad c_W = \frac{\tilde{\phi}_{WC} e^{-2Bk} - \tilde{\phi}_{SW}}{e^{-2Bk} - 1} \quad \text{for} \quad -k \leq y \leq 0 \quad (9.18)$$

and

$$a_W = \frac{\tilde{\phi}_{NW} - \tilde{\phi}_{WC}}{e^{2Bk} - 1} , \quad c_W = \frac{\tilde{\phi}_{WC} e^{2Bk} - \tilde{\phi}_{NW}}{e^{2Bk} - 1} \quad \text{for} \quad 0 \leq y \leq k \quad (9.19)$$

have been found to produce more accurate FA coefficients when convection is very strong (i.e. cell Reynolds number $2Ah$ or $2Bk$ greater than about 400). If the FA equation development in this section is repeated using boundary functions of the form of Equation (9.17) instead of Equation (9.6), the FA coefficients obtained will be identical to those given in Equations (9.13) and (9.14), except that E, EA, EB, E_2 and E_2' are now defined as

$$
\begin{aligned}
E &= \frac{1}{4\cosh(Ah)\cosh(Bk)} - \frac{Ah}{2\sinh(Ah)} E_2 - \frac{Bk}{2\sinh(bk)} E_2' , \\
EA &= AhE_2\coth(Ah) , \\
EB &= BkE_2'\coth(Bk) , \\
E_2 &= \sum_{m=1}^{\infty} \frac{1}{[(Ah)^2 + (\lambda_m h)^2]\cosh(\mu_m k)} \qquad \text{and} \\
E_2' &= \frac{h^2\cosh(Bk)}{k^2\cosh(Ah)} E_2 + \frac{Ak \cdot \tanh(Bk) - Bh \cdot \tanh(Ah)}{2AkBk \cdot \cosh(Ah)} ,
\end{aligned}
$$

with

$$
\begin{aligned}
\mu_m &= \sqrt{A^2 + B^2 + \lambda_m^2} \qquad \text{and} \\
\lambda_m &= \frac{(2m-1)\pi}{2h} .
\end{aligned}
$$

The coefficients developed earlier in this section using exponential and linear boundary functions such as Equation (9.6)) are more accurate than the piecewise-exponential coefficients for the case of pure diffusion (i.e. no fluid flow). For moderate cell Reynolds numbers, both formulations produce comparable coefficients. Further discussion of this matter can be found in [26].

9.3 The Poisson Equation

Equation (9.1) reduces to the Poisson equation when $\phi = \psi$ and $R = 0$.

$$\psi_{xx} + \psi_{yy} = -f \qquad (9.20)$$

The finite analytic solution of Equation (9.20) for node $P(i, j)$ on the $2h \times 2k$ FA element is easily obtained from Equations (9.12) to (9.14) as

$$\psi_P = \sum_{j=NE}^{NC} C'_j \psi_j + C'_P R f_P , \qquad (9.21)$$

where C'_j and C'_P are the FA coefficients given in Equations (9.12) and (9.16) with $A = B = 0$ (making them invariant because $A = B = 0$ always). For example, with $h = k$, the FA coefficients are

$$C'_{EC} = C'_{WC} = C'_{NC} = C'_{SC} = 0.205315,$$

$$C'_{NE} = C'_{NW} = C'_{SE} = C'_{SW} = 0.044685$$

and

$$C'_P = 0.294685h^2 .$$

Laplace's equation results from Equation (9.20) when $f = 0$. The FA solution for Laplace's equation, represented by Equation (9.21), compares very closely to the fourth order 9-point FD solution [114], which gives

$$C'_{EC} = C'_{WC} = C'_{NC} = C'_{SC} = 0.2$$

and

$$C'_{NE} = C'_{NW} = C'_{SE} = C'_{SW} = 0.05.$$

9.4 FA Solution for Nonuniform Grids

The derivation of the FA solution for the two-dimensional transport equation in Section 9.2 was accomplished on a uniform grid, where the spacing between grid lines (and nodes) is constant over the entire domain. In some applications it is advantageous to use nonuniform grid spacing (see Section 5.1). The FA formula for uniform grid applications is adapted to nonuniform grids in this section.

Consider the case $h_E < h_W$ and $h_N < h_S$ shown in Figure 9.3 as an example. Note that h_e in Figure 9.3 corresponds to h_E in the text. Similarly, $h_w = h_W$,

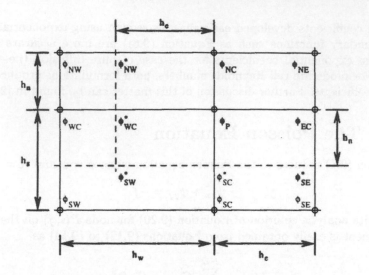

Figure 9.3: Nonuniform FA element.

$h_n = h_N$ and $h_s = h_S$. A smaller rectangular element of width $2h_E$, height $2h_N$ and with the interior point P located at the center is drawn as shown. The FA formula, Equation (9.12), derived previously for ϕ_P, can be written in terms of nodal values ϕ^*_{NW}, ϕ^*_{WC}, etc. on the smaller rectangular element as follows:

$$\phi_P = C_{NE}\phi_{NE} + C_{NW}\phi^*_{NW} + C_{SE}\phi^*_{SE} + C_{SW}\phi^*_{SW} +$$
$$C_{EC}\phi_{EC} + C_{WC}\phi^*_{WC} + C_{NC}\phi_{NC} + C_{SC}\phi^*_{SC} + C_P R f_P \,, \quad (9.22)$$

where the FA coefficients C_{NE}, C_{NW}, etc. are as defined in Equations (9.13)-(9.14) with grid sizes $h = h_E$ and $k = h_N$.

If suitable interpolation functions are employed to approximate the unknown nodal values (ϕ^*_{NW}, ϕ^*_{SE}, etc.) in terms of the known values (ϕ_{NW}, ϕ_{SE}, etc.) at the 8 boundary nodes and interior node P, a FA formula for the nonuniform element can be obtained. For example, the north boundary condition can be approximated in the same manner as Equation (9.6) by the boundary function

$$\phi_N(x) = a_N(e^{2Ax} - 1) + b_N x + c_N \,, \quad (9.23)$$

where

$$a_N = \frac{h_W\phi_{NE} + h_E\phi_{NW} - (h_E + h_W)\phi_{NC}}{h_W(e^{2Ah_E} - 1) + h_E(e^{-2Ah_W-1})} \,, \quad (9.24)$$

$$b_N = \frac{(e^{-2Ah_W} - 1)(\phi_{NE} - \phi_{NC}) - (e^{2Ah_E} - 1)(\phi_{NW} - \phi_{NC})}{h_W(e^{2Ah_E} - 1) + h_E(e^{-2Ah_W-1})} \quad (9.25)$$

and

$$c_N = \phi_N \,. \quad (9.26)$$

Note that the expressions in Equations (9.24) - (9.25) are in terms of the un-equally spaced nodal values ϕ_{NE}, ϕ_{NC} and ϕ_{NW} on the element boundary. The interpolated nodal value ϕ^*_{NE} is obtained by simply evaluating the boundary function Equation (9.23) at $x = -h_E$, which gives

$$\phi^*_{NW} = \phi_N(x = -h_E) = (s-1)\phi_{NE} + \bar{s}\phi_{NW} + (2-s-\bar{s})\phi_{NC} , \qquad (9.27)$$

where

$$s = \frac{h_W(e^{2Ah_E} + e^{-2Ah_E} - 2)}{h_W(e^{2Ah_E} - 1) + h_E(e^{-2Ah_W} - 1)}$$

and

$$\bar{s} = s\frac{h_E}{h_W} .$$

The other unknown nodal values (ϕ^*_{WC}, ϕ^*_{SC}, etc.) are found using similar linear-exponential boundary functions. The 9-point FA formula for a nonuniform element is obtained by substituting these expressions for the unknown nodal values into Equation (9.22). Letting

$$\begin{aligned}
\bar{\phi}^m_P &= \frac{1}{G}(b_{NE}\phi^m_{NE} + b_{NW}\phi^m_{NW} + b_{SE}\phi^m_{SE} + b_{SW}\phi^m_{SW} \\
&\quad + b_{EC}\phi^m_{EC} + b_{WC}\phi^m_{WC} + b_{NC}\phi^m_{NC} + b_{SC}\phi^m_{SC} - C_P f_P) \quad (9.28)
\end{aligned}$$

at the mth time step, the FA solution for ϕ at node P becomes

$$\phi^m_P = \left[\bar{\phi}^m_P + \frac{RC_P}{\tau}\phi^{m-1}_P\right]\left[G + \frac{C_P R}{\tau}\right]^{-1} , \qquad (9.29)$$

where

$$G = 1 - (2-s-\bar{s})C_{WC} - (2-t-\bar{t})C_{SC} - (2-s-\bar{s})(2-t-\bar{t})C_{SW}$$

and the coefficients b_{nb} are given as

$$\begin{aligned}
b_{NE} &= C_{NE} + (s-1)C_{NW} + (t-1)C_{SE} + (s-1)(t-1)C_{SW} , \\
b_{NW} &= \bar{s}C_{NW} + \bar{s}(t-1)C_{SW} , \\
b_{SE} &= \bar{t}C_{SE} + \bar{t}(s-1)C_{SW} , \\
b_{SW} &= \bar{s}\bar{t}C_{SW} , \\
b_{EC} &= C_{EC} + (s-1)C_{WC} + (2-t-\bar{t})C_{SE} + (s-1)(2-t-\bar{t})C_{SW} , \\
b_{WC} &= \bar{s}C_{WC} + \bar{s}(2-t-\bar{t})C_{SW} , \\
b_{NC} &= C_{NC} + (t-1)C_{SC} + (2-s-\bar{s})C_{NW} + (t-1)(2-s-\bar{s})C_{SW} , \\
b_{SC} &= \bar{t}C_{SC} + \bar{t}(2-s-\bar{s})C_{SW} , \qquad (9.30)
\end{aligned}$$

where

$$t = \frac{h_S(e^{2Bh_N} + e^{-2Bh_N} - 2)}{h_S(e^{2Bh_N} - 1) + h_N(e^{-2Bh_S} - 1)}$$

and

$$\bar{t} = t\frac{h_N}{h_S} .$$

Note the FA coefficients in Equations (9.30) are obtained from the small element with node P at the center and with grid sizes, h and k, equal to the smaller spacing (e.g. $h = h_E$ and $k = h_N$ in Figure 9.3). For the cases $h_E > h_W$ or $h_N > h_S$, the finite analytic algebraic equation given above can still be applied by simply reversing the flow direction and renaming the nodal points. It can be carried out easily through the change of signs and indices in numerical calculation.

9.5   Heat Transfer in a Driven Cavity

The FA numerical method for the two-dimensional transport equation will be used to solve for the flow and heat transfer in a two-dimensional lid-driven cavity. Consider a square cavity of depth and width L. The bottom wall, at temperature T_P, moves with speed U_P in the positive x direction. The other walls are at rest and at an isothermal temperature T_W. The flow is assumed to be two-dimensional, steady, incompressible, and laminar with constant transport properties.

Introducing the dimensionless variables $x = X/L$, $y = Y/L$, $u = U/U_P$, $v = V/U_P$, and $\theta = (T - T_W)/(T_P - T_W)$, the dimensionless Navier-Stokes equations in vorticity-stream function form are

$$(Re \cdot u)\omega_x + (Re \cdot v)\omega_y = \omega_{xx} + \omega_{yy} , \qquad (9.31)$$

$$\psi_{xx} + \psi_{yy} = -\omega , \qquad (9.32)$$

with

$$u = \psi_y \qquad \text{and} \qquad v = -\psi_x . \qquad (9.33)$$

The Reynolds number $Re = U_P L/\nu$ is based on the velocity of the moving wall U_P and the cavity depth L. The dimensionless energy equation is given by

$$(Pe \cdot u)\theta_x + (Pe \cdot v)\theta_y = \theta_{xx} + \theta_{yy} , \qquad (9.34)$$

where $Pe = Re \cdot Pr$ is the Peclet number and Pr is the Prandtl number. Equations (9.31) - (9.34) are five equations governing the variables ω, ψ, u, v and θ.

The boundary conditions to be satisfied are the no-slip and impermeable conditions, as well as an isothermal wall condition.

Bottom wall $(y = 0)$:

$$\psi = 0, \quad \psi_y = 1, \quad \theta = 1$$

and

$$\omega(x,0) = \frac{-2\psi(x,\Delta y)}{\Delta y^2} + \frac{2}{\Delta y} . \qquad (9.35)$$

Side walls ($x = 0$ and $x = 1$):

$$\psi = 0, \quad \psi_x = 0, \quad \theta = 0$$

and

$$\omega(0, y) = \frac{-2\psi(\Delta x, y)}{\Delta x^2}, \quad \omega(1, y) = -\frac{2\psi(1 - \Delta x, y)}{\Delta x^2}. \tag{9.36}$$

Top wall ($y = 1$):

$$\psi = 0, \quad \psi_y = 0, \quad \theta = 0$$

and

$$\omega(x, 1) = -\frac{2\psi(x, 1 - \Delta y)}{\Delta y^2}. \tag{9.37}$$

The vorticity boundary conditions for Equation (9.31) are derived from the Taylor series expansion of the stream function from the wall to an interior point on a line perpendicular to the wall. For example, on the bottom wall

$$\psi(x, \Delta y) = \psi(x, 0) + \psi_y(x, 0)\Delta y + \psi_{yy}(x, 0)\frac{\Delta y^2}{2} + \cdots .$$

With $\psi(x, 0) = 0$, $\psi_y(x, 0) = 1$ and $\psi_{yy}(x, 0) = -\omega(x, 0)$, one obtains the boundary condition Equation (9.35). Boundary conditions at the corners are specified as

$$\omega(1, 1) = \omega(0, 1) = 0 \quad \text{and} \quad \omega(0, 0) = \omega(1, 0) = \frac{2}{\Delta y}.$$

The cavity flow region is uniformly discretized with 51 × 51 nodes. Calculations are made for Reynolds numbers of 100 and 1000. Thermal fields were calculated for Prandtl numbers of 0.1, 1.0 and 10 for both Reynolds numbers. The vorticity equation [Equation (9.31)], which can be represented by a system of FA algebraic equations [Equation (9.12)], is solved first with an assumed velocity distribution and boundary conditions. This vorticity solution ω is substituted into Equation (9.32), which is solved for the stream function with boundary conditions $\psi = 0$. The velocity components u and v are obtained from the solution of the stream function by a difference expression for ψ_y and ψ_x. Their integral values u_P and v_P, averaged over the subregion, are used as the linearized constants in Equation (9.3). The improved, iterated solution for the vorticity is thus obtained from Equation (9.31). This completes one cycle through the iterative procedure. Once the flow field is solved, the temperature profile can likewise be solved from Equation (9.34). If needed, the pressure distribution in the flow can be obtained once the velocity variables are determined. This is achieved by solving the pressure equation, which is derived by taking the divergence of the momentum equation, yielding

$$\nabla^2 p = 2(u_x v_y - u_y v_x).$$

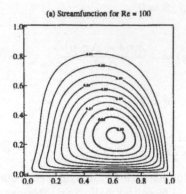
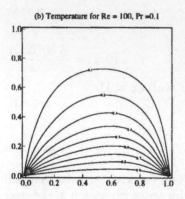
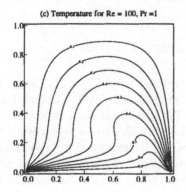
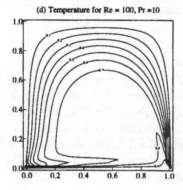

Figure 9.4: Stream function and temperature contours for $Re = 100$. (a) stream-function (b) temperature for $Pr = 0.1$ (c) temperature for $Pr = 1.0$ (d) temperature for $Pr = 10$.

Contour plots of stream function and temperature solutions for $R = 100$ are shown in Figure 9.4. The stream function plot is shown in Figure 9.4 (a). At this Reynolds number, the convective terms in the momentum equation are "stronger" than the diffusion terms, so that the resulting flow shows a strong circulation cell, centered approximately at $(x, y) = (0.7, 0.3)$. Parts (b) through (d) of Figure 9.4 are temperature contour plots each for a different Prandtl number, which is a measure of the relative strengths of kinematic viscosity and the thermal diffusivity. Perhaps in simplistic terms, a large Prandtl number usually implies weak thermal diffusivity relative to the diffusion of momentum. This is evident in Figure 9.4 where the thermal contours gradual approach the stream function contours as the Prandtl number increases. In Figure 9.4(b), the Prandtl number is 0.1, and the thermal contours appear as they would if there were no circulation.

The contours presented in Figure 9.5 are similar to those in Figure 9.4 except for a Reynolds number of 1000. Comparing the stream function contour in part (a) of Figure 9.5 to that in Figure 9.4, the increased relative strength of convective terms allows the flow to "carry" the momentum imparted to the fluid at the wall further into the cell. This is evident in the center of the circulation located closer to the center of the cell.

The strength of this convection is evident in the thermal contour shown in parts (b) through (d) of Figure 9.5. The circulation pattern in this case is suggested in the thermal pattern even in part (b), where the Prandtl number is 0.1. Note that this pattern is very similar to that for the Prandtl number of 1.0 in Figure 9.4(b). The relative lack of thermal diffusion is especially prominent in parts (c) and (d) of Figure 9.5 where the tightly packed thermal contours closely follow the stream lines. Progressively, the interior of the cell is heated primarily by the circulation carrying the slightly warmer fluid near the bottom of the cell into the upper reaches.

Further details and discussion of the solution is given by Chen, Naseri-Neshat and Ho [40] and Chen et al. [37, 45].

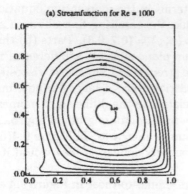
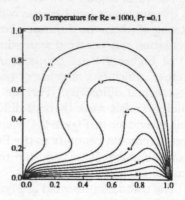

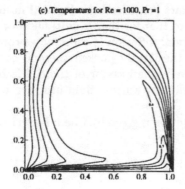
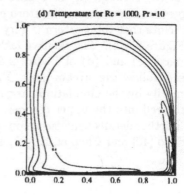

Figure 9.5: Stream function and temperature contours for $Re = 1000$. (a) streamfunction (b) temperature for $Pr = 0.1$ (c) temperature for $Pr = 1.0$ (d) temperature for $Pr = 10$.

Chapter 10

The Three-Dimensional Case

The development of the FA method for the three-dimensional transport equation is outlined in this chapter. The chapter begins by showing how the three-dimensional transport equation can be transformed into a linear, homogeneous, elliptic PDE on a small FA element. Then, several methods are given for formulating FA solutions of the linear, homogeneous transport equation on the element. The first method results in an algebraic formula that incorporates all 27 nodes of the three-dimensional FA element. The initial formulation is for uniform grid spacing. In the next section, the 27-point formulation is adapted for nonuniform grids. Then, two alternative FA formulations for the three-dimensional case are presented. One is a 19-node formula that is derived by treating the three-dimensional problem as three related two-dimensional problems. The next section is devoted to a comparative analysis of the 27- and 19-point schemes. A second alternative is an 11-point formula based on a hybrid FA-FD scheme, and its derivation is outlined in the next section. The chapter concludes with a presentation of 27- and 19-point formulation results for lid-driven cavity flow. The FA solutions are compared with computational results for the same problem found in earlier investigations using different discretization methods.

10.1   Three-Dimensional Transport Equation

Unsteady, three-dimensional, incompressible flow and heat transfer applications are governed by the Navier-Stokes and energy equations presented in Section 2.2.2. The governing equations for momentum (velocity) and energy can be expressed as a general nondimensional convective transport equation

$$R(\phi_t + u\phi_x + v\phi_y + w\phi_z) = \phi_{xx} + \phi_{yy} + \phi_{zz} + Rf , \qquad (10.1)$$

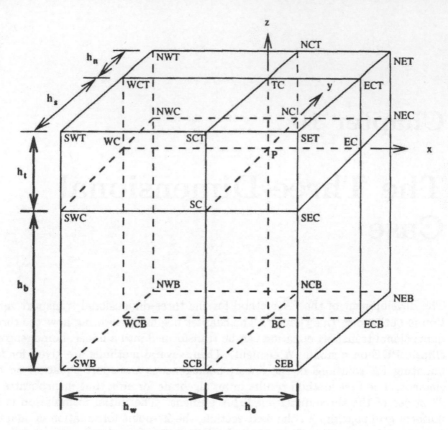

Figure 10.1: Three-dimensional FA subdomain with nonuniform spacing.

where ϕ may represent any one of the dimensionless velocity components or temperature. As stated previously, analytic solution methods for Equation (10.1) are very difficult (if not impossible) due to the coupling of variables, the nonlinearity of the governing equations, and complex geometry and boundary conditions.

To implement the FA method, the flow region is divided into a number of small, three-dimensional elements with nonuniform grid spacing h_E, h_W, h_N, h_S, h_T and h_B in the x, y and z directions, respectively (see Figure 10.1). A linearization scheme similar to that described for the two-dimensional case is employed to obtain a nominally linear convective transport equation from the nonlinear version in Equation (10.1).

In Equation (10.1), let $u = U + u'$, $v = V + v'$ and $w = W + w'$, where U, V, and W are representative constant values for the velocity components on the FA element. They may be the velocities at the interior point P of the FA element, or volume-averaged velocities over the FA element. The primed quantities represent deviations from the respective representative values, and may be thought of as higher-order correction terms.

Substituting these expressions into Equation (10.1), the convective transport equation may be written as

$$R\phi_t + RU\phi_x + RV\phi_y + RW\phi_z = \phi_{xx} + \phi_{yy} + \phi_{zz} + F(x, y, z, t, \phi_j) , \quad (10.2)$$

where F in Equation (10.2) is given by

$$F = Rf(x, y, z, t, \phi_j) - Ru'\phi_x - Rv'\phi_y - Rw'\phi_z . \quad (10.3)$$

In Equation (10.2), ϕ_j may be any dependent variable, including ϕ itself. The source term $f(x, y, z, t, \phi_j)$ and the higher-order correction term, $u'\phi_x + v'\phi_y + w'\phi_z$, of Equation (10.3) are approximated by known values either from the previous time step $t = m - 1$ or from a previous iteration in iterative steady-state formulation. In the case of the higher-order correction term, if the FA element is "sufficiently small", the higher-order correction terms are negligible, and $F \approx f(x, y, z, t, \phi_j)$.

To reduce the complexity in deriving the analytic solution for Equation (10.2), a finite difference expression is employed to approximate the time derivative term, and the nonhomogeneous term is approximated by a representative constant $F_P = f^{m-1}(x, y, z, \phi_j^{m-1})|_P$ on the FA element. With these substitutions, Equation (10.2) is simplified to a linear, elliptic PDE with a constant nonhomogeneous term

$$2A\phi_x + 2B\phi_y + 2C\phi_z - \phi_{xx} - \phi_{yy} - \phi_{zz} = g , \quad (10.4)$$

where

$$A = \frac{1}{2}RU, \quad B = \frac{1}{2}RV, \quad C = \frac{1}{2}RW \quad (10.5)$$

and

$$g = \frac{R}{\tau}(\phi_P^m - \phi_P^{m-1}) + F_P . \quad (10.6)$$

The FA element constants A, B, C and g may differ from one element to another, so the overall nonlinearity of Equation (10.1) is approximately preserved. Furthermore, the coupling of velocity and temperature equations through the convection coefficients A, B, C and source term g, may also be preserved by solving the coupled equations iteratively within each time step.

The last step in the transformation process is to introduce a change of variable to absorb the nonhomogeneous term in Equation (10.4). Let

$$\tilde{\phi} = \phi + \frac{g}{2(A^2 + B^2 + C^2)}(Ax + By + Cz) \quad (10.7)$$

so that $\tilde{\phi}$ satisfies the linear, homogeneous convective transport equation

$$\tilde{\phi}_{xx} + \tilde{\phi}_{yy} + \tilde{\phi}_{zz} = 2A\tilde{\phi}_x + 2B\tilde{\phi}_y + 2C\tilde{\phi}_z \quad (10.8)$$

in the FA element. With the boundary conditions properly specified, Equation (10.8) can be solved analytically by the method of separation of variables to provide the local analytic solution for each small FA element.

10.2   FA 27-Point Solution for Uniform Grids

This section outlines the FA method for uniform grids. In this case $h_e = h_w = h$, $h_n = h_s = k$ and $h_t = h_b = l$ in Figure 10.1. Specifying boundary conditions for the elliptic PDE, given by Equation (10.8), on all six surface boundaries of the FA element will make the problem well-posed. In the FA solution of one- and two-dimensional convective transport problems, several boundary function approximations have been investigated. Among those, a function composed of exponential and linear terms is relatively simple and gives the correct asymptotic behavior for both diffusion and convection dominated cases. Thus, in the FA formulation of unsteady, three-dimensional problems, a generalized linear-exponential boundary function will be employed to approximate the six boundary conditions in terms of the 26 boundary nodes.

As an example, the boundary function on the top surface may be written as

$$\begin{aligned}
\tilde{\phi}_T(x,y) \quad = \quad & a_{T1} + a_{T2}(e^{2Ax} - 1)(e^{2By} - 1) + a_{T3}(e^{2Ax} - 1)y \\
& + a_{T4}(e^{2By} - 1)x + a_{T5}(e^{2Ax} - 1) + a_{T6}(e^{2By} - 1) \\
& + a_{T7}x + a_{T8}y + a_{T9}xy ,
\end{aligned} \tag{10.9}$$

where the coefficients a_{T1}, a_{T2}, ..., a_{T9} are expressed in terms of the nine boundary nodes on the top boundary surface. The boundary functions for the bottom, east, west, north and south sides ($\tilde{\phi}_B(x,y)$, $\tilde{\phi}_E(x,y)$, $\tilde{\phi}_W(x,y)$, $\tilde{\phi}_N(x,y)$ and $\tilde{\phi}_S(x,y)$) can be similarly approximated by exponential and linear functions in terms of the nine nodal points available on each respective boundary surface.

The linearized, homogeneous transport equation (Equation (10.8)), with boundary conditions $\tilde{\phi}_B(x,y)$, $\tilde{\phi}_E(x,y)$, $\tilde{\phi}_W(x,y)$, $\tilde{\phi}_N(x,y)$ and $\tilde{\phi}_S(x,y)$, can be solved analytically by the method of separation of variables and the principle of superposition. A finite analytic algebraic equation is then obtained by evaluating the local analytic solution at the interior point P, which gives the 27-point FA formula

$$\tilde{\phi}_P = \sum_{nb=1}^{26} C_{nb}\tilde{\phi}_{nb}, \tag{10.10}$$

where the subscript "nb" denotes the neighboring nodal points of the interior point P. The FA coefficients C_{nb} in Equation (10.10) are given by

$$
\begin{aligned}
C_{NET} &= Pe^{-Ah-Bk-Cl}, & C_{NWT} &= Pe^{Ah-Bk-Cl}, \\
C_{SET} &= Pe^{-Ah+Bk-Cl}, & C_{SWT} &= Pe^{Ah+Bk-Cl}, \\
C_{NEB} &= Pe^{-Ah-Bk+Cl}, & C_{NWB} &= Pe^{Ah-Bk+Cl}, \\
C_{SEB} &= Pe^{-Ah+Bk+Cl}, & C_{SWB} &= Pe^{Ah+Bk+Cl}, \\
C_{NCT} &= QAe^{-Bk-Cl}, & C_{SCT} &= QAe^{Bk-Cl}, \\
C_{NCB} &= QAe^{-Bk+Cl}, & C_{SCB} &= QAe^{Bk+Cl}, \\
C_{ECT} &= QBe^{-Ah-Cl}, & C_{WCT} &= QBe^{Ah-Cl}, \\
C_{ECB} &= QBe^{-Ah+Cl}, & C_{WCB} &= QBe^{Ah+Cl}, \\
C_{NEC} &= QCe^{-Ah-Bk}, & C_{NWC} &= QCe^{Ah-Bk}, \\
C_{SEC} &= QCe^{-Ah+Bk}, & C_{SWC} &= QCe^{Ah+Bk}, \\
C_{EC} &= RAe^{-Ah}, & C_{WC} &= RAe^{Ah}, \\
C_{NC} &= RBe^{-Bk}, & C_{SC} &= RBe^{Bk}, \\
C_{TC} &= RCe^{-Cl}, & C_{BC} &= RCe^{Cl},
\end{aligned}
\tag{10.11}
$$

where

$$
\begin{aligned}
P &= \frac{1}{\substack{8cosh(Ah)cosh(Bk)cosh(Cl)\\-FA-FB-FC+GA+GB+BC}}, \\
QA &= 2cosh(Ah)(FA-GB-GC), \\
QB &= 2cosh(Bk)(FB-GA-GC), \\
QC &= 2cosh(Cl)(FC-GA-GB), \\
RA &= 4cosh(Bk)cosh(Cl)GA, \\
RB &= 4cosh(Ah)cosh(Cl)GB \quad \text{and} \\
RC &= 4cosh(Ah)cosh(Bk)GC.
\end{aligned}
\tag{10.12}
$$

Values for FA, FB and FC are defined by

$$
\begin{aligned}
FA &= 2Ahcoth(Ah)EA, \\
FB &= 2Bkcoth(Bk)EB \quad \text{and} \\
FC &= 2Clcoth(Cl)EC,
\end{aligned}
\tag{10.13}
$$

with

$$
\begin{aligned}
EA &= E_{12}^{y} + E_{21}^{z}, \\
EB &= E_{21}^{x} + E_{12}^{z}, \\
&= \left(\frac{h}{k}\right)^2 (EA) + \frac{1}{16cosh(Ah)cosh(Bk)cosh(Cl)} \\
&\quad \left[\frac{tanh(Bk)}{Bk} - \left(\frac{h}{k}\right)^2 \frac{tanh(Ah)}{Ah}\right] \quad \text{and} \\
EC &= E_{12}^{x} + E_{21}^{y}
\end{aligned}
$$

$$= \left(\frac{h}{l}\right)^2 (EA) + \frac{1}{16\cosh(Ah)\cosh(Bk)\cosh(Cl)}$$
$$\left[\frac{\tanh(Cl)}{Cl} - \left(\frac{h}{l}\right)^2 \frac{\tanh(Ah)}{Ah}\right]. \tag{10.14}$$

Values for GA, GB and GC are defined by

$$
\begin{aligned}
GA &= 4BkCl \cdot \coth(Bk) \cdot \coth(Cl) \cdot E_{22}^x , \\
GB &= 4AhCl \cdot \coth(Ah) \cdot \coth(Cl) \cdot E_{22}^y \quad \text{and} \\
GC &= 4AhBk \cdot \coth(Ah) \cdot \coth(Bk) \cdot E_{22}^z , \tag{10.15}
\end{aligned}
$$

where E_{ij}^x, E_{ij}^y and E_{ij}^z are double series summations

$$E_{ij}^x = \sum_{q=1}^{\infty}\sum_{r=1}^{\infty} \frac{(-1)^{q+r}(\mu_q k)(\delta_r l)}{2[(Bk)^2 + (\mu_q k)^2]^i[(Cl)^2 + (\delta_r l)^2]^j \cosh(\gamma_{gr} h)} ,$$

$$E_{ij}^y = \sum_{p=1}^{\infty}\sum_{r=1}^{\infty} \frac{(-1)^{p+r}(\lambda_p h)(\delta_r l)}{2[(Cl)^2 + (\delta_r l)^2]^i[(Ah)^2 + (\delta_p h)^2]^j \cosh(\gamma_{pr} k)} \quad \text{and}$$

$$E_{ij}^z = \sum_{p=1}^{\infty}\sum_{q=1}^{\infty} \frac{(-1)^{p+q}(\lambda_p h)(\mu_q k)}{2[(Ah)^2 + (\lambda_p h)^2]^i[(Bk)^2 + (\mu_q k)^2]^j \cosh(\gamma_{pq} l)} , \tag{10.16}$$

where $i, j = 1, 2$, and

$$
\begin{aligned}
\gamma_{pr} &= \sqrt{A^2 + B^2 + C^2 + \lambda_p^2 + \delta_r^2} , \\
\gamma_{qr} &= \sqrt{A^2 + B^2 + C^2 + \mu_p^2 + \delta_r^2} \quad \text{and} \\
\gamma_{pq} &= \sqrt{A^2 + B^2 + C^2 + \lambda_p^2 + \mu_q^2} ,
\end{aligned}
$$

with

$$
\begin{aligned}
\lambda_p h &= \left(p - \frac{1}{2}\right)\pi , \\
\mu_q k &= \left(q - \frac{1}{2}\right)\pi \quad \text{and} \\
\delta_r l &= \left(r - \frac{1}{2}\right)\pi . \tag{10.17}
\end{aligned}
$$

An algebraic formula for ϕ_P is obtained by substituting Equation (10.7) into Equation (10.10) for each of the nodal points. This gives the expression

$$\phi_P = \sum_{nb=1}^{26} C_{nb}\left[\phi_{nb} + \frac{g}{2(A^2 + B^2 + C^2)}(Ax_{nb} + By_{nb} + Cz_{nb})\right] , \tag{10.18}$$

where (x_{nb}, y_{nb}, z_{nb}) is the Cartesian coordinate position of the given neighboring node. Also, g may include the unsteady term, source function and the higher-order correction term used to compensate for the assumption of constant convective velocities on the FA element.

Substituting g of Equation (10.6) into Equation (10.18), a 27-point FA formula for the unsteady, three-dimensional convective transport equation is obtained for the FA element on a uniform grid.

$$\phi_P = \frac{1}{1 + \frac{R}{\tau}C_P} \left[\sum_{nb=1}^{26} C_{nb}\phi_{nb} + \frac{R}{\tau}C_P\phi_P^{n-1} - C_P F_P \right] \tag{10.19}$$

In Equation (10.19),

$$
\begin{aligned}
C_P &= -\sum_{nb=1}^{26} \frac{(Ax_{nb} + By_{nb} + Cz_{nb})C_{nb}}{2(A^2 + B^2 + C^2)} \\
&= \frac{1}{2(A^2 + B^2 + C^2)}(CP_1 - CP_2 \cdot CP_3)
\end{aligned}
$$

with

$$
\begin{aligned}
CP_1 &= Ah \cdot tanh(Ah) + Bk \cdot tanh(Bk) + Cl \cdot tanh(Cl) , \\
CP_2 &= 16cosh(Ah)cosh(Bk)cosh(Cl) , \\
CP_3 &= (Ah)^2(EA) + (Bk)^2(EB) + (Cl)^2(EC)
\end{aligned}
$$

and

$$F_P = f(x, y, z, t, \phi_j) + R[(u'\phi)_x + (v'\phi)_y + (w'\phi)_z]\Big|_P .$$

The nodal values without a superscript in Equation (10.19) denote values evaluated at the nth time step, while ϕ_p^{n-1} and F_P are, respectively, the nodal value and source function (including the higher order correction term) at the interior point P evaluated at the $(n-1)^{th}$ time step.

In the limiting case $Rh^2/\tau \to 0$, the steady-state solution given by Equation (10.18) with $g = F_p$ will be recovered. It can be written as

$$\phi_P = \sum_{nb=1}^{26} C_{nb}\phi_{nb} - C_p F_p . \tag{10.20}$$

The same steady-state solution can also be obtained by equating ϕ_p^{n-1} with ϕ_p^n in Equation (10.19).

This section concludes with examples of FA coefficient values for three different convective flow situations for various Reynolds numbers. Table 10.2 lists the coefficients for $h = k = l$, $Ah = Bk = 0$ and $Cl = 1$, which represents the scenario of flow from the bottom of the cell. Each column of coefficients is for a different Reynolds number, either 0, 1, 10 or 100. When $Re = 0$ (column one), there is no flow, so the nodes closest to node P (those at the center of each of the six element faces) should have equal, and greatest, influence on the value at node P. Note that each of these nodes have coefficient values of 0.1136 in

	0	1	10	100
C_{BC}	0.11363	0.25008	0.81127	0.98010
$C_{ECB} = C_{WCB} = C_{SCB} = C_{NCB}$	0.02394	0.04966	0.04380	0.00495
$C_{SEB} = C_{SWB} = C_{NEB} = C_{NWB}$	0.00386	0.00780	0.00257	0.00003
$C_{EC} = C_{WC} = C_{NC} = C_{SC}$	0.11363	0.09419	0.00072	10^{-16}
$C_{NEC} = C_{NWC} = C_{SEC} = C_{SWC}$	0.02394	0.01960	0.00008	10^{-17}
$C_{ECT} = C_{WCT} = C_{NCT} = C_{SCT}$	0.02394	0.00672	10^{-10}	10^{-89}
$C_{NET} = C_{NWT} = C_{SET} = C_{SWT}$	0.00386	0.00106	10^{-11}	10^{-92}
C_{TC}	0.11363	0.03384	10^{-11}	10^{-87}

Table 10.1: FA coefficients for $h = k = l$ and $Ah = Bk = 0$.

column one. For the case $Re = 1$, the influence of the nodes on the bottom of the FA cell, denoted with a "B" in their subscript, increases and the influence of all other nodes decrease because of the slight convective nature of the flow. Yet even in this case, the facial center nodes still have coefficient value 0.09419, greater than all bottom nodes except the one at the center.

The stronger convective nature of the flow results in very little influence on node P by any node not on the bottom face. This is reflected in the coefficients listed in columns three and four. In fact, in the case $Re = 100$, only the center node on the bottom face has any significant influence on node P.

Tables 10.2 and 10.2 are similar in structure to Table 10.2, but for different flow scenarios. The scenario for Table 10.2 ($h = k = l$, $Ah = 0$ and $Bk = Cl$) is flow from the bottom front edge of the FA element (see Figure 10.1). Note that for the case $Re = 0$, the coefficient values for the element nodes are identical to those in Table 10.2. As in the previous scenario, when the Reynolds number of the flow increases, the position of a node relative to the direction of flow becomes more important than its distance from node P. The coefficients listed in Table 10.2 are for the scenario $h = k = l$ and $Ah = Bk = Cl$, which represents convection from the vertex located at the bottom, front, left corner of the FA element depicted in Figure 10.1.

10.3   FA Formulation on Nonuniform Grids

There are engineering applications for which the use of nonuniform grid spacing is essential (see Section 5.1). Therefore, an analytic solution for a FA element in a nonuniform grid is derived in this section. As an example, consider the case $h_E < h_W$, $h_N < h_S$ and $h_T < h_B$ shown in Figure 10.1. Note that h_e in Figure 10.1 corresponds to h_E in the text. Similarly, $h_w = h_W$, $h_n = h_N$, $h_s = h_S$, $h_t = h_T$ and $h_b = h_B$. The adaptation of formulas from the uniform case to the nonuniform case in three dimensions is similar to the two-dimensional adaptation presented in Section 9.4.

Figure 10.2 shows how a smaller uniform cell of width $2h_E$, depth $2h_N$ and height $2h_T$ is embedded in the original nonuniform element. The algebraic

	0	1	10	20	50
C_{SBC}	0.0239	0.1044	0.5981	0.7181	0.8246
C_{NCT}	0.0239	0.0019	10^{-18}	10^{-35}	10^{-87}
$C_{BC} = C_{SC}$	0.1136	0.2106	0.1596	0.1191	0.0785
$C_{NC} = C_{TC}$	0.1136	0.0285	10^{-10}	10^{-18}	10^{-45}
$C_{EC} = C_{WC}$	0.1136	0.0793	0.0001	10^{-7}	10^{-8}
$C_{NCB} = C_{SET}$	0.0239	0.0141	10^{-9}	10^{-18}	10^{-44}
$C_{SEB} = C_{SWB}$	0.0039	0.016	0.0275	0.0167	0.0080
$C_{NET} = C_{NWT}$	0.0039	0.0003	10^{-19}	10^{-37}	10^{-89}
$C_{ECB} = C_{WCB} =$ $C_{SEC} = C_{SWC}$	0.0239	0.0413	0.0069	0.0026	0.0006
$C_{NEB} = C_{NWB} =$ $C_{SET} = C_{SWT}$	0.0039	0.0022	10^{-10}	10^{-19}	10^{-46}
$C_{NEC} = C_{ECT} =$ $C_{NWC} = C_{WCT}$	0.0239	0.0056	10^{-11}	10^{-20}	10^{-47}

Table 10.2: FA coefficients for $h = k = l$, $Ah = 0$ and $Bk = Cl$.

	0	5	30
C_{SWB}	0.003862	0.255536	0.539682
$C_{SCB} = C_{WCB} = C_{SWC}$	0.023943	0.163256	0.127785
$C_{SEB} = C_{NWB} = C_{SWT}$	0.003862	0.000012	10^{-18}
$C_{SC} = C_{WC} = C_{BC}$	0.113631	0.084868	0.025654
$C_{SEC} = C_{NWC} = C_{SCT}$	0.023943	0.000007	10^{-18}
$C_{WCT} = C_{ECB} = C_{NCB}$	0.023943	0.000007	10^{-18}
$C_{NEB} = C_{SET} = C_{NWT}$	0.003862	0.000007	10^{-18}
$C_{EC} = C_{BC} = C_{TC}$	0.113631	0.000004	10^{-19}
$C_{NEC} = C_{ECT} = C_{NCT}$	0.023943	10^{-10}	10^{-36}
C_{NET}	0.003862	10^{-14}	10^{-53}

Table 10.3: FA coefficients for $h = k = l$ and $Ah = Bk = Cl$.

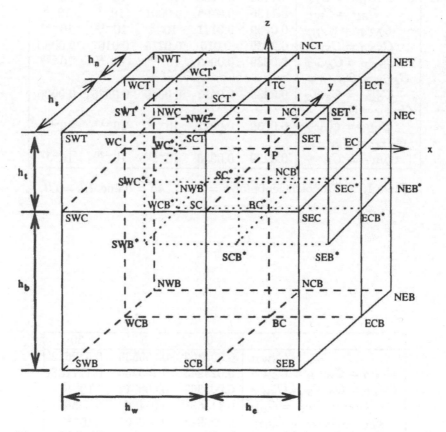

Figure 10.2: Three-dimensional FA element with inscribed uniform cell.

formula for the dependent quantity ϕ_P is

$$\phi_P = \phi_P^* = \frac{1}{1 + \frac{R}{\tau}C_P}\left[\sum_{nb=1}^{26} C_{nb}\phi_{nb}^* + \frac{R}{\tau}C_P\phi_P^{n-1} - C_P F_P\right], \qquad (10.21)$$

where the unknown nodal values ϕ_{nb}^* on the boundaries of the uniform element may be approximated by simple interpolation formulas using the known nodal values ϕ_{nb} on the larger nonuniform grid. To minimize interpolation error, a function consisting of exponential and linear terms will be employed to obtain the interpolated nodal value ϕ_{nb}^*. For example, the unknown nodal value ϕ_{NWT}^* may be approximated using the values ϕ_{NWT}, ϕ_{NCT} and ϕ_{NET} on the boundary of the larger nonuniform element in the formula

$$\phi_{NWT}^* = (s-1)\phi_{NET} + \bar{s}\phi_{NWT} + (2 - s - \bar{s})\phi_{NCT}, \qquad (10.22)$$

where

$$s = \frac{h_W(e^{2Ah_E} + e^{-2Ah_E} - 2)}{h_W(e^{2Ah_E} - 1) + h_E(e^{-2Ah_W} - 1)} \quad \text{and} \quad \bar{s} = s\frac{h_E}{h_W}. \qquad (10.23)$$

Similar formulations may be employed to obtain other interpolated nodal values.

Substituting these interpolated nodal values into Equation (10.21), the following 27-point FA formula for the unsteady, three-dimensional convective transport equation on the nonuniform element is obtained

$$\phi_P = \frac{1}{G + \frac{R}{\tau}b_P}\left[\sum_{nb=1}^{26} b_{nb}\phi_{nb} + \frac{R}{\tau}b_P\phi_P^{n-1} - b_P f_P)\right], \qquad (10.24)$$

where

$$
\begin{aligned}
G &= 1 - (2 - s - \bar{s})C_{WC} - (2 - t - \bar{t})C_{SC} - (2 - r - \bar{r})C_{BC} \\
&\quad -(2 - s - \bar{s})(2 - t - \bar{t})C_{SWC} - (2 - s - \bar{s})(2 - r - \bar{r})C_{WCB} \\
&\quad -(2 - t - \bar{t})(2 - r - \bar{r})C_{SCB} - (2 - s - \bar{s})(2 - t - \bar{t})(2 - r - \bar{r})C_{SWB}, \\
b_{NET} &= C_{NET} + (s-1)C_{NWT} + (t-1)C_{SET} + (r-1)C_{NEB} \\
&\quad +(s-1)(t-1)C_{SWT} + (t-1)(r-1)C_{SEB} + (s-1)(r-1)C_{NWB} \\
&\quad +(s-1)(t-1)(r-1)C_{SWB}, \\
b_{ECT} &= C_{ECT} + (s-1)C_{WCT} + (2 - t - \bar{t})C_{SET} + (r-1)C_{ECB} \\
&\quad +(s-1)(r-1)C_{WCB} + (2 - t - \bar{t})(r-1)C_{SEB} \\
&\quad +(s-1)(2 - t - \bar{t})C_{SWT} + (s-1)(2 - t - \bar{t})(r-1)C_{SWB}, \\
b_{NCT} &= C_{NCT} + (2 - s - \bar{s})C_{NWT} + (t-1)C_{SCT} + (r-1)C_{NCB} \\
&\quad +(2 - s - \bar{s})(t-1)C_{SWT} + (2 - s - \bar{s})(r-1)C_{NWB} \\
&\quad +(t-1)(r-1)C_{SCB} + (2 - s - \bar{s})(t-1)(r-1)C_{SWB}, \\
b_{NEC} &= C_{NEC} + (s-1)C_{NWC} + (t-1)C_{SEC} + (2 - r - \bar{r})C_{NEB} \\
&\quad +(s-1)(t-1)C_{SWC} + (s-1)(2 - r - \bar{r})C_{NWB}
\end{aligned}
$$

$$+(t-1)(2-r-\bar{r})C_{SEB}+(s-1)(t-1)(2-r-\bar{r})C_{SWB},$$

$$b_{EC} = C_{EC}+(s-1)C_{WC}+(2-t-\bar{t})C_{SEC}+(2-r-\bar{r})C_{ECB}$$
$$+(s-1)(2-t-\bar{t})C_{SWC}+(s-1)(2-r-\bar{r})C_{WCB}$$
$$+(2-t-\bar{t})(2-r-\bar{r})C_{SEB}+(s-1)(2-t-\bar{t})(2-r-\bar{r})C_{SWB},$$

$$b_{NC} = C_{NC}+(t-1)C_{SC}+(2-s-\bar{s})C_{NWC}+(2-r-\bar{r})C_{NCB}$$
$$+(2-s-\bar{s})(t-1)C_{SWC}+(t-1)(2-r-\bar{r})C_{SCB}$$
$$+(2-s-\bar{s})(2-r-\bar{r})C_{NWB}+(2-s-\bar{s})(t-1)(2-r-\bar{r})C_{SWB},$$

$$b_{TC} = C_{TC}+(r-1)C_{BC}+(2-s-\bar{s})C_{WCT}+(2-t-\bar{t})C_{SCT}$$
$$+(2-s-\bar{s})(r-1)C_{WCB}+(2-t-\bar{t})(r-1)C_{SCB}$$
$$+(2-s-\bar{s})(2-t-\bar{t})C_{SWT}+(2-s-\bar{s})(r-1)(2-t-\bar{t})C_{SWB},$$

$$b_{NWT} = \bar{s}[C_{NWT}+(t-1)C_{SWT}+(r-1)C_{NWB}+(t-1)(r-1)C_{SWB}],$$

$$b_{NWC} = \bar{s}[C_{NWC}+(t-1)C_{SWC}+(2-r-\bar{r})C_{NWB}+(t-1)(2-r-\bar{r})C_{SWB}],$$

$$b_{WCT} = \bar{s}[C_{WCT}+(2-t-\bar{t})C_{SWT}+(r-1)C_{WCB}+(2-t-\bar{t})(r-1)C_{SWB}],$$

$$b_{WC} = \bar{s}[C_{WC}+(2-t-\bar{t})C_{SWC}+(2-r-\bar{r})C_{WCB}$$
$$+(2-t-\bar{t})(2-r-\bar{r})C_{SWB}],$$

$$b_{SET} = \bar{t}[C_{SET}+(s-1)C_{SWT}+(r-1)C_{SEB}+(s-1)(r-1)C_{SWB}],$$

$$b_{SEC} = \bar{t}[C_{SEC}+(s-1)C_{SWC}+(2-r-\bar{r})C_{SEB}+(s-1)(2-r-\bar{r})C_{SWB}],$$

$$b_{SCT} = \bar{t}[C_{SCT}+(2-s-\bar{s})C_{SWT}+(r-1)C_{SCB}+(2-s-\bar{s})(r-1)C_{SWB}],$$

$$b_{SC} = \bar{t}[C_{SC}+(2-s-\bar{s})C_{SWC}+(2-r-\bar{r})C_{SCB}$$
$$+(2-s-\bar{s})(2-r-\bar{r})C_{SWB}],$$

$$b_{NEB} = \bar{r}[C_{NEB}+(s-1)C_{NWR}+(t-1)C_{SEB}+(s-1)(t-1)C_{SWB}],$$

$$b_{ECB} = \bar{r}[C_{ECB}+(s-1)C_{WCB}+(2-t-\bar{t})C_{SEB}$$
$$+(s-1)(2-t-\bar{t})C_{SWB}],$$

$$b_{NCB} = \bar{r}[C_{NCB}+(2-s-\bar{s})C_{NWB}+(t-1)C_{SCB}$$
$$+(2-s-\bar{s})(t-1)C_{SWB}],$$

$$b_{BC} = \bar{r}[C_{BC}+(2-s-\bar{s})C_{WCB}+(2-t-\bar{t})C_{SCB}$$
$$+(2-s-\bar{s})(2-t-\bar{t})C_{SWB}],$$

$$b_{SEB} = \bar{r}\bar{t}[C_{SEB}+(s-1)C_{SWB}],$$

$$b_{SCB} = \bar{r}\bar{t}[C_{SCB}+(2-s-\bar{s})C_{SWB}],$$

$$b_{NWB} = \bar{r}\bar{s}[C_{NWB}+(t-1)C_{SWB}],$$

$$b_{WCB} = \bar{r}\bar{s}[C_{WCB}+(2-t-\bar{t})C_{SWB}],$$

$$b_{SWT} = \bar{t}\bar{s}[C_{SWT}+(r-1)C_{SWB}],$$

$$b_{SWC} = \bar{t}\bar{s}[C_{SWC}+(2-r-\bar{r})C_{SWB}],$$

$$b_{SWB} = \bar{t}\bar{s}\bar{r}C_{SWB},$$

$$b_p = C_p, \tag{10.}$$

where s and $\bar{s}$ are defined in Equation (10.23), and t, $\bar{t}$, r and $\bar{r}$ are similarly defined as

$$t = \frac{h_S(e^{2Bh_N} + e^{-2Bh_N} - 2)}{h_S(e^{2Bh_N} - 1) + h_N(e^{-2Bh_S} - 1)} \; ,$$

$$\bar{t} = t\frac{h_N}{h_S} \; ,$$

$$r = \frac{h_B(e^{2Ch_T} + e^{-2Ch_T} - 2)}{h_B(e^{2Ch_T} - 1) + h_T(e^{-2Ch_B} - 1)} \quad \text{and}$$

$$\bar{r} = r\frac{h_T}{h_B} \; .$$

The coefficients C_{nb} are given in Equations (10.11) with $h = h_E$, $k = h_N$ and $l = h_T$. For the cases $h_E > h_W$, $h_N > h_S$, etc., the FA solution given by Equation (10.24) can still be used by reversing the flow direction and renaming the indices of neighboring nodes.

10.4   The 19-Point FA Formula

This section presents a method for deriving a FA solution for the three-dimensional transport equation using a 19 node element. Consider the linearized version of the three-dimensional transport equation

$$2A\phi_x + 2B\phi_y + 2C\phi_z - \phi_{xx} - \phi_{yy} - \phi_{zz} = g \; , \tag{10.26}$$

where $2A = RU$, $2B = RV$, $2C = RW$ and R is either the Reynolds or Peclet number. The constants U, V and W are representative velocities and g has possible components of a source term, a correction term for the constant velocities, and a finite difference term in the event that the problem has an unsteady solution.

Equation (10.26) can be transformed to the homogeneous form

$$\psi_{xx} + \psi_{yy} + \psi_{zz} = 2A_P\psi_x + 2B_P\psi_y + 2C_P\psi_z \tag{10.27}$$

by defining

$$\psi = \phi + \frac{g \cdot (Ax + By + Cz)}{2(A^2 + B^2 + C^2)} \; , \tag{10.28}$$

where

$$g = (R\phi_t - F_P) \tag{10.29}$$

and

$$\phi_t \approx \frac{(\phi_P^m - \phi_P^{m-1})}{\Delta t} \; . \tag{10.30}$$

The 19-point FA scheme, proposed by Chen et al. [35, 38], is based on the superposition principle for linear, homogeneous PDEs. Equation (10.27) is

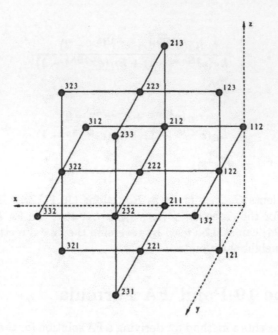

Figure 10.3: FA element for the 19-point formula.

written in an equivalent form as three two-dimensional PDEs.

$$\psi_{xx} + \psi_{yy} = 2A\psi_x + 2B\psi_y + G_z , \tag{10.31}$$

$$\psi_{yy} + \psi_{zz} = 2B\psi_y + 2C\psi_z + G_x \quad \text{and} \tag{10.32}$$

$$\psi_{zz} + \psi_{xx} = 2C\psi_z + 2A\psi_x + G_y , \tag{10.33}$$

where

$$G_z = 2C\psi_z - \psi_{zz} , \tag{10.34}$$

$$G_x = 2A\psi_x - \psi_{xx} \quad \text{and} \tag{10.35}$$

$$G_y = 2B\psi_y - \psi_{yy} . \tag{10.36}$$

Equations (10.31) through (10.33) are similar to the two-dimensional transport equation solved in Chapter 9 except for the source terms. In fact, they represent the two-dimensional flows on the xy, yz and zx planes, respectively, if G_z, G_x and G_y are considered constant over each planar element shown in Figure 10.3. Therefore, one may use the 9-point algebraic formula given by Equation (9.12) to give the following discretized forms for Equations (10.31) - (10.33):

$$\psi_P = \sum_{nb=1}^{8} c_{nb}^z \psi_{nb}^z - c_P^z G_z, \tag{10.37}$$

$$\psi_P = \sum_{nb=1}^{8} c_{nb}^x \psi_{nb}^x - c_P^x G_x, \tag{10.38}$$

$$\psi_P = \sum_{nb=1}^{8} c_{nb}^y \psi_{nb}^y - c_P^y G_y. \tag{10.39}$$

From Equations (10.27), (10.34), (10.35) and (10.36), it follows that

$$G_x + G_y + G_z = 0. \tag{10.40}$$

Combining Equations (10.37) through (10.40), the following algebraic equation for the three-dimensional transport equation is obtained:

$$\psi_P = \sum_{nb=1}^{8} H_{nb}^x \psi_{nb}^x + \sum_{nb=1}^{8} H_{nb}^y \psi_{nb}^y + \sum_{nb=1}^{8} H_{nb}^z \psi_{nb}^z, \tag{10.41}$$

where

$$H_{nb}^x = c_P \frac{c_{nb}^x}{c_P^x},$$

$$H_{nb}^y = c_P \frac{c_{nb}^y}{c_P^y},$$

$$H_{nb}^z = c_P \frac{c_{nb}^z}{c_P^z}$$

and

$$\frac{1}{c_P} = \frac{1}{c_P^x} + \frac{1}{c_P^y} + \frac{1}{c_P^z}. \tag{10.42}$$

Using Equations (10.28) to (10.30), Equation (10.41) can be transformed to the following equation in the original variable ϕ_P:

$$\phi_P = \sum_{nb=1}^{8} H_{nb}^x \phi_{nb}^x + \sum_{nb=1}^{8} H_{nb}^y \phi_{nb}^y + \sum_{nb=1}^{8} H_{nb}^z \phi_{nb}^z + H_P g_P, \tag{10.43}$$

where

$$H_P = -\frac{1}{2(A^2 + B^2 + C^2)} \left[\sum_{nb=1}^{8} H_{nb}^x (Ax + By + Cz)_{nb}^x \right.$$

$$+ \sum_{nb=1}^{8} H_{nb}^y (Ax + By + Cz)_{nb}^y$$

$$\left. + \sum_{nb=1}^{8} H_{nb}^z (Ax + By + Cz)_{nb}^z \right]. \tag{10.44}$$

Equation (10.44) relates the value at the central point P to the values at the 18 neighboring points. Further details concerning the 19-point method are given in Chen et al. [35].

10.5   Analysis of 19- and 27-point FA Schemes

This section analyzes the accuracy of the 19-point finite analytic discretization and other numerical schemes for the steady three-dimensional convection-diffusion equation in a cubic element $2h \times 2h \times 2h$ with $h = 1$. The analysis centers on analyzing the variation of the nodal value with respect to the velocity vector and Reynolds number. The finite analytic algebraic representation of the steady Navier-Stokes equations for an element can now be written as

$$\phi_P = \sum_{i,j,k} C_{ijk}\phi_{ijk} + C_P g_P , \qquad (10.45)$$

where P denotes the center node of the element and coefficient C_{ijk} denotes the finite analytic coefficient associated with the nodal value at (i, j, k). Depending on the scheme used, these coefficients may have a positive or zero value. Values Ah, Bk and Cl (or A, B and C, since $h = k = l = 1$) can be considered the *element Reynolds numbers* in the x, y and z directions, respectively.

Table 10.4 shows the finite analytic coefficients for the 19-point finite analytic scheme for various element Reynolds numbers A, B and C. These values, indexed as C_{ijk}, are obtained from the local analytic solution expressed in Equations (10.43) and (10.44). For example, with the origin located at the bottom northeast corner of the cubic element, C_{123} denotes the nodal coefficient at the first x node, the second y node, and the third z node. The first three columns of Table 10.4 give the x, y and z element Reynolds numbers Ah, Bh, Ch with $h = 1$. The values of C_{ijk} given in the table are the FA coefficients multiplied by 10^5. Note that FA coefficients are all less than one. Table 10.4 also illustrates the gradual variation of coefficients with the velocity vector, or equivalently, the element Reynolds number. This means that the convection and diffusion terms in Equation (10.27) are properly simulated. For example, $A = 30$ and $B = C = 0$ implies that the flow is in the positive x direction. In this case, the node "upwind" from node P is indexed (122) and its corresponding coefficient C_{122} is 0.94008. This may be interpreted to mean that when the element Reynolds number is 30, the boundary node at (122) contributes approximately 94% of the new value at node P, indicating a very strong up-winding effect.

In order to verify the proper representation of the 19-point FA discretization, Figure 10.4 compares the 19-point FA coefficients with the 27-point FA coefficients for six different convective vectors. The value in each block is the corresponding C_{ijk} value. The 27-point FA coefficients are printed at the lower line of the block. The value shown in the block for node P is the corresponding C_P value for the nonhomogeneous or source term. The comparison shows that both the 27-point and the 19-point FA coefficients exhibit the same proper trend in the numerical value variation. It should be mentioned that if a 7-point discretization is used, as is often done in the finite difference and the finite volume methods, the boundary representation can not have the resolution of the 19-point or 27-point FA methods. The 19-point FA scheme has a definite advantage over the more accurate 27-point FA scheme since the 19-point FA coefficients require only a single series summation. Figure 10.5 illustrates the

C233	C123	C223	C323	C213	C132	C232	C332	C122	C222	C322	C112	C212	C312	C231	C121	C221
1489	1489	13686	1489	1489	1489	13686	1489	13689	19645	13686	1489	13689	1489	1489	1489	13689
1363	3024	11825	409	1363	3024	11825	409	29508	17988	3993	3025	11827	409	1363	3025	11827
378	2799	10291	378	2799	843	3496	114	25837	16649	3496	6230	25837	843	378	2800	10293
106	787	3079	106	787	787	3079	106	22753	15549	3079	5819	22753	787	787	5819	22753
633	3608	3478	0	633	3608	3478	0	69112	8354	3	3609	3479	0	633	3609	3479
185	3522	3075	0	1371	1089	1092	0	65379	8156	2	8025	8069	0	185	3523	3076
54	1066	973	0	404	1066	973	0	61926	7981	2	7880	7190	0	404	7880	7190
0	2564	807	0	2564	0	1	0	33864	5938	1	20389	33864	0	0	2565	807
0	788	270	0	788	0	1	0	32124	5898	1	20254	32124	0	0	5823	1996
0	0	0	0	0	0	0	0	15902	5075	0	17428	15902	0	0	17428	15902
122	810	564	0	122	810	564	0	94008	1620	0	810	564	0	122	810	564
36	807	499	0	271	252	179	0	93146	1614	0	1867	1324	0	36	807	499
11	251	159	0	81	251	159	0	92324	1609	0	1861	1177	0	81	1861	1177
0	767	104	0	663	0	0	0	82902	1535	0	7676	6350	0	0	767	104
0	240	35	0	204	0	0	0	82350	1533	0	7669	5950	0	0	1774	259
0	0	0	0	0	0	0	0	75110	1502	0	7511	22353	0	0	7511	2353
0	555	0	0	555	0	0	0	32222	1111	0	33333	32222	0	0	555	0
0	173	0	0	173	0	0	0	31874	1111	0	33333	31874	0	0	1285	0
0	0	0	0	0	0	0	0	27777	1111	0	33333	27777	0	0	5555	0
0	0	0	0	0	0	0	0	0	1111	0	33333	0	0	0	33333	0

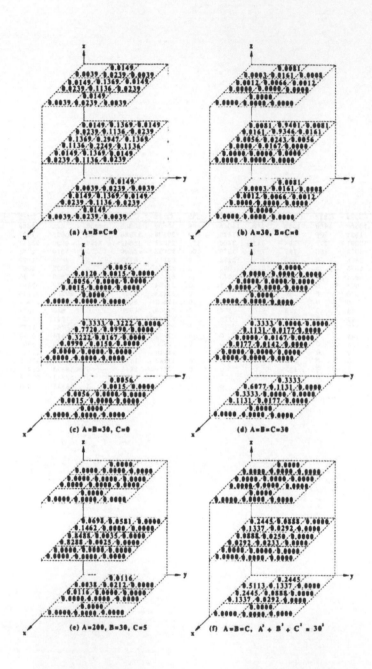

Figure 10.4: Comparison of 19- and 27-point FA Coefficients.

variation of the 19-point FA coefficients for six different convection scenarios as functions of the x element Reynolds number Ah, with $h = 1$.

10.6 11-Point FA Formula

An 11-point FA formula for the three-dimensional transport equation can be derived using a hybrid FA-FD scheme [37, 52]. The derivation of the formula and its subsequent use in calculations is much simpler than the 27- and 19-point formulations.

As in the previous formulations, the derivation begins with the linearized version of the three-dimensional transport equation presented in Equation (10.4). For the sake of reference, it is presented again here.

$$2A\phi_x + 2B\phi_y + 2C\phi_z - \phi_{xx} - \phi_{yy} - \phi_{zz} = g. \tag{10.46}$$

Recall that the nonhomogeneous term g in Equation (10.46) includes a finite difference representation of the time derivative and the constant term F_P, which includes the nonhomogeneous term of the original transport equation and possibly a correction term for the constant velocity approximations in A, B and C.

Re-arrangement of Equation (10.46) gives

$$2A\phi_x + 2B\phi_y - \phi_{xx} - \phi_{yy} = \phi_{zz} - 2C\phi_z + g . \tag{10.47}$$

Referring to Figure 10.6, the derivative terms on the right-hand side of Equation (10.47) are replaced by constant values approximated using finite differences formed in the z direction across the xy plane on the t^{n-1} time step. Once this is done, Equation (10.47) is cast in the form of Equation (9.3) with g in the former two-dimensional equation equal to the right-hand side of Equation (10.47). Therefore, the FA algebraic formula given in Equation (9.16) can be used to give an algebraic solution for Equation (10.47)

$$\phi_P = \frac{\sum_{nb=1}^{8} C_{nb}\phi_{nb} + C_P(b^2\phi_P^d + (Cb + b^2)\phi_P^u + F_P + R\phi_P^{m-1}/\tau}{l_P + C_P(R/\tau + Ce + 2b^2)} , \tag{10.48}$$

where $b = \frac{1}{\delta_z}$, and C_{nb}, $nb = NC, NE, \cdots, NW$, are the FA coefficients for the two-dimensional transport equation given in Equations (9.13) and (9.14). Equation (10.48) is an eleven point algebraic formula for the three-dimensional, unsteady transport equation, with eight nodal values of ϕ_{nb} on the xy plane, one "upstream" value ϕ_P^u, one "downstream" value ϕ_P^d, and the "initial" value of ϕ at node P at time $t = m - 1$, ϕ_P^{m-1}. It should be remarked that the algebraic solution to Equation (10.47) is obtained with upwind differencing for $C\phi_z$, and is fully elliptic if all variables other than ϕ_P^{m-1} are evaluated at time t^m. If ϕ_P^d is neglected, the FA solution given by Equation (10.48) may be thought of as a partially parabolic formula. Keeping ϕ_P^d in Equation (10.48), but evaluating it at the $(m-1)^{st}$ time step, results in an formula that may be thought of as semi-elliptic.

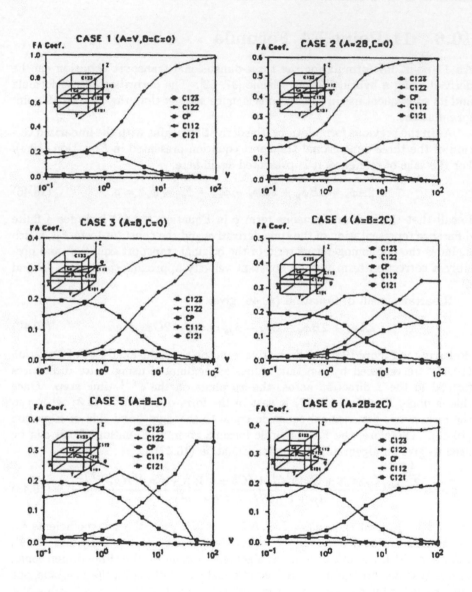

Figure 10.5: FA coefficients for 19-point formula for various convective scenarios.

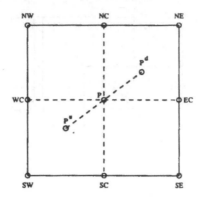

Figure 10.6: FA element for the 11-point formulation.

10.7 FA Solution of 3D Cavity Flow

Lid-driven flow in a three-dimensional cavity has served as a benchmark flow model for the testing of numerical procedures. Chen et al. [35] used both the 27- and 19-point FA solution techniques to solve for the flow in a cubic cavity. This problem is particularly attractive because of its simple geometry and straight-forward boundary conditions. Yet, flow within the cavity has some interesting complexities.

Figure 10.7 depicts the coordinate system and geometry of the three-dimensional cavity problem. The cavity has dimensionless size $1 \times 1 \times 1$. The fluid within the cavity is assumed to be incompressible, having constant transport properties. The governing equations for the problem are the dimensionless Navier-Stokes equations given as Equations (2.7) - (2.8) in Section 2.2.2. The following boundary and initial conditions were used:

Bottom Boundary:
$$u = 1.0, \text{ and } v = w = 0.$$

All Other Boundaries:
$$u = v = w = 0.$$

Initial Conditions:
$$u = v = w = p = 0.$$

Pressure Boundary Condition:
$$p(1, 1, 1) = 0.$$

The momentum equations were solved using both the 27- and 19-point FA formulas. Both uniform and nonuniform grids were used in the study. The continuity equation was used to generate pressure and pressure correction equations by the SIMPLER procedure described by Patankar [145]. Dependent variable values at the nodal locations were calculated by generating systems of equations

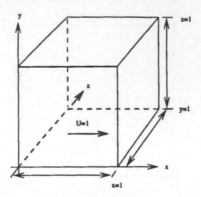

Figure 10.7: Three-dimensional cavity.

and using a line-by-line ADI method with correction schemes similar to those outlined by Kelkar and Pantankar [102]. The computations were made with a time marching procedure with a time-step of $\tau = 0.3$. The flow was assumed to be steady-state at a low Reynolds number, so the numerical solution was deemed time-independent when the error in the mass source (see Patankar [145]) was less than 0.004. Calculations were done for Reynolds numbers of 100, 400 and 1000.

The authors compared their solution for the cubic cavity with numerical results found independently by Goda [73], Cazalbou et al. [31], Hwang and Huynh [85], Ku et al. [108] and Iwatsu et al. [90, 89] for the same problem.

The solutions of the 19-point method for $Re = 100$ were obtained with uniform grids of $5 \times 5 \times 5$, $10 \times 10 \times 10$, and a finer grid of $25 \times 25 \times 25$. The profile of the u velocity component obtained with the 19-point FA method with a $25 \times 25 \times 25$ grid is plotted along the central vertical line in Figure 10.8(a). This profile is compared with the solution obtained with the 27-point FA method and the one obtained by Iwatsu et al. [90, 89], which was accomplished with a modified MAC (Marker and Cell) method on a nonuniform grid of $81 \times 81 \times 81$. There is no appreciable difference among all these solutions. Similar conclusions can be derived from Figure 10.8(b). In this figure, the w velocity component along the horizontal center line on the same symmetry plane of the cavity is illustrated.

Figures 10.9(a) and 10.9(b) show the same profiles of velocities u and w along the vertical and horizontal lines at the center of the corresponding symmetry planes for $Re = 400$. The computational grid in this case also had 25 nodes in each direction. The numerical solution of $Re = 400$ obtained by different researchers, usually with different numerical methods, do not agree as well as the results for $Re = 100$. The FA solution obtained with the 19-point FA method is very close to the more accurate FA solution attained by the 27-point FA method. The solutions of Ku et al. [108] and Iwatsu et al. [90, 89] show a higher velocity near the upper wall when compared with the present FA solution. However, their results predicted a smaller velocity near the center of the cavity.

The program used to compute the 19-point and 27-point FA solutions was the same except for the subroutine that evaluates the FA coefficients. The time per iteration for the 27-point FA method on a $25 \times 25 \times 25$ grid was 1024 seconds. The time per iteration for the 19-point FA method on the same grid was only 400 seconds. Therefore, a greater reduction in computational time can be achieved with the 19-point FA scheme. Nevertheless, the 27-point solution may still serve as a numerical benchmark solution.

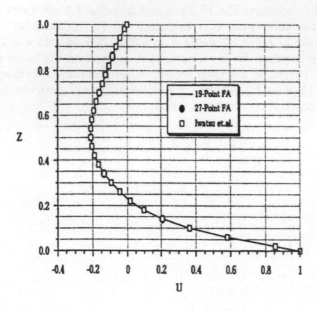

(a) *u* velocity

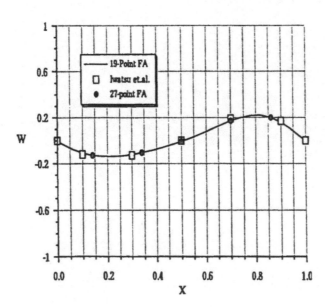

(b) *w* velocity

Figure 10.8: Comparison of velocities at the center planes of symmetry for *Re* = 100

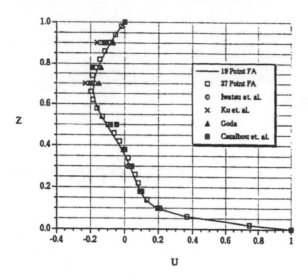

(a) u velocity

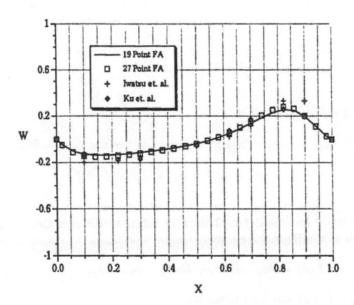

(b) w velocity

Figure 10.9: Comparison of velocities at the center planes of symmetry for Re = 400.

Chapter 11

Stability and Convergence

Stability and convergence properties of the finite analytic numerical method for the two-dimensional transport equation are examined in this chapter. First, three operators that are used to established the desired results are defined. One way to show that the discretized equation generated from a numerical method converges to the "true" solution is to first show that the discretized equation is consistent with the original PDE. Consequently, it is shown that the discretized equation generated with the FA method is, indeed, consistent with the governing PDE. Next, it is necessary to establish that the FA method is stable. This is done through an examination of the matrix that is generated using the FA coefficients. The desired stability result is accomplished by first proving some important properties about the FA coefficients. Finally, convergence is shown by combining the consistency and stability results.

11.1 Three Operators

The two-dimensional convective transport equation has the general form

$$R(\phi_t + u\phi_x + v\phi_y) = \phi_{xx} + \phi_{yy} + f \,, \tag{11.1}$$

where ϕ is the dependent variable, u is the velocity in the x direction, v is the velocity in the y direction, f is the source term, and R is a dimensionless parameter such as the Reynolds number. Define a nonlinear operator F as

$$F(\phi) = R\phi_t + Ru\phi_x + Rv\phi_y - \phi_{xx} - \phi_{yy} \tag{11.2}$$

so that Equation (11.1) may be written as

$$F(\phi) = f \,. \tag{11.3}$$

Suppose the objective is to solve Equation (11.1) on a rectangular computational domain using the FA method. The domain is discretized using I equally spaced computational nodes in the x direction and J equally spaced computational nodes in the y direction.

Next, the linearized version of F as the linear operator L is defined as

$$L(\phi) = R\phi_t + 2A\phi_x + 2B\phi_y - \phi_{xx} - \phi_{yy} , \qquad (11.4)$$

where $2A = RU$ and $2B = RV$. The terms U and V are representative constant velocities for u and v, respectively, on small FA elements. Using the linearized version of F, Equation (11.1) becomes

$$L(\phi) = f . \qquad (11.5)$$

In the FA method, the linear PDE given in Equation (11.5) is solved analytically on small FA elements. The source term on the right-hand side of Equation (11.5) may include a finite difference approximation for the time derivative ϕ_t, and possibly a higher-order correction term for the velocity approximations in the linear operator L.

As outlined in Section 9.2, the FA solution of Equation (11.5) on a small FA element resulted in an algebraic equation of the form

$$\phi_P^m = \sum_{nb=1}^{8} C_{nb} \cdot \phi_{nb}^m + \frac{R}{\tau}C_P \cdot \phi_P^{m-1} - C_P F_P , \qquad (11.6)$$

where ϕ_P^m represents the value, at time $t = m$, of ϕ at the center node of the FA element shown in Figure 9.1. When FA solutions are found for all interior nodes of the problem domain, the resulting algebraic equations form a system of linear equations. Let A be the coefficient matrix for this system, the elements of which are the FA coefficients of Equation (11.6). The matrix A is a square matrix of size $n = (I \times J)$, where n is the number of interior nodes of the problem domain. Define the discrete linear operator L_n as

$$L_n(\vec{\phi_n}) = A \cdot \vec{\phi_n} , \qquad (11.7)$$

where $\vec{\phi_n}$ solves the matrix equation $L_n(\vec{\phi_n}) = \vec{f_n}$, and is the approximate solution to ϕ, at the discrete nodal locations, of Equation (11.4).

These three operators have been defined as a way of establishing the stability and convergence of the FA method in the case of the two-dimensional transport equation. In the next section, it is shown that the linear operator L_n is a *consistent*, or "proper," representation of the nonlinear operator F.

11.2 Consistency of the FA Solution

In this section, the consistency of the discrete linear operator L_n with the nonlinear operator F is established. To do so, it is necessary to define the norm of the difference $L_n(\vec{\phi_n}) - F(\phi)$. As it stands, this is not possible because $L_n(\vec{\phi_n})$ is an element of $\mathbb{R}^n$, while $F(\phi)$ is an element of $\mathbb{Z}$, the space of twice-differentiable functions on the domain of the problem. Unfortunately, $\mathbb{R}^n$ is not a subspace of $\mathbf{Z}$. To compare outputs of these operators, the linear *restriction* operator $r_n : \mathbf{Z} \mapsto \mathbb{R}^n$ is introduced such that

$$r_n(\phi) = \vec{\phi}_n = (\phi(t_i)) = \begin{bmatrix} \phi(t_1) \\ \phi(t_2) \\ \vdots \\ \phi(t_n) \end{bmatrix}, \tag{11.8}$$

where t_i represents the nodal locations in the discretized domain. Next, we define what it means for two operators to be consistent .

Definition 11.1 *The approximation* $\{L_n, r_n\}$ *is said to be* consistent *with* F *if, for every* $\phi \in \mathbf{Z}$

$$\lim_{n \to \infty, \tau \to 0} \|r_n F(\phi) - L_n(r_n \phi)\| = 0. \tag{11.9}$$

The norm used in Definition 11.1 is the *vector norm* defined as

$$\|\vec{x}\| = \max_i |x_i| \, ,$$

where the maximum is taken over the entire set of FA cells defined by two consecutive time planes and the discretization of the two-dimensional spatial planes. In the case of uniform nodal spacing, as n goes to infinity, the spacing h and k go to zero. The following theorem establishes the consistency of L_n and F.

Theorem 11.1 *(Consistency) The discrete linear operator* L_n, *defined in Equation (11.7), is consistent with the operator* F *given in Equation (11.2).*

Proof: Begin by adding and subtracting the restriction $r_n L$ of the linear operator L inside the norm expression, and use the triangle inequality to write

$$\begin{aligned} \|r_n F(\phi) - L_n(r_n \phi)\| &= \|r_n F(\phi) - r_n L(\phi) + r_n L(\phi) - L_n(r_n \phi)\| \\ &\leq \|r_n F(\phi) - r_n L(\phi)\| + \|r_n L(\phi) - L_n(r_n \phi)\| \, . \end{aligned}$$

The first term on the right-hand side is the vector norm of the difference between the restricted nonlinear and restricted linear operators. The second term is the vector norm of the difference between the restricted linear operator and the discrete linear operator of the FA method. Each of these terms will be analyzed individually. The first term simplifies to

$$\begin{aligned} \|r_n F(\phi) - r_n L(\phi)\| &= \|r_n(F(\phi) - L(\phi))\| \\ &= \|r_n(R(U - u)\phi_x + R(V - v)\phi_y)\| \\ &\leq R\|r_n((U - u)\phi_x)\| + R\|r_n((V - v)\phi_y)\| \\ &= R\max_i |[(U - u)\phi_x]_i| + R\max_i |[(V - v)\phi_y]_i| \, , \end{aligned}$$

where we use the fact that operator r_n is linear, the triangle inequality, the definition of r_n, and the definition of the vector norm. Note that the time derivative term in both operators is the same, so that the difference in the operators depends only on n. The two terms on the right-hand side of the last expression can be estimated by focusing on the FA element.

The differences $(U - u)$ and $(V - v)$ need to be bounded on the FA element. This is accomplished using a Taylor series expansion. Assuming $u(x, y)$ is sufficiently differentiable, and knowing $U = u_P$ (the value of u at node P), it follows that

$$
\begin{aligned}
u - U &= u - u_P \\
&= \left.\frac{\partial u}{\partial x}\right|_P (x - x_P) + \left.\frac{\partial u}{\partial y}\right|_P (y - y_P) + \left.\frac{\partial^2 u}{\partial x^2}\right|_P \frac{(x - x_P)^2}{2} + \\
&\quad \left.\frac{\partial^2 u}{\partial y^2}\right|_P \frac{(y - y_P)^2}{2} + \left.\frac{\partial^2 u}{\partial x \partial y}\right|_P \frac{(x - x_P)(y - y_P)}{2} + \cdots
\end{aligned}
$$

Because the partial derivatives of u are continuous on the closed and bounded FA element, there exist numbers M_{i1}^x and M_{i1}^y such that

$$
\left|\frac{\partial u}{\partial x}\right| \le M_{i1}^x \quad \text{and} \quad \left|\frac{\partial u}{\partial y}\right| \le M_{i1}^y .
$$

This leads to

$$
|u - U| \le M_{i1}^x h + M_{i1}^y k + O(h^2, k^2, hk) ,
$$

where h and k are the length and width of the FA element, respectively. In a similar way, assert the existence of numbers M_{i2}^x and M_{i2}^y such that

$$
|v - V| \le M_{i2}^x h + M_{i2}^y k + O(h^2, k^2, hk) ,
$$

and M_{i3}^x and M_{i3}^y such that

$$
\phi_x \le M_{i3}^x \quad \text{and} \quad \phi_y \le M_{i3}^y .
$$

Using these bounds, write

$$
\begin{aligned}
\|r_n F(\phi) - r_n L(\phi)\| &\le R \max_i [(M_{i1}^x h + M_{i1}^y k + O(h^2, k^2, hk)) M_{i3}^x] \\
&\quad + R \max_i [(M_{i2}^x h + M_{i2}^y k + O(h^2, k^2, hk)) M_{i3}^y] .
\end{aligned}
$$

Because there are a finite number of FA elements, it is possible to find numbers $M_j^x = \max_i(M_{i,j}^x)$ and $M_j^y = \max_i(M_{i,j}^y)$ for $j = 1, 2$ and 3. Therefore,

$$
\begin{aligned}
\|r_n F(\phi) - r_n L(\phi)\| &\le R(M_1^x h + M_1^y k + M_3^x O(h^2, k^2, hk)) \\
&\quad + R(M_2^x h + M_2^y k + M_3^y O(h^2, k^2, hk)) .
\end{aligned}
$$

Now, as $n \to \infty$ the values of h and k go to zero, which gives

$$
\lim_{n \to \infty} \|r_n F(\phi) - r_n L(\phi)\| = 0 . \tag{11.10}
$$

Next, consider the second term $\|r_n L(\phi) - L_n(r_n \phi)\|$, the difference between the linear operator L and its discrete approximation L_n. As with any numerical approximation, this term includes errors due to truncation, machine round-off, and accumulation. The definition of the vector norm leads to

$$\|r_d L(\phi) - L_d(r_d \phi)\| = \max_i | (L(\phi))_i - L_d(\phi_i) | \, ,$$

where the difference on the right-hand side is that associated with an arbitrary FA element. In terms of the sources of error on the FA element, we will consider only those due to truncation, believing that the round-off error is significantly less than any associated with truncation error, and ignore the accumulation effects. Truncation error in the FA method includes the interpolation error in approximating the boundary functions on the FA element boundary and the use of a finite difference approximation for the time derivative ϕ_t. The fact that the separation of variables technique provides an analytic solution for ϕ on the FA element means, theoretically, that there are no sources of truncation error other than the two already mentioned. However, the actual calculation of the FA coefficients requires the approximation of an infinite series, so there is some truncation error in reality, but this error can be as small as one wishes by including a sufficient number of terms in the partial sum approximation of the series. Therefore, the difference in the linear operator and discrete linear operator on an arbitrary FA element obeys

$$| (L(\phi))_i - L_d(\phi_i) | \leq E_{bnd} + E_{fd} \, ,$$

where E_{bnd} is the error due to boundary function interpolation, and E_{fd} is that incurred by using a finite difference approximation for ϕ_t.

The latter error is bounded using a Taylor series expansion. That is,

$$\phi^{m-1} = \phi^m + \frac{\partial \phi}{\partial t}(-\tau) + \frac{\partial^2 \phi}{\partial t^2}\frac{(-\tau)^2}{2} + \cdots . \qquad (11.11)$$

Solving for the first derivative gives

$$\frac{\partial \phi}{\partial t} = \frac{\phi^m - \phi^{m-1}}{\tau} + O(\tau^2) \, , \qquad (11.12)$$

and so

$$E_{fd} = \left| \frac{\partial \phi}{\partial t} - \left[\frac{\phi^m - \phi^{m-1}}{\tau} \right] \right| \leq O(\tau^2) \, . \qquad (11.13)$$

The last inequality shows that $\lim_{\tau \to 0} E_{fd} = 0$.

Next, a bound can be constructed for the error E_{bnd} due to the boundary interpolation . To do so, consider the northern boundary of the FA element. Let $\phi^N(x)$ be the actual function for ϕ on the boundary, and let $\phi_a^N(x)$ represent the approximation constructed by the means outlined in Section 9.2. Referring to Appendix B, the true solution for Problem I, where the nonhomogeneous boundary condition is taken along the northern boundary, is

$$w^N(x, y) = \sum_{n=1}^{\infty} A_n sinh(\mu_n[y + k]) sin(\lambda_n[x + h]) \, , \qquad (11.14)$$

where

$$A_n = \frac{1}{h \cdot sinh(2\mu_n k}\int_{-h}^{h} \phi^N(x)sin(\lambda_n[x+h])dx .$$ (11.15)

In a similar way, the approximate solution for this problem can be represented as

$$w_a^N(x,y) = \sum_{n=1}^{\infty} B_n sinh(\mu_n[y+k])sin(\lambda_n[x+h]) ,$$ (11.16)

where

$$B_n = \frac{1}{h \cdot sinh(2\mu_n k)}\int_{-h}^{h} \phi_a^N(x)sin(\lambda_n[x+h])dx .$$ (11.17)

The difference between the actual and approximate solution is found by subtracting Equation (11.16) from Equation (11.14), giving

$$w^N(x,y) - w_a^N(x,y) = \sum_{n=1}^{\infty} (A_n - B_n) sinh(\mu_n[y+k])sin(\lambda_n[x+h]) .$$ (11.18)

The sine and hyperbolic sine are both bounded on the interval $(-h, h)$. Therefore,

$$\left| w^N(x,y) - w_a^N(x,y) \right| \le B \sum_{n=1}^{\infty} |A_n - B_n| ,$$ (11.19)

where B is a bound for both functions on the interval. Inequality (11.19) indicates the need to examine the difference $A_n - B_n$. Using the definitions for these coefficients gives

$$A_n - B_n = \frac{1}{h \cdot sinh(2\mu_n k)}\int_{-h}^{h} \left(\phi^N(x) - \phi_a^N(x) \right) sin(\lambda_n[x+h])dx .$$ (11.20)

Equation (11.20) indicates that the difference between the true boundary function and its approximation must be examined next.

Let $e^N(x) = \phi^N(x) - \phi_a^N(x)$. If we let $x = x_i$ at the center of the boundary, then Taylor's theorem can be used to express $e^N(x)$ along the boundary,

$$e^N(x) = e^N(x_i) + e_x^N(x_i)(x - x_i) + e_{xx}^N(x_i)\frac{(x - x_i)^2}{2} + O((x - x_i)^3) .$$ (11.21)

It is true that $e^N(x_i) = 0$, since ϕ_a^N matches ϕ^N at three nodes on the boundary, including the middle node x_i. The continuity of $e^N(x)$ on the finite boundary assures the existence of a constant M_i^N such that $| e^N(x) | \le M_i^N$ for all x on the boundary. Using this result, the fact that $| x - x_i | < h$, and Equation (11.21) yields

$$|e^N(x)| \le M_i^N \cdot h + O(h^2) .$$ (11.22)

Using Equation (11.22) in Equation (11.20), and the fact that $sinh(2\mu_n k) \geq n$ for $n \geq 1$, results in

$$
\begin{aligned}
|A_n - B_n| &\leq \frac{M_i^N h}{h \cdot sinh(2\mu_n k)} \left| \int_{-h}^{h} sin(\lambda_n[x+h]) dx \right| \\
&= \frac{M_i^N}{sinh(2\mu_n k)} \frac{4h}{n\pi} \\
&\leq \frac{M_i^N 4h}{\pi n^2} \\
&= \frac{K_i^N h}{n^2} ,
\end{aligned}
\tag{11.23}
$$

where $K_i^N = 4M_i^N/\pi$.

Substituting the result in Equation (11.23) into Equation (11.19) gives

$$
\begin{aligned}
\left| w^N(x,y) - w_a^N(x,y) \right| &\leq B \sum_{n=1}^{\infty} \frac{K_i^N h}{n^2} \\
&= BK_i^N h \sum_{n=1}^{\infty} \frac{1}{n^2} .
\end{aligned}
\tag{11.24}
$$

The infinite series in the last expression converges so that the error in the approximation is bounded by a constant times h. Consequently,

$$
lim_{h \to 0} E_{bnd} = lim_{h \to 0} \left| w^N(x,y) - w_a^N(x,y) \right| = 0 .
$$

This same analysis may be applied to the other three sub-problems presented in Appendix B. Consequently, the error on the arbitrary element due to the boundary function approximation will go to zero as h and k tend to zero.

Since there are a finite number of FA elements in the discretization of the problem domain, one can assert the existence of $M = \max_i(B \cdot M_i)$ so that

$$
\|r_d L(\phi) - L_d(r_d \phi)\| \leq M \cdot \max(h,k) + O(\tau^2) ,
$$

and consider the limit of the right-hand side as $n \to \infty$ (that is, $\max(h,k) \to 0$) and $\tau \to 0$. The conclusion is

$$
\lim_{n \to 0, \tau \to 0} \|r_d L(\phi) - L_d(r_d \phi)\| = 0 .
\tag{11.25}
$$

Combining Equations (11.10) and (11.25) yields

$$
\lim_{n \to \infty, \tau \to 0} \|r_n F(\phi) - L_n(r_n \phi)\| = 0 ,
$$

and thus L_n is consistent with F. $\square$

11.3 Stability and Convergence

11.3.1 Stability

In this section, the stability of the FA numerical method is established. A numerical method is *stable* if the resulting solution depends, in a continuous way, on the parameters of the problem. Problem parameters include coefficients in the operator and boundary or initial conditions.

The FA method for solving the unsteady, two-dimensional transport equation results in a linear system of equations was represented in the previous section as $L_n(\vec{\phi}_n) = \vec{f}_n$. In order to examine the effect small perturbations in the problem parameters have on the solution $\vec{\phi}_n$, consider the equation

$$(L_n + \Delta L_n)(\vec{\phi}_n + \Delta\vec{\phi}_n) = \vec{f}_n + \Delta\vec{f}_n , \qquad (11.26)$$

where ΔL_n represents perturbations in the discrete operator L_n, and $\Delta \vec{f}_n$ represents perturbations in the constant vector. In the case of the FA method, the discrete operator is a matrix whose terms are the FA coefficients that depend on the flow parameters, such as the Reynolds number Re, and the interpolated boundary functions. Perturbations in the vector $\vec{f}$ are given by such things as the initial conditions and source terms in the original equation.

The objective is to use Equation (11.26) to gauge the size of $\Delta\vec{\phi}_n$ relative to $\vec{\phi}_n$ in terms of the other perturbations. The desired relationship is given in a theorem from numerical analysis (see Linz [123]).

Theorem 11.2 *If* $L : X \to Y$*, where* X *and* Y *are Banach spaces, has a bounded inverse on* Y*, and if the perturbation* $\Delta L : X \to Y$ *satisfies*

$$\|\Delta L\| < \frac{1}{\|L^{-1}\|},$$

then

$$\frac{\|\Delta x\|}{\|x\|} \leq \frac{k}{1 - k\|\Delta L\|/\|L\|} \left[\frac{\|\Delta y\|}{\|y\|} + \frac{\|\Delta L\|}{\|L\|} \right] , \qquad (11.27)$$

where $k = \|L\|\|L^{-1}\|$.

A *Banach space* is a complete, normed, linear space. In our case, both X and Y are $\mathbb{R}^n$ with finite dimension n. $\mathbb{R}^n$ is a linear space with the usual definitions for vector addition and scalar multiplication. The norm is the vector norm described in the previous section. A space is *complete* if the limit of an arbitrary convergent sequence is an element of the space. This is true for every finite dimensional space.

The inequality stated in the conclusion of Theorem 11.2 establishes the continuous dependence on parameters, which is what is meant by "stability." Recall that changes in the problem parameters result in perturbations in either the operator L or the source term y, represented by ΔL and Δy respectively. Inequality (11.27) may be rearranged to give

$$\|\Delta x\| \leq \left(\frac{k}{1 - k\|\Delta L\|/\|L\|} \left[\frac{\|\Delta y\|}{\|y\|} + \frac{\|\Delta L\|}{\|L\|} \right] \right) \|x\| , \qquad (11.28)$$

which shows that $\|\Delta x\|$ is bounded by the sum of the two perturbations ΔL and Δy. This implies that as these perturbations go to zero, the perturbation in x goes to zero as well. That is, there is no abrupt change in x given an infinitesimal change in problem parameters.

The key condition for stability is the boundedness of the inverse of L. For if either L or L^{-1} is unbounded, then k is not finite, and the inequality given by (11.28) provides no bound on Δx. The norm of an operator , such as L, is defined as

$$\|L\| = \sup_{\|\vec{x}\|=1} \|L\vec{x}\| .$$

In the case of the FA method, it is the boundedness of the norm of the inverse of the coefficient matrix, $\|A^{-1}\|$, that needs to be established. This is done through examination of A. The nature of A will reveal if A^{-1} exists and, consequently, if its norm is bounded.

Let ϕ_P^m represent the value of the dependent variable at an arbitrary interior node of the computational domain at time step $t = m$. As shown in Section 9.2, the FA numerical method results in the following algebraic version of Equation (11.1) for $\phi_{i,j}^m$ in the small FA element

$$\left(1 + \frac{C_P R}{\tau}\right) \phi_P^m - C_{NC}\phi_{NC}^m - C_{NE}\phi_{NE}^m$$
$$-C_{EC}\phi_{EC}^m - C_{SE}\phi_{SE}^m - C_{SC}\phi_{SC}^m$$
$$-C_{SW}\phi_{SW}^m - C_{WC}\phi_{WC}^m - C_{NW}\phi_{NW}^m \;=\; -C_P F_P + \frac{C_P R \phi_P^{m-1}}{\tau} .$$

Collecting similar equations for all the interior nodes results in a linear system of equations $A\vec{\phi}^m = \vec{g}$. The matrix A is sparse, with a given row having at most nine nonzero terms. Formulas for the FA coefficients in Equation (11.29) were given in Section 9.2 and are repeated here for the sake of reference.

$$C_{EC} = EBe^{-Ah} , \qquad C_{NE} = Ee^{-Ah-Bk} ,$$
$$C_{WC} = EBe^{Ah} , \qquad C_{NW} = Ee^{Ah-Bk} ,$$
$$C_{SC} = EAe^{Bk} , \qquad C_{SE} = Ee^{-Ah+Bk} ,$$
$$C_{NC} = EAe^{-Bk} , \qquad C_{SW} = Ee^{Ah+Bk}$$

and

$$C_P =$$
$$\frac{Ah}{2(A^2 + B^2)}[C_{NW} + C_{WC} + C_{SW} - C_{NE} - C_{EC} - C_{SE}]$$
$$+\frac{Bk}{2(A^2 + B^2)}[C_{SW} + C_{SC} + C_{SE} - C_{NW} - C_{NC} - C_{NE}] , \quad (11.29)$$

where

$$E = \frac{1}{4cosh(Ah)cosh(Bk)} - AhE_2coth(Ah) - BkE_2'coth(Bk) ,$$

$$EA = 2Ah\frac{cosh^2 Ah}{sinh(Ah)}E_2, \quad EB = 2Bk\frac{cosh^2 Bk}{sinh(Bk)}E_2',$$

$$E_2 = \sum_{m=1}^{\infty}\frac{-(-1)^m(\lambda_m h)}{[(Ah)^2 + (\lambda_m h)^2]^2 cosh(\mu_m k)},$$

$$E_2' = \sum_{m=1}^{\infty}\frac{-(-1)^m(\lambda_m' k)}{[(Bk)^2 + (\lambda_m' k)^2]^2 cosh(\mu_m' h)},$$

$$\mu_m = \sqrt{A^2 + B^2 + \lambda_m^2}, \quad\quad \lambda_m = \frac{(2m-1)\pi}{2h},$$

$$\mu_m' = \sqrt{A^2 + B^2 + (\lambda_m')^2} \quad \text{and} \quad \lambda_m' = \frac{(2m-1)\pi}{2k}.$$

The following relationship between E_2 and E_2' exists:

$$E_2' = \left(\frac{h}{k}\right)^2 E_2 + \frac{Ak \cdot tanh(Bk) - Bh \cdot tanh(Ah)}{4AkBk \cdot cosh(Ah)cosh(Bk)}.$$

Next, the following important properties concerning the FA coefficients are established.

Property 11.1 *The FA coefficients C_{nb} given in Equations (11.29) are such that $C_{nb} \geq 0$, $nb = NC, NE, \cdots, NW$.*

Proof: The key to establishing this result is in the definitions for EA, EB, and E. These three values are defined in terms of the infinite series represented by E_2 and E_2'. Both series are defined by sequences whose terms alternate in sign and decrease in absolute value. Both series converge by the alternating series test, and the sign of the sum is determined by the sign (positive or negative) of the first term in the sequence. In both cases the first term is positive, so each series is positive. Now we can claim that EA and EB are both positive. Both are defined as quotients, and the hyperbolic cosine term in both cases is positive. The only terms in either quotient that may be negative are simultaneously negative; Ah and $sinh(Ah)$ in the definition of EA, Bk and $sinh(Bk)$ in the definition of EB. Next, notice that E is positive through the definition of E in terms of E_2, E_2' and the relationship between E_2 and E_2'.

Since E, EA and EB are all positive, we know that each of the coefficients C_{EC} through C_{SW} are positive. Finally, the definition of C_P includes Ah and Bk. If $Ah \geq 0$ then $C_{NW} - C_{NE} > 0$, $C_{WC} - C_{EC} > 0$ and $C_{SW} - C_{SE} > 0$. If $Ah < 0$, then each of these differences is negative, and the first term in the definition of C_P is still positive. A similar result is true for the second term in the definition of C_P, which includes Bk. Therefore, each FA coefficient is nonnegative for any value of Ah and Bk. $\square$

The next property states that each FA coefficient is bounded by one, and the sum of the eight neighboring coefficients is equal to one. The fact that their sum is one is crucial to establishing that $\|A^{-1}\|$ is bounded.

Property 11.2 *The finite analytic coefficients C_{nb}, $nb = NC, NE, \cdots, NW$ are bounded by one (that is, $C_{nb} \leq 1$), and satisfy*

$$\sum_{nb=1}^{8} C_{nb} = 1 . \tag{11.30}$$

Proof: The proof begins by showing that the sum of the neighboring coefficients is one.

$$
\begin{aligned}
\sum_{nb=1}^{8} C_{nb} &= EBe^{Ah} + EBe^{-Ah} + EAe^{Bk} + EAe^{-Bk} \\
&\quad + E(e^{Ah+Bk} + e^{-Ah+Bk} + e^{Ah-Bk} + e^{-Ah-Bk}) \\
&= 2EBcosh(Ah) + 2EAcosh(Bk) + E\left[e^{Bk}(e^{Ah} + e^{-Ah})\right. \\
&\quad \left. + e^{-Bk}(e^{Ah} + e^{-Ah})\right] \\
&= 2EBcosh(Ah) + 2EAcosh(Bk) + E\left[e^{Bk}2cosh(Ah)\right. \\
&\quad \left. + e^{-Bk}2cosh(Ah)\right] \\
&= 2EBcosh(Ah) + 2EAcosh(Bk) + 4Ecosh(Bk)cosh(Ah) .
\end{aligned}
$$

Substituting for EA, EB and E gives

$$
\begin{aligned}
\sum_{nb=1}^{8} C_{nb} &= 4Bkcosh(Ah)\frac{cosh^2(Bk)}{sinh(Bk)}E_2' \\
&\quad + 4Ahcosh(Bk)\frac{cosh^2(Ah)}{sinh(Ah)}E_2 \\
&\quad + 4\left[\frac{1}{4cosh(Ah)cosh(Bk)} - AhE_2coth(Ah)\right. \\
&\quad \left. - BkE_2'coth(Bk)\right] cosh(Ah)cosh(Bk) .
\end{aligned}
$$

The terms involving E_2 and E_2' cancel in the last equation, leaving

$$\sum_{nb=1}^{8} C_{nb} = \frac{4cosh(Ah)cosh(Bk)}{4cosh(Ah)cosh(Bk)} = 1 .$$

Now that it has been shown that the sum of the neighboring coefficients is one, and that each coefficient is nonnegative (Property 11.1), it follows that no coefficient can be greater than one. This completes the proof. □

These important FA coefficient properties will be used to establish the following theorem.

Theorem 11.3 *The coefficient matrix A of the linear operator L_n associated with the FA method is diagonally dominant.*

Proof: A matrix is *diagonally dominant* if the absolute value of each diagonal term is greater than the sum of the absolute values of all other terms in the row of the diagonal element. For an arbitrary row in A, the diagonal element is $1 + \frac{C_P}{\tau}$. Since C_P and τ are positive, the diagonal element is greater than one. There are eight remaining elements in the given row, and they are the FA coefficients for the eight neighboring nodes of the node represented by the diagonal element. Properties 11.1 and 11.2 guarantee that since each coefficient is positive and their sum is one, the sum of the absolute values of the remaining row elements is less than $1 + \frac{C_P}{\tau}$. Therefore, A is diagonally dominant. $\square$

Because A is diagonally dominant, it is assured that A^{-1} exists (see, for example, Lay [112]). Since the Banach spaces in the problem are finite dimensional, we know that $\|A^{-1}\|$ is bounded. Knowing this, Theorem 11.2 guarantees the stability for the FA method, provided the perturbation ΔL_n is sufficiently small. The preceding arguments and Theorem 11.2 establish the validity of the following stability theorem for the FA method.

Theorem 11.4 *(Stability) The approximation $\{\ L_n, r_n\ \}$, where L_n is the discrete operator defined by matrix A, and r_n is the restriction operator, is a stable approximation to F, the nonlinear operator associated with the unsteady, two-dimensional transport equation.*

11.3.2 Convergence

This section demonstrates that convergence is an immediate consequence of consistency and stability. This result is stated in the following theorem (see Linz [123]).

Theorem 11.5 *If $\{L_n, r_n\}$ is a stable and consistent approximation to L, then the sequence $\{\vec{x}_n\}$ converges to x, where x solves $L(x) = y$ and $\vec{x}_n$ solves $L_n \vec{x}_n = r_n y = \vec{y}_n$.*

Theorem 11.5 is now restated in the context of the FA method for the two-dimensional transport equation.

Theorem 11.6 *(Convergence) The sequence of discrete FA approximations $(\vec{\phi}_n)$ converges to the true solution ϕ of $F(\phi) = f$ as $n \to \infty$ and $\tau \to 0$.*

The contents of this chapter show that the FA method is a stable method producing a discrete solution that converges to the true solution in the case of the unsteady, two-dimensional transport equation. Similar arguments may be used to establish the same conclusions for the one- and three-dimensional cases. An alternate method for establishing the stability and convergence of the FA method for the two-dimensional transport equation is offered by Zeng and Li [183].

Chapter 12

Hyperbolic PDEs

In this chapter, the finite analytic techniques for hyperbolic PDEs are presented. Hyperbolic PDEs are commonly associated with compressible, supersonic flow. The chapter begins by considering an analytical method that incorporates the use of characteristics for solving a hyperbolic PDE. However, many realistic engineering problems are too complex for analytical techniques. Therefore, the FA numerical method for solving hyperbolic equations is introduced. The FA solution is based on the method of characteristics presented initially. The chapter ends with an example of how the FA method is applied to a supersonic flow in an asymmetric expanding channel.

12.1 Hyperbolic Equations

As shown in Chapter 3, PDEs may be classified by order, linearity and characteristics. Hyperbolic PDEs are those that have two families of characteristics. Recall that a characteristic family is a set of curves in the xy plane on which given initial data do not guarantee a unique solution to the problem.

As an example, consider the following two equation system of PDEs that governs hydraulic channel flows:

$$\frac{\partial a}{\partial t} + u\frac{\partial a}{\partial x} = -a\frac{\partial u}{\partial x} \tag{12.1}$$

$$\frac{\partial u}{\partial t} + u\frac{\partial u}{\partial x} = g\frac{1}{f(a)}\frac{\partial a}{\partial x} + fc(u) \tag{12.2}$$

In this system, a represents the cross-section velocity and u is the average velocity in the channel. Equation (12.1) is a form of the continuity equation, and Equation (12.2) is the momentum equation in the x direction. The general form of the system shown above is

$$a_1 u_x + b_1 u_y + c_1 v_x + d_1 v_y = q_1, \tag{12.3}$$

$$a_2 u_x + b_2 u_y + c_2 v_x + d_2 v_y = q_2, \tag{12.4}$$

where a_i, b_i, c_i, d_i and q_i may be functions of u, v, x or y.

An example of a second order hyperbolic PDE is

$$(a^2 - \phi_x^2)\phi_{xx} - 2\phi_x\phi_y\phi_{xy} + (a^2 - \phi_y^2)\phi_{yy} = f(\phi_x, \phi_y, \phi, x, y) , \qquad (12.5)$$

where $a^2 = a_0^2 - \frac{\gamma-1}{2}(\phi_x^2 + \phi_y^2)$ is the local sound velocity and ϕ is the velocity potential related to u and v by $u = \phi_x$ and $v = \phi_y$. The general form of Equation (12.5) is

$$Au_{xx} + Bu_{xy} + Cu_{yy} = F(u_x, u_y, u, x, y) , \qquad (12.6)$$

where A, B, C, and F are functions of u, u_x, u_y, x and y.

The general forms of the two examples cited above are hyperbolic if

$$B^2 - 4AC > 0 , \qquad (12.7)$$

where A, B and C correspond to the same symbols in Equation (12.6). In the case of the general first order system given in Equations (12.3) and (12.4),

$$A = a_1c_2 - a_2c_1 , \qquad (12.8)$$

$$B = -a_1d_2 + a_2d_1 - b_1c_2 + b_2c_2 \qquad (12.9)$$

and

$$C = b_1a_2 - b_2d_1 . \qquad (12.10)$$

The families of characteristics are curves $\Gamma(x, y) = constant$ for which

$$\frac{dy}{dx} = \frac{-B \pm \sqrt{B^2 - 4AC}}{2A} . \qquad (12.11)$$

In physical terms, the system example given in Equations (12.1) and (12.2) is hyperbolic when the hydraulic flow is super-critical , or when is has a hydraulic jump . Equation (12.5) is hyperbolic if $\phi_x > a^2$.

12.2 Method of Characteristics

The families of characteristics defined in the previous section provide an analytic method for solving hyperbolic PDEs. Let $\Gamma(x, y) = 0$ be an arbitrary curve in the xy plane. Suppose the Cauchy data u, u_x and u_y are given along Γ. Knowing these data along the curve $\Gamma(x, y) = 0$, the following system of equations may be set up in an effort to determine the second derivatives u_{xx}, u_{xy} and u_{yy} along Γ.

$$\begin{aligned}
Au_{xx} + Bu_{xy} + Cu_{yy} &= F(x, y, u, u_x, u_y) \\
dxu_{xx} + dyu_{xy} + 0 &= du_x \\
0 + dxu_{xy} + dyu_{yy} &= du_y
\end{aligned} \qquad (12.12)$$

Application of Cramer's Rule yields

$$u_{xx} = \frac{\begin{vmatrix} F & B & C \\ du_x & dy & 0 \\ du_y & dx & dy \end{vmatrix}}{\begin{vmatrix} A & B & C \\ dx & dy & 0 \\ 0 & dx & dy \end{vmatrix}} \, , \qquad u_{xy} = \frac{\begin{vmatrix} A & F & C \\ dx & du_x & 0 \\ 0 & du_y & dy \end{vmatrix}}{\begin{vmatrix} A & B & C \\ dx & dy & 0 \\ 0 & dx & dy \end{vmatrix}}$$

and

$$u_{yy} = \frac{\begin{vmatrix} A & B & F \\ dx & dy & du_x \\ 0 & dx & du_y \end{vmatrix}}{\begin{vmatrix} A & B & C \\ dx & dy & 0 \\ 0 & dx & dy \end{vmatrix}} \, . \tag{12.13}$$

The formulas for the second derivatives in Equation (12.13) fail if

$$\begin{vmatrix} A & B & C \\ dx & dy & 0 \\ 0 & dx & dy \end{vmatrix} = 0 \, . \tag{12.14}$$

In order to determine what Γ curves may result in a zero determinant, expand the left-hand side of Equation (12.14). This results in a quadratic equation in $\frac{dy}{dx}$,

$$A \left(\frac{dy}{dx} \right)^2 + B \frac{dy}{dx} + C = 0 \, . \tag{12.15}$$

Solving Equation (12.15) for the derivative gives

$$\frac{dy}{dx} = \frac{-B \pm \sqrt{B^2 - 4AC}}{2A} \, . \tag{12.16}$$

The hyperbolic condition given in Equation (12.7) implies that there are two distinct families of curves, called characteristics, along which the Cauchy data do not guarantee a unique solution for u_{xx}, u_{xy} and u_{yy}. The 'α' family of characteristics corresponds to the case

$$\frac{dy}{dx} = \frac{-B + \sqrt{B^2 - 4AC}}{2A} = \alpha \, , \tag{12.17}$$

and the 'β' family to the case

$$\frac{dy}{dx} = \frac{-B - \sqrt{B^2 - 4AC}}{2A} = \beta \, . \tag{12.18}$$

The only hope for u_{xx}, u_{xy} and u_{yy} to be defined uniquely along a characteristic is for

$$\begin{vmatrix} F & B & C \\ du_x & dy & 0 \\ du_y & dx & dy \end{vmatrix} = \begin{vmatrix} A & F & C \\ dx & du_x & 0 \\ 0 & du_y & dy \end{vmatrix} = \begin{vmatrix} A & B & F \\ dx & dy & du_x \\ 0 & dx & du_y \end{vmatrix} \, . \tag{12.19}$$

The zero determinant conditions given in Equation (12.19) are called *compatibility conditions*. They determine how u, u_x and u_y must vary along the α and β characteristics so that u_{xx}, u_{xy} and u_{yy} may be determined.

As an example, the second determinant in Equation (12.19) is expanded to give

$$A\frac{dy}{dx}du_x + Cdu_y - Fdy = 0 . \tag{12.20}$$

Substituting for $\frac{dy}{dx}$ in Equation (12.20) with α first, then β, gives

$$A\alpha du_x + Cdu_y - Fdy = 0 \tag{12.21}$$

and

$$A\beta du_x + Cdu_y - Fdy = 0 . \tag{12.22}$$

Equations (12.21) and (12.22) can be integrated to determine u_x and u_y along the two characteristic lines α and β. Having done this, u_x and u_y may be used in Equation (12.23) to calculate u along α and β by integrating

$$du = u_x dx + u_y dy . \tag{12.23}$$

Two characteristics pass through every point in the interior of the problem domain. These characteristics may be determined using the slope formulas given in Equations (12.17) and (12.18). In theory, one could follow one of the characteristics until it intersects the curve on which the Cauchy data for the problem is given. Next, the compatibility conditions could be used to determine u_x and u_y on the characteristic. Once this is accomplished, Equation (12.23) could be used to determine u at the arbitrary interior node.

The theoretical method of characteristics provides a solution when the coefficients A, B, C and F are variables, even so far as to make the original PDE quasilinear. In reality, complex forms of A, B, C and F appearing in many engineering problems prevent the method of characteristics from being a practical method for numerical computation.

12.3 FA Method

Analytic solution techniques, such as the method of characteristics outlined in the previous section or Fourier series methods, may be used to solve hyperbolic PDEs. However, the complexity of many engineering applications requires a numerical solution method to solve the governing hyperbolic PDE. Numerical methods include the explicit Lax-Wendroff and the implicit Crank-Nicholson finite difference methods. In this section, the classical method of characteristics will be modified to provide an alternative numerical procedure.

Consider the second-order, hyperbolic PDE

$$A\phi_{xx} + B\phi_{xy} + C\phi_{yy} = F , \tag{12.24}$$

where the coefficients A, B and C, and nonhomogeneous term F may be functions of x, y, ϕ and derivatives of ϕ. The computational domain is discretized

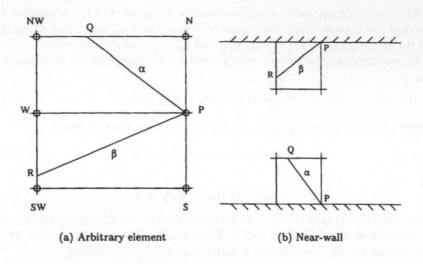

(a) Arbitrary element   (b) Near-wall

Figure 12.1: Arbitrary and near-wall FA elements.

into small FA elements. If Equation (12.24) is not linear, it is made linear on each FA element, as depicted if Figure 12.1(a), by assigning representative constant values $A(P)$, $B(P)$, $C(P)$ and $F(P)$ for the coefficients and nonhomogeneous term on the FA element.

As stated in the previous section, there are two distinct characteristics passing through the node P. These curves are defined by the slope formulas given in Equations (12.17) and (12.18). On the small FA element, these formulas may be accurately approximated by constant values $\alpha(P)$ and $\beta(P)$, which are given by

$$\alpha(P) = \frac{dy}{dx} = \frac{1}{2A(P)} \left[B(P) + \sqrt{B^2(P) - 4A(P)C(P)} \right] \qquad (12.25)$$

and

$$\beta(P) = \frac{dy}{dx} = \frac{1}{2A(P)} \left[B(P) - \sqrt{B(P)^2 - 4A(P)C(P)} \right] . \qquad (12.26)$$

These constant slopes define two straight lines that originate from node P and intersect one of the boundaries of the FA element, as depicted in Figure 12.1(a).

Suppose the straight-line characteristic with slope $\alpha(P)$ intersects the FA element boundary at point Q, and the characteristic with slope $\beta(P)$ intersects the boundary at point R. In general, these points will not coincide with the discrete nodal locations labeled N, NW, WC, SW or S. In this case, values on the boundary must be interpolated from values at the nodes to use the compatibility conditions given in Equations (12.21) and (12.22) to determine ϕ_x and ϕ_y along the characteristics, and more specifically, at node P.

Once these values are determined, the compatibility equations

$$A\alpha d\phi_x + C d\phi_y - F dy = 0 \qquad (12.27)$$

and

$$A\beta d\phi_x + Cd\phi_y - Fdy = 0 \qquad (12.28)$$

are integrated along the characteristics to give

$$\frac{1}{2}[A(P)\alpha(P) + A(Q)\alpha(Q)][\phi_x(P) - \phi_x(Q)]$$

$$+\frac{1}{2}[C(P) + C(Q)][\phi_y(P) - \phi_y(Q)]$$

$$-\frac{1}{2}[F(P) + F(Q)][y(P) - y(Q)] = 0 \qquad (12.29)$$

and

$$\frac{1}{2}[A(P)\beta(P) + A(R)\beta(R)][\phi_x(P) - \phi_x(R)]$$

$$+\frac{1}{2}[C(P) + C(R)][\phi_y(P) - \phi_y(R)]$$

$$-\frac{1}{2}[F(P) + F(R)][y(P) - y(R)] = 0 , \qquad (12.30)$$

from which $\phi_x(P)$ and $\phi_y(P)$ are determined.

Now that $\phi_x(P)$ and $\phi_y(P)$ are known, $\phi(P)$ is obtained by integrating $d\phi = \phi_x dx + \phi_y dy$ along either characteristic to give

$$\phi(P) = \phi_P$$

$$= \phi(Q) + \frac{1}{2}[\phi_x(P) + \phi_x(Q)][x(P) - x(Q)]$$

$$+\frac{1}{2}[\phi_y(P) + \phi_y(Q)][y(P) - y(Q)] \qquad (12.31)$$

or

$$\phi(P) = \phi_P$$

$$= \phi(R) + \frac{1}{2}[\phi_x(P) + \phi_x(R)][x(P) - x(R)]$$

$$+\frac{1}{2}[\phi_y(P) + \phi_y(R)][y(P) - y(R)] . \qquad (12.32)$$

The formula in Equation (12.31) or (12.32) can be written for all interior nodes at a given step of Δx or x_i. The points Q and R, where the characteristic line intersect the FA element, will determine if the FA method for hyperbolic equations is implicit or strictly explicit. If Q and R lie on the *SW-NW* boundary in all the FA elements for the x_i row, then the interpolation procedure uses data from the x_{i-1} row, which are known from the previous step in x. Therefore, the method is explicit. On the other hand, if any of the characteristics intersect the *NW-N* or *SW-S* boundary in their respective element, the interpolation formula includes values for ϕ from the current x_i row, where data is unknown, so the method is implicit.

Figure 12.1(b) shows two typical near-wall FA elements where node P is on the boundary. Known boundary conditions will include ϕ at the x_{i-1} row and, typically, ϕ_y at the x_i row. Then, the solution of $\phi(P)$ on the north wall can be obtained from Equations (12.30) and (12.32) with the corresponding boundary conditions. The solution of $\phi(P)$ on the south wall is determined from Equations (12.29) and (12.31) with the corresponding boundary conditions.

12.4   Supersonic Flow in a 2D Channel

Supersonic flow in a two-dimensional expanding channel is solved using the FA method. The channel is symmetric with respect to the horizontal center line. As shown in Figure 12.2(a), the upper and lower channel wall geometries are described by $y_{max}(x) = N(x)$ and $y_{min}(x) = M(x)$, respectively. For the case of a symmetric channel, $y_{max}(x) = y_{min}(x)$. The fluid is an ideal gas. The inlet velocity profiles ϕ_x and ϕ_y are prescribed on the left boundary.

Under the above assumptions, the governing equation is the full Euler potential equation

$$(a^2 - \phi_x^2)\phi_{xx} + (a^2 - \phi_y^2)\phi_{yy} - 2\phi_x\phi_y\phi_{xy} = 0 , \tag{12.33}$$

with

$$a^2 = a_0^2 - \frac{\gamma - 1}{2}\left(\phi_x^2 + \phi_y^2\right) , \tag{12.34}$$

$$u = \phi_x \quad \text{and} \tag{12.35}$$

$$v = \phi_y . \tag{12.36}$$

The dependent variable ϕ is the velocity potential, a is the local speed of sound and γ is the ratio of specific heats ($\gamma = \frac{c_p}{c_v} = 1.4$). The subscript "0" represents the stagnation state .

The channel inlet is on the left-hand side. Boundary conditions at the inlet may be thought of as "initial" conditions because the computational method calculates values for ϕ in a marching procedure starting with values at the channel inlet and stepping downstream. At the inlet, velocity potential can be specified from the given velocity distribution in x and y components. That is,

$$\phi_x(0, y) = H(y) \tag{12.37}$$

and

$$\phi_y(0, y) = G(y) . \tag{12.38}$$

At this point, the problem is converted from the physical coordinate system to the boundary-fitted coordinate system shown in Figure12.2(b). Define

$$\xi = x \tag{12.39}$$

and

$$\eta = \frac{y - M(x)}{K(x)} , \tag{12.40}$$

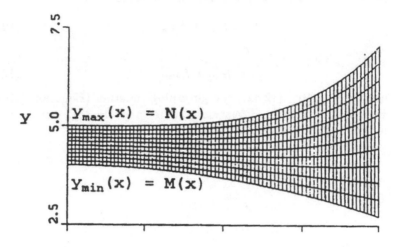

(a) Physical domain

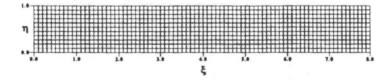

(b) Computational domain

Figure 12.2: Physical and computational domains.

where $K(x) = N(x) - M(x)$. If f is a dependent variable, then

$$f_x = f_\xi \xi_x + f_\eta \eta_x \tag{12.41}$$

and

$$f_y = f_\xi \xi_y + f_\eta \eta_y \,. \tag{12.42}$$

Using Equations (12.39) - (12.42), the governing equation (Equation (12.33)) and the boundary conditions (Equations (12.37) and (12.38)) can be transformed onto the ξ-η computational domain, yielding

$$A\phi_{\xi\xi} + B\phi_{\xi\eta} + C\phi_{\eta\eta} = F \,, \tag{12.43}$$

where

$$A = a^2 - \phi_x^2 \,,$$

$$B = -2\left[\frac{(a^2 - \phi_x^2)(M'(\xi) + \eta K'(\xi)) + \phi_x \phi_y}{K(\xi)}\right] \,,$$

$$\begin{aligned}
C &= \frac{1}{K^2(\xi)}\left[(a^2 - \phi_y^2) + (a^2 - \phi_x^2)(M'(\xi) + \eta K'(\xi))^2\right. \\
&\quad \left. + 2\phi_x \phi_y (M'(\xi) + \eta K'(\xi))\right] \,,
\end{aligned}$$

$$\begin{aligned}
F &= \frac{2\phi_y}{K^2(\xi)}\left[\phi_x \phi_y K'(\xi) + (a^2 - \phi_x^2)[M'(\xi)\right. \\
&\quad \left. + \eta K'(\xi)) - \frac{1}{2}(M''(\xi) + \eta K''(\xi))K(\xi)]\right]
\end{aligned}$$

$$u = \phi_x = \phi_\xi - \frac{M'(\xi) + \eta K'(\xi)}{K(\xi)}\phi_\eta \qquad \text{and}$$

$$v = \phi_y = \frac{\phi_\eta}{K(\xi)} \,.$$

The initial conditions in $\xi - \eta$ coordinates are

$$\phi_\eta(0, \eta) = K(0)g(0, \eta) \tag{12.44}$$

and

$$\phi_\xi(0, \eta) = h(0, \eta) + [M'(0) + \eta K'(0)]g(0, \eta) \,. \tag{12.45}$$

where $g(0, \eta)$ and $h(0, \eta)$ are the transformation of $G(y)$ and $H(y)$ evaluated at $\xi = 0$.

The boundary conditions in $\xi - \eta$ coordinates are

$$\phi_\eta(\xi, 1) = \phi_\xi(\xi, 1)\frac{N'(\xi)K(\xi)}{1 + [M'(\xi) + K'(\xi)]N'(\xi)} \tag{12.46}$$

and

$$\phi_\eta(\xi, 0) = \phi_\xi(\xi, 0) \frac{M'(\xi)K(\xi)}{1 + (M'(\xi))^2}. \tag{12.47}$$

The channel appears simply as a rectangle in boundary-fitted coordinates. The transformed governing equations are solved by the finite analytic numerical procedure described in the previous section on the grid shown in Figure 12.2(b).

Once the FA solution for the velocity components ϕ_ξ and ϕ_η is obtained, the speed of sound a can be computed from Equation (12.34). Then, other flow properties, such as density, pressure, temperature, pressure coefficient and Mach number, can be evaluated using the following isentropic relations:

$$\frac{\rho}{\rho_R} = \left(\frac{a}{a_R}\right)^{\frac{2}{\gamma-1}},$$

$$\frac{T}{T_R} = \left(\frac{a}{a_R}\right)^2,$$

$$\frac{p}{p_R} = \left(\frac{a}{a_R}\right)^{\frac{2\gamma}{\gamma-1}},$$

$$C_P = \frac{\frac{p}{p_R} - 1}{\frac{\gamma M_R^2}{2}} \quad \text{and}$$

$$M = \sqrt{\frac{\phi_x^2 + \phi_y^2}{a^2}},$$

where subscript "R" refers to the reference state.

Since the governing equation is quasi-linear, iteration is needed during computation. In general, only three or four iterations are required for convergence of the solution in each time step.

Figure 12.3 shows contour plots of the local Mach numbers (Figure 12.3(a)), relative Mach numbers (Figure 12.3(b)) and the local speed of sound (Figure 12.3(c)) for an inlet Mach number of 1.5. The symmetric, parabolic, divergent channel has wall geometries given by $y_{max}(x) = 1 + 0.02x^2$ and $y_{min}(x) = -1 - 0.02x^2$. It is seen from the plot of the solution contour that there exist two families of simple wave regions (only upper half shown) near the inlet issuing from the upper and lower edge of the channel. These waves are then reflected at the symmetry plane.

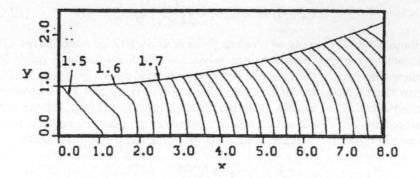

(a) Local Mach number contour (increment of 0.05)

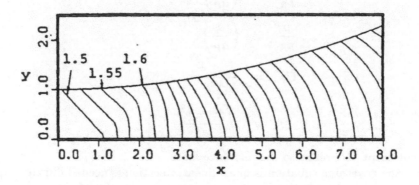

(b) Relative Mach number contour (increment of 0.025)

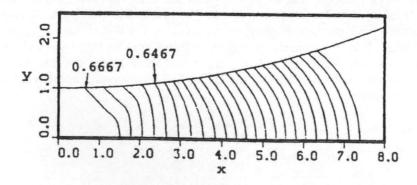

(c) Sound speed contour (increment of 0.005)

Figure 12.3: FA solution of a symmetric channel flow with inlet Mach number of 1.5.

Chapter 13

Explicit Finite Analytic Method

The derivation of an explicit finite analytic method for the two-dimensional transport equation for convection dominated flows is presented in this chapter. The FA method developed in Chapters 7 - 10 is implicit in that the algebraic formulation for the transport equation results in an equation incorporating unknown quantities of the dependent variable at neighboring nodal locations. These implicit formulations are somewhat complicated in their derivation and implementation.

The explicit method outlined in this chapter reduces some of the complication. It is a numerical method based on an analytic solution using characteristics. The chapter begins by describing how the two-dimensional transport equation is reduced to a linear equation when the flow is dominated by convection. Next, it is shown how the homogeneous version of this linear equation is solved analytically, and the how the concept of characteristics is used to devise a numerical method. Once the explicit FA method is described, the results using the new method are compared with analytic and other numerical solutions for a single convective transport equation.

13.1 Convection Dominated Transport Equation

Development begins with the unsteady, two-dimensional transport equation

$$\phi_t + u\phi_x + v\phi_y = \frac{1}{R}(\phi_{xx} + \phi_{yy}) + f , \qquad (13.1)$$

where ϕ is the dependent variable, x and y are the independent spatial variables, t is time, and f is a source term that may depend on x, y, t or ϕ. The convective terms in Equation (13.1) are those containing the velocities u and v in the x and y directions, respectively. The symbol R represents a flow parameter such as the Reynolds number or Peclet number.

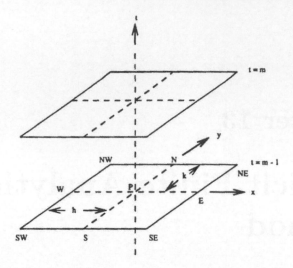

Figure 13.1: Arbitrary cell for the explicit FA method.

The nonlinear nature of Equation (13.1) makes analytic solution methods unlikely for realistic flow situations. Similar to the development of the numerical method presented in previous chapters, the explicit development begins by discretizing the problem domain into many small elements. An example of such an element is shown if Figure 13.1, where time increases along the vertical axis. The small cell includes rectangular planes from time planes labeled $t = m - 1$ and $t = m$.

The next step is to linearize Equation (13.1) on the small element using methods outlined in Chapter 9. The result is

$$\phi_t + U\phi_x + V\phi_y = \frac{1}{R}(\phi_{xx} + \phi_{yy}) + F , \qquad (13.2)$$

where U and V are representative constant velocities for the element, and F is a representative constant source term for the element.

Highly convective flows are characterized by large values of the parameter R. In this case, the diffusion term in Equation (13.2) becomes negligible because of the factor $\frac{1}{R}$. Instead of omitting the diffusion term, assume

$$\frac{1}{R}(\phi_{xx} + \phi_{yy}) \approx \frac{1}{R}\left(\delta_x^2\phi_P^{m-1} + \delta_y^2\phi_P^{m-1}\right) , \qquad (13.3)$$

where

$$\delta_x^2\phi_P^{m-1} = \frac{1}{\Delta x^2}(\phi_{EC}^{m-1} - 2\phi_P^{m-1} + \phi_{WC}^{m-1}) \qquad (13.4)$$

and

$$\delta_y^2\phi_P^{m-1} = \frac{1}{\Delta y^2}(\phi_{NC}^{m-1} - 2\phi_P^{m-1} + \phi_{SC}^{m-1}) . \qquad (13.5)$$

Note that this approximation uses values of ϕ from the $m - 1$ time plane, which is consistent with the goal of formulating an explicit method. If the requirement is made that the representative source term F also be taken from the $m-1$ time plane, a constant g can be defined as

$$g = \frac{1}{R}(\delta_x^2 \phi_P^{m-1} + \delta_y^2 \phi_P^{m-1}) + F_P^{n-1} . \tag{13.6}$$

Substituting g into Equation (13.2) gives the following linear, first-order PDE:

$$\phi_t + U\phi_x + V\phi_y = g \tag{13.7}$$

Note that the constants U, V and g may differ from one element to another, thus the formulation is "nonlinear" on a global basis.

13.2 Analytic Solution

The explicit FA numerical method will be based on the analytical solution to the following initial value problem (IVP), where U and V are constant velocities.

$$\begin{cases} \phi_t + U\phi_x + V\phi_y &= 0 \\ \phi(x,y,0) &= \phi_0(x,y) \end{cases} \tag{13.8}$$

An analytic solution to this IVP is

$$\phi(x,y,t) = \phi_0(x - U \cdot t, y - V \cdot t) . \tag{13.9}$$

A closer look at Equation (13.9) reveals that the solution ϕ at any time t is a copy of the initial condition $\phi_0(x,y)$ translated a distance $U \cdot t$ in the x direction and $V \cdot t$ in the y direction. For example, when U and V are positive, the initial condition translates from the southwest to the northeast as shown in Figure 13.2. In other words, the analytic solution is expressed in terms of the planes $\lambda = x - U \cdot t$ and $\eta = y - V \cdot t$ in the (x,y,t) coordinate system. The line of intersection between these planes is called a *characteristic line*. An example how a characteristic through node P on the time plane $t = m$ looks in relation to the FA cell is shown in Figure 13.3(a). It is evident from the figure that $|\frac{U\Delta t}{\Delta x}| \leq 1$ and $|\frac{V\Delta t}{\Delta y}| \leq 1$ must be satisfied in order to have the characteristic remain within the local element. This condition is known as the *Courant-Friedrichs-Lewy* (CFL) condition.

13.3 FA Solution

The details of the explicit FA method are presented in this section. The process has two main steps. The first is approximating the initial condition on the $m-1$ time plain. The second step uses this approximate initial condition in the analytic solution to Equation (13.7) as a way to calculate ϕ at node P on the $t = m$ plane.

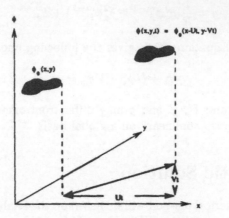

Figure 13.2: Translation of the initial condition.

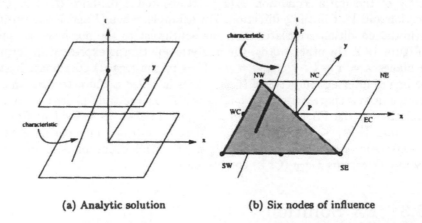

(a) Analytic solution   (b) Six nodes of influence

Figure 13.3: Characteristic and FA solution.

13.3.1 Approximating the Initial Condition

The explicit FA method for Equation (13.7) begins by approximating the initial condition, or in general, the value for ϕ on the $m-1$ time plane. A typical element involves nine nodes. Thus, the initial condition for the linearized convective transport Equation (13.7) may be approximately specified by these nine nodal values. The approximation is accomplished by fitting the second order polynomial

$$\phi_n(x, y, 0) = C_0 + C_1 x + C_2 y + C_3 x^2 + C_4 y^2 + C_5 xy \qquad (13.10)$$

to the known ϕ quantities on the $m-1$ plane. There are six unknowns $C_i (i = 0, 1, ..., 5)$ in Equation (13.10) and nine data points (nodes) on the plane. If all nine were used, the problem of finding the coefficients would be over-determined. However, the location of the characteristic line within the FA element indicates which of these nine nodal values are more important in determining ϕ_P^m.

As an example, the location of the characteristic in the FA cell shown in Figure 13.3(b) suggests that the six nodal values at P, SW, SC, WC, SE and NW are more important than the other nodal values when U and V are positive.

Using only these six data points results in the following expressions for C_i:

$$
\begin{aligned}
C_0 &= \phi_P^{m-1}, \\
C_1 &= \frac{1}{2h}(2\phi_P^{m-1} + \phi_{SW}^{m-1} + \phi_{SE}^{m-1} - 2\phi_{SC}^{m-1} - 2\phi_{WC}^{m-1}), \\
C_2 &= \frac{1}{2h}(2\phi_P^{m-1} + \phi_{SW}^{m-1} + \phi_{NW}^{m-1} - 2\phi_{SC}^{m-1} - 2\phi_{WC}^{m-1}), \\
C_3 &= \frac{1}{2h^2}(\phi_{SW}^{m-1} + \phi_{SE}^{m-1} - 2\phi_{SC}^{m-1}), \\
C_4 &= \frac{1}{2h^2}(\phi_{SW}^{m-1} + \phi_{NW}^{m-1} - 2\phi_{WC}^{m-1}) \quad \text{and} \\
C_5 &= \frac{1}{h^2}(\phi_P^{m-1} + \phi_{SW}^{m-1} - \phi_{WC}^{m-1} - \phi_{SC}^{m-1}).
\end{aligned} \qquad (13.11)
$$

The constants may be determined in a similar way for other flow scenarios. When $U < 0$ and $V > 0$, they are

$$
\begin{aligned}
C_0 &= \phi_P^{m-1}, \\
C_1 &= -\frac{1}{2h}(2\phi_P^{m-1} + \phi_{SW}^{m-1} + \phi_{SE}^{m-1} - 2\phi_{SC}^{m-1} - 2\phi_{EC}^{m-1}), \\
C_2 &= -\frac{1}{2h}(2\phi_P^{m-1} + \phi_{SE}^{m-1} + \phi_{NE}^{m-1} - 2\phi_{SC}^{m-1} - 2\phi_{EC}^{m-1}), \\
C_3 &= \frac{1}{2h^2}(\phi_{SW}^{m-1} + \phi_{SE}^{m-1} - 2\phi_{SC}^{m-1}), \\
C_4 &= \frac{1}{2h^2}(\phi_{SE}^{m-1} + \phi_{NE}^{m-1} - 2\phi_{EC}^{m-1}) \quad \text{and} \\
C_5 &= -\frac{1}{h^2}(\phi_P^{m-1} + \phi_{SE}^{m-1} - \phi_{EC}^{m-1} - \phi_{SC}^{n-1}).
\end{aligned} \qquad (13.12)
$$

When $U > 0$ and $V < 0$, they are given by

$$C_0 = \phi_P^{m-1},$$

$$C_1 = \frac{1}{2h}(2\phi_P^{m-1} + \phi_{NE}^{m-1} + \phi_{NW}^{m-1} - 2\phi_{NC}^{m-1} - 2\phi_{WC}^{m-1}),$$

$$C_2 = -\frac{1}{2h}(2\phi_P^{m-1} + \phi_{SW}^{m-1} + \phi_{NW}^{m-1} - 2\phi_{NC}^{m-1} - 2\phi_{WC}^{m-1}),$$

$$C_3 = \frac{1}{2h^2}(\phi_{NW}^{m-1} + \phi_{NE}^{m-1} - 2\phi_{NC}^{m-1}),$$

$$C_4 = \frac{1}{2h^2}(\phi_{SW}^{m-1} + \phi_{NW}^{m-1} - 2\phi_{WC}^{m-1}) \quad \text{and}$$

$$C_5 = -\frac{1}{h^2}(\phi_P^{m-1} + \phi_{NW}^{m-1} - \phi_{WC}^{m-1} - \phi_{NC}^{m-1}). \tag{13.13}$$

Finally, when both U and V are negative the constants are given by

$$C_0 = \phi_P^{m-1},$$

$$C_1 = -\frac{1}{2h}(2\phi_P^{m-1} + \phi_{NE}^{m-1} + \phi_{NW}^{m-1} - 2\phi_{NC}^{m-1} - 2\phi_{EC}^{m-1}),$$

$$C_2 = -\frac{1}{2h}(2\phi_P^{m-1} + \phi_{SE}^{m-1} + \phi_{NE}^{m-1} - 2\phi_{NC}^{m-1} - 2\phi_{EC}^{m-1}),$$

$$C_3 = \frac{1}{2h^2}(\phi_{NW}^{m-1} + \phi_{NE}^{m-1} - 2\phi_{NC}^{m-1}),$$

$$C_4 = \frac{1}{2h^2}(\phi_{SE}^{m-1} + \phi_{NE}^{m-1} - 2\phi_{EC}^{m-1}) \quad \text{and}$$

$$C_5 = \frac{1}{h^2}(\phi_P^{m-1} + \phi_{NE}^{m-1} - \phi_{EC}^{m-1} - \phi_{NC}^{m-1}). \tag{13.14}$$

The derivation of the above coefficients is given in greater detail by Dai and Chen [60]. Coefficient determination based on the direction of the characteristic is similar to the upwind finite difference scheme used to solve the one-dimensional hyperbolic equation $\phi_t + \bar{u}\phi_x = 0$ (see the book by Anderson et al. [8]). In the one-dimensional case, the explicit finite analytic scheme is identical to the first-order upwind scheme.

13.3.2   Explicit Finite Analytic Scheme

Once the initial condition has been approximated on the $t = m - 1$ plane, the FA solution is obtained by evaluating Equation (13.9) at node P on the $t = m$ plane. That is,

$$\phi_P^m = \phi_{m-1}(-U\Delta t, -V\Delta t) + \Delta t g \tag{13.15}$$

or

$$\phi_P^m = C_0 - C_1 U\Delta t - C_2 V\Delta t + C_3(U\Delta t)^2 + C_4(V\Delta t)^2 + C_5 U V \Delta t^2 + \Delta t g, \tag{13.16}$$

where constants the C_i are given in the previous section.

If the constants U and V are chosen as the velocities u_P^{m-1} and v_P^{m-1} from the previous time step at the central point P, then Equation (13.16) may be written

$$
\begin{aligned}
\phi_P^m \;=\;& C_0 - C_1 u_P^{m-1}\Delta t - C_2 v_P^{m-1}\Delta t + C_3(u_P^{m-1}\Delta t)^2 + C_4(v_P^{m-1}\Delta t)^2 \\
& + C_5(u_P^{m-1} v_P^{m-1}\Delta t^2) + \Delta t g ,
\end{aligned}
\tag{13.17}
$$

where

$$
g = \frac{1}{R}(\delta_x^2 \phi_P^{m-1} + \delta_y^2 \phi_P^{m-1}) + f_P^{m-1} .
$$

The formulas for C_i in Equation (13.17) involve ϕ values from the $t = m - 1$ plane only, making Equation (13.17) explicit. The numerical method is called the *Explicit Finite Analytic Scheme* (EFAS) for this reason.

As mentioned before, the characteristic through node P on the $t = m$ plane must remain within the FA element. This is true if $|\frac{U\Delta t}{\Delta x}| \leq 1$ and $|\frac{V\Delta t}{\Delta y}| \leq 1$. Indeed, the stability of the EFAS is lost if this requirement is not met.

13.4 EFAS Solution for a Single Equation

As an example, consider the linear convective transport equation for a scalar u. If $U = V = 1$, $F = 0$ and $\frac{1}{Re} = \mu$, Equation (13.2) becomes

$$
u_t + u_x + u_y = \mu(u_{xx} + u_{yy}) .
\tag{13.18}
$$

Suppose the initial condition is given by

$$
u(x,y,0) = sin(\pi x) + sin(\pi y) ,
\tag{13.19}
$$

and the boundary conditions are

$$
\begin{aligned}
u(0,y,t) &= e^{-\mu\pi^2 t}(sin(-\pi t) + sin\pi(y - t)) , \\
u(1,y,t) &= e^{-\mu\pi^2 t}(sin(1 - t) + sin\pi(y - t)) , \\
u(x,0,t) &= e^{-\mu\pi^2 t}(sin(x - t) + sin(-\pi t)) \qquad \text{and} \\
u(x,1,t) &= e^{-\mu\pi^2 t}(sin\pi(x - t) + sin\pi(1 - t)) .
\end{aligned}
\tag{13.20}
$$

The exact solution to this problem is

$$
u(x,y,t) = e^{-\mu\pi^2 t}(sin\pi(x - t) + sin\pi(y - t)) .
\tag{13.21}
$$

Since the solution is periodic, only the exact solutions for $t = 1$ and $t = 2$ will be discussed. Numerical solutions for the cases $\mu = 0.001$ and $\mu = 0.0000001$ were computed using the EFAS, an upwind finite difference scheme, and a forward-time, central-space (FTCS) finite difference scheme on a uniform 11×11 grid with a time step of 0.01. A comparison of the numerical results along the x and y centerlines obtained using the EFAS, FTCS and upwind schemes is shown in Figure 13.4. These selected results are indicative the general case

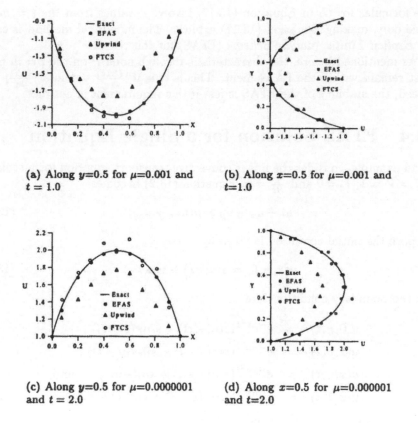

(a) Along $y=0.5$ for $\mu=0.001$ and $t=1.0$

(b) Along $x=0.5$ for $\mu=0.001$ and $t=1.0$

(c) Along $y=0.5$ for $\mu=0.0000001$ and $t=2.0$

(d) Along $x=0.5$ for $\mu=0.000001$ and $t=2.0$

Figure 13.4: Comparison of EFAS and other results.

showing the EFAS results are consistently closer to the exact solution than those calculated using the upwind and FTCS methods for both μ cases.

It should be noted that the basic principles used in developing the explicit finite analytic method for convection dominated two-dimensional flows may be applied to three-dimensional flows as well. The reader is referred to the paper by Dai and Chen [60] for details of the development.

13.5   Exercises for Part II

1. Explain how the FA method includes an "automatic upwinding" effect.

2. Outline the numerical procedure for the method of characteristics to solve the partial differential equation

$$\phi_{tt} - A\phi_{zz} = B ,$$

   where A and B are positive constants. The initial and boundary conditions are

$$
\begin{array}{llll}
\text{I.C.:} & t = 0, & 0 \le x \le 1, & \phi(0, x) = x(1 - x), \\
\text{B.C.:} & t \ge 0, & x = 0, & \phi(t, 0) = 0, \\
\text{B.C.:} & t \ge 0, & x = 1, & \phi(t, 1) = 0.
\end{array}
$$

3. Describe the FA method for deriving the algebraic approximation for

$$(1 + x^2)\phi_x = \phi_{xx} + \phi_{yy} .$$

4. (a) Use the method of separation of variables to solve the following parabolic PDE with the given initial and boundary conditions:

$$\phi_t = \phi_{xx},$$

$$
\begin{array}{llll}
\text{I.C.:} & t = 0, & -h \le x \le h, & \phi(0, x) = \phi_0(x), \\
\text{B.C.:} & x = -h, & -k \le t \le k, & \phi(t, -h) = \phi_w(t), \\
\text{B.C.:} & x = h, & -k \le t \le k, & \phi(t, h) = \phi_e(t).
\end{array}
$$

   (b) If $\phi_w = 0$, $\phi_0 = c_0 =$constant and $\phi_e(t) = a_0 + a_1 t$, find an expression for $\phi(0, k)$ in terms of h, k, c_0, a_0 and a_1.

5. (a) Use the method of separation of variables to solve the following elliptic PDE with the given boundary conditions:

$$\phi_{xx} + \phi_{yy} = 0,$$

$$
\begin{array}{lll}
x = -h, & -k \le y \le k, & \phi(-h, y) = \phi_w(y), \\
x = h, & -k \le y \le k, & \phi(h, y) = \phi_e(y), \\
-h \le x \le h, & y = -k, & \phi(x, -k) = \phi_s(x), \\
-h \le x \le h, & y = k, & \phi(x, k) = \phi_n(x).
\end{array}
$$

   (b) Show that the analytic solution to Burgers' equation is that given in Equation (8.17).

   (c) If $\phi_w = \phi_n = \phi_s = 0$ and $\phi_e(y) = a_0 + a_1 y + a_2 y^2$, find an expression for $\phi(0, 0)$ in terms of h, k, a_0, a_1 and a_2.

6. Derive the boundary coefficient formulas given in Equations 8.7.

7. Show that the functions $\phi(x,y) = 1$, $\phi(x,y) = Ay - Bx$ (A and B constant) and $\phi(x,y) = e^{2Ax+2By}$ (A and B constant) solve the homogeneous, linear form of the second-order transport equation shown in Equation (9.5).

8. The vorticity boundary condition used in the cavity flow simulation of section 9.5 is of the first order in Δx and Δy. Use the Taylor series expansion to derive a second order boundary condition for ω.

9. Outline the numerical procedure for the method of characteristics to solve the following partial differential equation

$$\phi_{tt} - A\phi_{xx} = B ,$$

where A and B are positive constants. Use the following initial and boundary conditions:

$$
\begin{array}{llll}
\text{I.C.:} & t = 0, & 0 \leq x \leq 1, & \phi(0,x) = x(1-x), \\
\text{B.C.:} & t \geq 0, & x = 0, & \phi(t,0) = 0, \\
\text{B.C.:} & t \geq 0, & x = 0, & \phi(t,1) = 0.
\end{array}
$$

10. (Computer Exercise) Write a computer program to solve Burger's equation using the FA 1D method. Do the same for the FA hybrid method and compare your results.

11. (Computer Exercise) This exercise is a continuation of exercise 15 of part I. Consider the following PDE for vorticity ω

$$\frac{\partial \omega}{\partial t} + u \cdot \frac{\partial \omega}{\partial x} = \nu \frac{\partial \omega}{\partial x^2} ,$$

with initial and boundary conditions

$$
\begin{array}{lll}
\text{I.C.:} & t = 0, & \omega = \sin\frac{\pi x}{L}, \\
\text{B.C.:} & x = 0, & \omega = 0, \\
\text{B.C.:} & x = L, & \omega = 0.
\end{array}
$$

The kinematic viscosity ν is 10^{-2} $ft^2 sec$, the reference velocity U of 10 $ft \cdot sec^{-1}$ and length L of 1 give a Reynolds number of $Re = \frac{LU}{\nu}$.

(a) Find an analytic solution for this PDE using the method of separation of variables.

(b) Solve the problem using the finite analytic method. Compute the solution for ω with a uniform spatial increment of $\Delta x = 0.1$ and temporal increment $\Delta t \leq 0.5$.

(c) Plot analytic and numerical results for $t = 0$, 0.1, 0.2, 0.3, 0.4 and 0.5 on the same set of axes. If you completed exercise 15 in part I, include the finite difference solution on the plot as well.

(d) Comment on the accuracy of the numerical results relative to the analytic solution.

Part III

Numerical Grid Generation

The material in Part III of the book is devoted to numerical methods for grid generation. Chapter 14 begins with an introduction and an overview of the three general methods discussed in the following chapters. Chapter 15 gives an introduction to boundary fitted coordinate techniques, which is an example of differential coordinate generation. Chapter 16 is devoted to the diagonal Cartesian method for coordinate generation on complex geometries. The last chapter of Part III concerns the use of the finite analytic method on complex geometries. In this chapter we show how the methodology developed in Part II of the text is adapted to complex geometry applications.

Chapter 14

Introduction to Grid Generation

Advances in high performance computing have made numerical simulation of fluid flows and heat transfer problems possible. Numerical techniques provide excellent results for many problems on rectangular domains. Applications involving cylindrical or spherical boundaries are successful with the proper transformation of the governing PDEs to cylindrical or spherical coordinates. However, simulations involving complex geometries remain a challenging task.

Accurate approximation of the boundary shape and conditions is essential for accurate numerical simulation of fluid flows and heat transfer. Consequently, the goal in grid generation is a transformation of the complex physical domain to a simple rectangular, preferably square, domain on which the crucial boundary conditions may be more accurately approximated.

Grid generation comes with costs. Some generation techniques require a moderate amount of computer coding and solutions may require significant computer resources. Moreover, the governing differential equations must be expressed in the new coordinate variables, complicating the discretization process. Creating acceptable grids in a timely manner has become one of the major obstacles in bringing computational fluid dynamics (CFD) technology out of specialized research group environments and into the design and project engineering areas [130].

The key idea to many grid generation efforts is to assure a boundary in the computational domain corresponds to a boundary in the physical domain. This goal may be referred to as *geometric adaptation*. Additionally, it may be desired to include greater nodal resolution in the physical domain where a dependent variable changes quickly. This process may be referred to as *solution adaptation*. It may be desirable to increase grid resolution in regions where a dependent variable has a large gradient change during the solution process. This process may be referred to as *automatic* or *dynamic* solution adaptation. This technique may also be fruitful for moving boundary problems.

The chapter begins by presenting some general objectives in grid generation and principles one would follow in order to attain these objectives. In the next section, we introduce a general mathematical frame work and terminology for purposes of description and discussion of grid generation methods. This general framework is used in the next section to introduce some classic, or standard, coordinate systems such as polar, cylindrical and spherical. There are several broad categories for more specialized grid generation including algebraic and differential methods systems. We provide a brief overview of these methods in section 14.4. The method of boundary fitted coordinates uses elliptic PDEs, and is of special interest to many researchers and practitioners of computational fluid dynamics. Therefore, we devote Chapter 15 to this topic. The last section of this chapter is devoted to adaptive techniques such as nonuniform grids and multilevel techniques.

This chapter is not intended to be a comprehensive, detailed introduction to grid generation methods. Indeed, the subject has been the focus of much research, over the last twenty years or so, which has lead to a wealth journal articles, conference proceedings, and monographs. If the reader is interested in more extensive survey of coordinate generation techniques, we suggest the book by Thompson et al. [168] The reader is referred to the book by Knupp and Steinberg [106] for a solid mathematical and practical look at the fundamentals of grid generation. The chapter on vector calculus and differential geometry provides important background for a strong fundamental understanding of planar grid generation. For additional mathematical theory on grid generation, we suggest the book edited by Castillo [30].

14.1   Objectives and Principles

The construction of a computational grid should be done in such a way as to attain the following objectives.

A. **Minimize numerical error.** Grid resolution and orientation with respect to flow direction may impact sources of numerical error such as round-off and truncation error.

B. **Provide numerical stability.** In some methods, such the explicit FD and FA methods, the stability of the method depends on the size of the discretization element.

C. **Provide computational economy.** Obviously, more computation is required as the number of grid nodes increases.

D. **Provide ease in handling boundary conditions.** In some cases, boundary conditions may involve normal derivatives. Consequently, it is advantageous for some grid lines to adjoin the boundary in a normal fashion.

Note that some objectives in the list above are at odds with each other. For example, objective C suggests the need for fewer nodes while objectives A and B seem to require more nodes. Indeed, the tension between too few and too many nodes is often at the center of the grid generation issue. The overall objective is the *optimal* grid; the most sparse grid system that provides the desired stability and accuracy.

The principles outlined below may be used to attain one or more of the stated objectives. The objectives addressed by a given principle are listed in parentheses.

1. The problem geometry aligns with the coordinate system. (A, C, and D)

2. If possible, flow and heat flux vectors should run parallel to the coordinate lines. (A)

3. In the case of nonuniform grid spacing, the ratio (larger to smaller) of spacing for two adjacent cells should be less than 2. (A and B)

4. The coordinate system should be orthogonal or nearly orthogonal whenever possible. (A and B)

5. Node density should be proportional to the gradient of a dependent variable. (A and B)

It is known that grid spacing alone affects the amount of error in an approximation. Item number 3 indicates that the rate at which grid spacing changes from one cell to another is another consideration in approximation error. Additionally, the lack of orthogonality in grid lines is another source of error (Mastin [127]), as stated in item number 4.

14.2 Mathematical Framework

The process of generating a computational grid for three-dimensional domains is considered in this section. The problem domain will be referred to as the *physical space*. It is the region where the flow or heat transfer problem is defined by the physical boundaries and corresponding boundary conditions. In general, the shape of the problem domain is not rectangular. The physical domain is *convex* if any two arbitrary points in the domain can be connected by a single line segment that lies entirely within the domain. As we shall see, grid generation on convex domains is often less problematic.

The *logical*, or *indexed*, space is a unit cube with equally spaced coordinate lines represented by ξ, η and ζ. Typically, $0 \leq \xi, \eta, \zeta \leq 1$. The governing differential equations, sometimes called the *hosted equations*, are discretized and solved in the logical space as functions of ξ, η and ζ.

Grid generation may be thought of as a *transformation* , or *mapping*, from the logical space to the physical space. Each node in the logical space, including boundary nodes, is mapped to node in the physical space. The x, y and z coordinates in the physical space of a point with (ξ, η, ζ) coordinates in the logical space are given by *coordinate functions* $x(\xi, \eta, \zeta)$, $y(\xi, \eta, \zeta)$ and

$z(\xi, \eta, \zeta)$, respectively. In the case of a three-dimensional logical space being mapped to a three-dimensional physical space, the transformation is given by three coordinate functions, each a function of the three variables ξ, η and ζ. In vector notation, let $\vec{x} = (x, y, z)$ and $\vec{\xi} = (\xi, \eta, \zeta)$ represents all three coordinate transformation functions as $\vec{x}(\vec{\xi})$.

There may be many different transformations for mapping the logical space to the physical space. However, we require certain characteristics for our transformation. For example, we require that the boundary of the logical space be mapped to the boundary of the physical domain. Such a coordinate map is called *boundary conforming* or *boundary fitted*. Additionally, each point of the logical space must be mapped to a unique point in the physical space. Such a transformation is called *one-to-one*. Finally, each point in the physical domain has a point in the logical domain, called its *pre-image*, that is mapped to it. In this case, the map is said to be *onto*. A transformation that is both one-to-one and onto is called a *homeomorphism*.

Many possible grid transformations remain even under the restrictions outlined in the previous paragraph. However, we can continue to restrict the possibilities by requiring the map to be *smooth*. That is, we may require the coordinate functions $x(\xi, \eta, \zeta)$, $y(\xi, \eta, \zeta)$ and $z(\xi, \eta, \zeta)$ to be differentiable with respect to ξ, η and ζ. A transformation that is one-to-one, onto and smooth is called a *diffeomorphism*.

When the coordinate maps of a grid transformation are smooth, their partial derivatives with respect to ξ, η and ζ are defined and continuous. In this case the *Jacobian matrix* $\mathcal{J}$ becomes a very useful tool in coordinate generation. For the three-dimensional to three-dimensional map, the Jacobian matrix $\mathcal{J}$ is defined as

$$\mathcal{J} = \begin{bmatrix} x_\xi & x_\eta & x_\zeta \\ y_\xi & y_\eta & y_\zeta \\ z_\xi & z_\eta & z_\zeta \end{bmatrix}. \tag{14.1}$$

Using vector notation, the element of $\mathcal{J}$ in row i and column j is expressed as $\vec{x}_{i\xi_j}$.

The Jacobian matrix may be used to determine if a transformation is one-to-one. The *Inverse Mapping Theorem* assures us that if the determinant J of the Jacobian matrix $\mathcal{J}$ is never zero on the interior of the logical space, then the transformation is one-to-one. This result is very important, for in the case of homeomorphic transformations, we know an *inverse mapping* from the physical space to the logical space exists. This inverse map is used to define the coordinate transformation in many applications.

Another very important tool in grid generation theory is the *metric matrix* or *metric tensor* $\mathcal{G}$. It is defined in terms of the Jacobian matrix $\mathcal{J}$ as

$$\mathcal{G} = \mathcal{J}^T \mathcal{J}, \tag{14.2}$$

where $\mathcal{J}^T$ is the transpose of $\mathcal{J}$. To understand one significance of the metric tensor, its individual terms are written in an indexed fashion. That is, if $g_{i,j}$ is the element of $\mathcal{G}$ in row i and column j, then

$$g_{i,j} = \sum_{l=1}^{3} \frac{\partial x_l}{\partial \xi_i} \frac{\partial x_l}{\partial \xi_j}, \quad 1 \le i, j \le 3. \tag{14.3}$$

The indexed notation implies that $x_1 = x$, $x_2 = y$, $x_3 = z$, $\xi_1 = \xi$, $\xi_2 = \eta$, and $\xi_3 = \zeta$. With the indexed definition, it is evident that $g_{i,j} = g_{j,i}$, so that $\mathcal{G}$ is symmetric. Also, the indexed notation shows the terms of $\mathcal{G}$ are actually the dot or *inner product* of the tangent vector $\vec{x}_{\xi_i}$ of the ith coordinate curve with the tangent vector $\vec{x}_{\xi_j}$ of the jth coordinate curve. If the two such curves are orthogonal, then $g_{i,j} = 0$, and conversely. Consequently, a newly generated grid in the physical space is orthogonal if and only if

$$g_{i,j} = 0, \qquad i \ne j.$$

In subsequent sections of this chapter, the framework and notation presented in this section is used to introduce various grid, or coordinate, systems. The next section introduces some standard transformations that result in familiar coordinate systems.

14.3 Standard Coordinate Systems

Coordinate systems such as cylindrical and spherical are commonly used in applications where objects in the physical space have circular boundaries or profiles. We use the mathematical framework for grid generation to introduce these coordinate systems. The coordinate functions in these cases involve trigonometric and other well-know functions.

Polar-Cylindrical Coordinates (r, θ, z)

Suppose the physical space is a sector of an solid annulus with inner radius r_0 and outer radius r_1 such that $0 \le r_i < r_o$, with angle limits θ_0 and θ_1 such that $0 \le \theta_0 < \theta_1 \le 2\pi$ and height limits $h_0 < h_1$. Figure 14.1(a) shows the orientation of the coordinate variables in the physical space. The coordinate functions for x, y and z for this case are

$$\begin{aligned}
x &= r \cdot cos(\theta), \\
y &= r \cdot sin(\theta), \\
z &= h,
\end{aligned}$$

with

$$r = r_i + (r_0 - r_1)\eta, \quad \theta = \theta_0 + (\theta_1 - \theta_0)\xi \quad \text{and} \quad h = h_0 + (h_1 - h_0)\zeta.$$

The definitions of r, θ and h in terms of ξ, η and ζ show why the logical space is sometimes referred to as the index space. The ξ, η and ζ intervals are discretized forming an index for purposes of coding or plotting. As an example, Figure 14.1(b) shows the images of three ζ levels ($\zeta = \frac{1}{3}, \frac{2}{3}$ and 1) for this transformation.

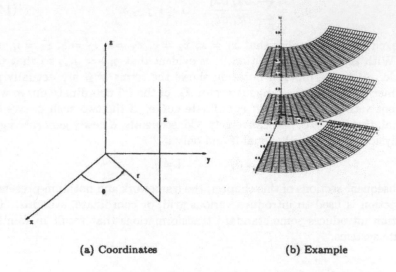

(a) Coordinates   (b) Example

Figure 14.1: Polar-cylindrical coordinates.

Parabolic-Cylindrical Coordinates (r, s, z)

Parabolic-cylindrical coordinates are useful in the event the physical space is defined by a parabolic boundary. The coordinate transformation functions are

$$
\begin{aligned}
x &= \frac{1}{2}(r^2 - s^2), \\
y &= r \cdot s, \\
z &= h,
\end{aligned}
$$

with

$$r = 1 + \xi, \quad s = 1 + \eta \text{ and } h = \zeta.$$

Spherical Coordinates (r, θ, ϕ)

Spherical coordinates are used when an object or boundary in the physical space has a spherical shape. The physical space is defined by the angles θ and ϕ, and the radius r, as shown in Figure 14.2(a). The coordinate transformations are defined as

$$
\begin{aligned}
x &= r \cdot sin\theta cos\phi, \\
y &= r \cdot sin\theta sin\phi, \\
z &= r \cdot cos\theta,
\end{aligned}
$$

with

$$\theta = \theta_0 + (\theta_1 - \theta_0)\xi, \quad \phi = \phi_0 + (\phi_1 - \phi_0)\eta, \quad \text{and} \quad r = r_0 + (r_1 - r_0)\zeta.$$

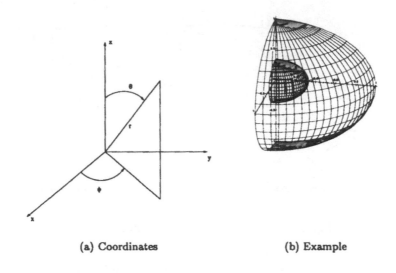

(a) Coordinates   (b) Example

Figure 14.2: Spherical coordinates.

14.4 Algebraic Methods

This section is devoted to coordinate transformations for which the coordinate functions are algebraic. The first type of algebraic map discussed is the *bilinear map*. This two-dimensional transformation is widely used when the physical space is a quadrilateral. The second transformation is the *transfinite interpolation* method, which is often used when the boundaries of a two-dimensional physical space are more general than line segments.

Algebraic methods are popular for their ease and speed in calculation. This is in contrast to the differential methods to be discussed in the next section. However, we will see that one of the disadvantages to algebraic methods is the lack of smoothness in the grid in some cases.

14.4.1 Bilinear Map

Suppose the physical domain is given as the quadrilateral with four corner points (x_1, y_1), (x_2, y_2), (x_3, y_3) and (x_4, y_4), as shown in Figure 14.3(a). The transformation will map the unit square in the logical space to the physical domain using the coordinate functions

$$x(\xi, \eta) = (1 - \xi)(1 - \eta)x_1 + (1 - \xi)\eta x_2 + \xi\eta x_3 + \xi(1 - \eta)x_4$$

and

$$y(\xi, \eta) = (1 - \xi)(1 - \eta)y_1 + (1 - \xi)\eta y_2 + \xi\eta y_3 + \xi(1 - \eta)y_4 \,.$$

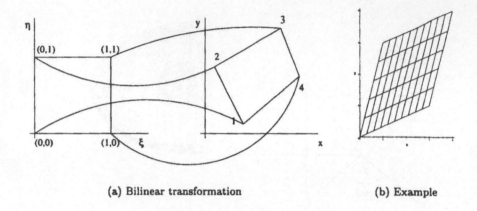

(a) Bilinear transformation   (b) Example

Figure 14.3: Bilinear mapping example.

Figure 14.3(b) shows the result of the bilinear transformation of 10 grid lines in ξ and 5 grid lines in η. Note that the spacing along any one of the quadrilateral boundaries is uniform.

14.4.2   Transfinite Interpolation

The transfinite interpolation method (TFI) is similar to bilinear maps except the boundary of our physical domain is expressed as functions, not necessarily linear, of the variables ξ and η. Figure 14.4(a) shows the relationships between corresponding boundaries in the physical and logical domains. For example, the left boundary of the logical domain is mapped to the left boundary of the physical domain through the function $\vec{x}_l(\eta)$. Note that this function is a vector relationship that specifies both the x and y coordinates of the image point. That is,

$$\vec{x}_l(\eta) = (x_l(\eta), y_l(\eta))$$

Similar functions for the top, right and bottom boundaries are given as well.

The coordinate functions for this transformation are

$$
\begin{aligned}
x(\xi, \eta) = {} & (1-\eta)x_b(\xi) + \eta x_t(\xi) + (1-\xi)x_l(\eta) + \xi x_r(\eta) \\
& -[\xi\eta x_t(1) + \xi(1-\eta)x_b(1) + \eta(1-\xi)x_t(0) \\
& +(1-\xi)(1-\eta)x_b(0)]
\end{aligned}
$$

and

$$
\begin{aligned}
y(\xi, \eta) = {} & (1-\eta)y_b(\xi) + \eta y_t(\xi) + (1-\xi)y_l(\eta) + \xi y_r(\eta) \\
& -[\xi\eta y_t(1) + \xi(1-\eta)y_b(1) + \eta(1-\xi)y_t(0) \\
& +(1-\xi)(1-\eta)y_b(0)] .
\end{aligned}
$$

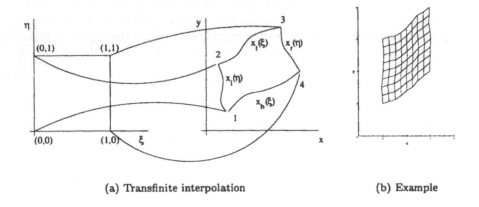

(a) Transfinite interpolation

(b) Example

Figure 14.4: Transfinite interpolation example.

An example of a TFI map is shown in Figure 14.4(b).

14.5 Differential Coordinate Systems

A differential coordinate system (DCS) is one in which the coordinate functions of the transformation are defined using ordinary or partial differential equations. Differential coordinate system transformations are particularly suitable for problems having physical domains with complex or irregular boundaries. They are sometimes preferred over algebraic methods because they may be tailored to ensure smooth and even orthogonal grid generation. Drawbacks include more sophisticated *solvers* (computer algorithms) and increased computer time.

The differential equations used in the transformation have nothing to do with the physical conservation laws that are the basis for the governing equations for fluid flow and heat transfer problems. Therefore, the methods associated with differential coordinate generation are applicable to a wide variety of physics or engineering applications.

Discussion of methods in this section will concern planar transformations. That is, transformations from the two-dimensional logical space to the two-dimensional physical space. This is due, in part, to the fact that much of the theory and application of differential grid generations is the result of *conformal mapping* theory, which is restricted to the two-dimensional case. However, much of the development resulting from conformal mapping, such as elliptical transformations, may be applied to three-dimensional mappings.

14.5.1   Conformal Maps

Let D be a simply connect domain in the physical space given in the xy plane. From the theory of functions of one complex variable, we know there exists a *conformal map* from region D onto a rectangular region R in the logical $\xi\eta$ plane. A conformal map is one that preserves angle measure and sense of direction. If the conformal function is represented by

$$F(z) = F(x + iy) = \xi(x, y) + i\eta(x, y),$$

then the real valued functions ξ and η are differentiable with respect to both x and y. Since the map is conformal we know these functions must satisfy the *Cauchy-Riemann* equations

$$\xi_x = \eta_y \quad \text{and} \quad \xi_y = -\eta_x.$$

As a consequence, we have

$$\xi_x \xi_y + \eta_x \eta_y = 0, \tag{14.4}$$
$$\xi_x \eta_y - \xi_y \eta_x = \xi_x^2 + \xi_y^2. \tag{14.5}$$

Equation 14.5 implies the map is orthogonal and Equation 14.5 implies the Jacobian of the map is strictly positive, which means the map is invertible. The inverse functions are the coordinate transformations $x(\xi, \eta)$ and $y(\xi, \eta)$ from the logical space to the physical space. These functions satisfy the Cauchy-Riemann equations as well.

Because the component functions are differentiable they each solve Laplace's equation. That is,

$$\xi_{xx} + \xi_{yy} = 0,$$
$$\eta_{xx} + \eta_{yy} = 0.$$

The same is true for the inverse transformation functions.

$$x_{\xi\xi} + x_{\eta\eta} = 0,$$
$$y_{\xi\xi} + y_{\eta\eta} = 0.$$

Functions that solve Laplace's equation are called *harmonic functions*.

Note that the transformations defined above are for a rectangular domain R in the logical space. The dimensions (aspect ratio $= \frac{length}{height}$) of this rectangular domain are unknown, making boundary function maps (boundary conditions) required to solve the system of Laplace's equations impossible. If we restrict the transformation to a square region in the logical domain, we need to introduce a change of variable

$$\nu = \frac{\eta}{M}, \tag{14.6}$$
$$\mu = \xi, \tag{14.7}$$

where the quantity M is known as the *conformal module*. Using the Cauchy-Riemann equations for ξ and η along with the definitions for ν and μ it follows

$$\mu_x = M\nu_y, \tag{14.8}$$
$$\mu_y = -M\nu_x. \tag{14.9}$$

We apply the chain rule to write

$$M x_\mu = y_\nu,$$
$$x_\nu = -M y_\mu.$$

Note that the last equations provide a way to calculate the module M. Squaring both sides of the equations, adding and solving for M gives

$$M^2 = \frac{x_\nu^2 + y_\nu^2}{x_\mu^2 + y_\mu^2}. \tag{14.10}$$

Equations 14.8 and 14.9 show that the composite maps $\mu(x, y)$ and $\nu(x, y)$ are not conformal. However, the functions $\mu(x, y)$ and $\nu(x, y)$ are harmonic, so that

$$\mu_{xx} + \mu_{yy} = 0,$$
$$\nu_{xx} + \nu_{yy} = 0.$$

Next, through the $\xi - \mu$, $\eta - \nu$ relationships and the chain rule we may write

$$M^2 x_{\mu\mu} + y_{\nu\nu} = 0, \tag{14.11}$$
$$M^2 y_{\mu\mu} + y_{\nu\nu} = 0. \tag{14.12}$$

The difficulty in solving the last set of equations is due to the conformal module M. The definition for M given in Equation 14.10 indicates it is domain dependent and makes the system of equations nonlinear.

Several methods for estimating M have been proposed. Barfield [13] used the following formulation for M

$$M = \int\int_S \left[\frac{x_\nu^2 + y_\nu^2}{x_\mu^2 + y_\mu^2}\right]^{\frac{1}{2}} d\mu d\nu, \tag{14.13}$$

where the M is a constant determined over the entire physical domain S. Because of the nonlinear nature of Equations 14.11 and 14.12, they must be solved in an iterative nature, so that the calculation of M given in Equation 14.13 provides an "average" aspect ratio determined from the grid in the previous iteration.

Conformal mapping provides orthogonal, angle preserving grid transformations; extremely desirable characteristics for grid generation. However, there are some aspects that make conformal mapping less desirable; the difficulty with M being one already cited. Others include the two-dimensional limitation, inflexible boundary point distribution, and a tendency for instability to be dependent on the shape of the physical domain [106].

14.5.2   Elliptic Grid Generation

One of the key components of the conformal map is the system of Laplace's equations. This elliptic equation is the basis for a robust differential method know as *elliptic grid generation*. The coordinate transformation functions $x(\xi, \eta)$ and $y(\xi, \eta)$ are defined in terms of the system

$$x_{\xi\xi} + x_{\eta\eta} = 0,$$
$$y_{\xi\xi} + y_{\eta\eta} = 0,$$

with boundary conditions specified as

$$x(\xi, 0) = x_b(\xi),$$
$$x(\xi, 1) = x_t(\xi),$$
$$x(0, \eta) = x_l(\eta),$$
$$x(1, \eta) = x_r(\eta),$$

for $x(\xi, \eta)$, and

$$y(\xi, 0) = y_b(\xi),$$
$$y(\xi, 1) = y_t(\xi),$$
$$y(0, \eta) = y_l(\eta),$$
$$y(1, \eta) = y_r(\eta),$$

for $y(\xi, \eta)$. Such a formulation is well-posed because boundary conditions are specified on all four boundaries.

Elliptic methods are stable with respect to small changes in the physical boundary. The interior grid lines are smooth, even if the boundary functions are not. The major drawback to such a formulation is the lack of a guaranteed one-to-one solution. That is, there may be regions in our physical domain where the generated grid is *folded*. This is especially true for the case of nonconvex physical domains.

The next chapter is devoted to elliptical grid generation. A method for avoiding grid folding will be shown there. Also, methods for controlling the interior placement of nodes and grid lines will be presented.

14.5.3   Other Differential Methods

In this section we mention a few alternative differential methods for generating grids. The first is a *hyperbolic* method used by Steger and Chaussee [163] to construct a grid for calculating flow around a two-dimensional air foil. The system of generating PDEs is

$$x_\xi x_\eta + y_\xi y_\eta = 0,$$
$$x_\xi y_\eta - x_\eta y_\xi = V(\xi, \eta),$$

where the nonhomogeneous term V is a user-defined control function. As these equations are nonlinear and, therefore, not easily classified. However, the linearized version of the given system is indeed hyperbolic.

The nature of this system of equations is such that they may be solved using a marching technique wherein the "initial" condition is specified on the inner boundary (the airfoil surface) of an unbounded domain. The grid is the constructed outward from the given boundary. The reader is referred to the article by Steger and Chaussee [163] for details of the numerical marching procedure used to solve the system.

Hyperbolic techniques have some limitations including the fact that they are limited to infinite domains. The boundary conditions on the interior boundary must be continuous and smooth. As mentioned before, the source term V is user defined; it cannot be found in an automatic way. Instead, some experimentation is necessary to construct the desired grid. Care is required in the solution process as shock-like behavior is possible. Truncation errors are rather large, so course grid resolution may be problematic.

The *parabolic* differential method introduced by Nakamura [137] is an attempt to avoid some of the problems of the hyperbolic method. The system of equations is

$$
\begin{aligned}
x_\eta &= A x_{\xi\xi} + S_x, \\
y_\eta &= A y_{\xi\xi} + S_y,
\end{aligned}
$$

where the coefficient A is a constant determined by experimentation, and the source terms S_x and S_y control grid spacing and orthogonality. The domain must be infinite, with boundary conditions specified on an interior boundary, as in the case of hyperbolic generation methods. Application is useful in the case of airfoils.

14.6 Adaptive and Multilevel Methods

Adaptive methods are those where the grid is tailored because of geometric considerations or solution characteristics. Geometric adaptation usually results in refining the grid near boundaries. It is a "static" adaptation because the refinement is usually done prior to actually solving the governing differential equations on the generated grid. Solution adaptive techniques may be static or "dynamic" in that the grid resolution may change as the solution evolves. An example would be where the grid refinement follows a moving boundary in a two-phase flow.

Another general category for grid generation is the so-called *multilevel* methods. These methods are generally categorized by the use of several grids, either in a sequential application or perhaps an *overlayment* process.

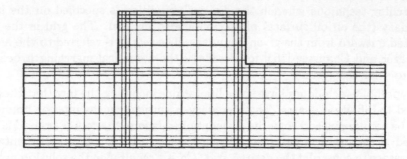

Figure 14.5: Nonuniform grid for a furnace cavity.

14.6.1   Adaptive Grids

We provide several examples, with brief descriptions and a figure, for several adaptive techniques.

Nonuniform Grids

Nonuniform coordinates are frequently used when there is need for increased node resolution in certain regions of the computational domain. Typically, these are near-wall regions or portions of the computational domain where steep gradients for the dependent variable are expected. Figure 14.5 shows how nonuniform grids may be used in the case for two-dimensional flow and heat transfer within a furnace cavity. Note the greater packing of horizontal grid lines near the walls, and the greater packing of vertical grid lines in the cavity region below the vertical walls. The later region is one where steep gradients are expected.

Regional Coordinates

The method of regional coordinates involves using different coordinates systems for different regions of the computational domain. An example of this method is pictured in Figure 14.6, where polar-cylindrical coordinates employed near the cylinder boundaries and Cartesian coordinates used in the open region between the cylinders.

Irregular Coordinates

In the irregular coordinates scheme, each node or element is individually determined by considering the geometric shape of a boundary. It is frequently used

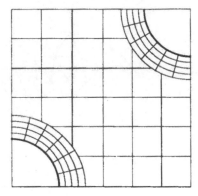

Figure 14.6: Regional coordinates for flow between cylinders.

in finite element methods. In this case, the triangular shape of the cell allows for easier alignment with irregular boundaries. Figure 14.7 indicates how the irregular placement of nodes, and the corresponding elements created by the nodes, makes for a fair approximation of a circular boundary.

Solution Adaptation

As an example of how the method is used, we consider the problem of frontal boundary moving through the physical domain of a time-dependent solution. The frontal boundary is often identified by a steep gradient in one or more of the dependent variables.

An example of such a problem would be the seabreeze flow associated with an land-water interface, such as a sea coast. The difference in surface temperatures between the land and water creates an on-shore flow of cool and moist air. The flow begins at the land-water boundary and moves inland creating a "front," where an abrupt change in wind speed, air temperature and humidity occur. It is important to refine the grid near the front to simulate and investigate the dynamics of this phenomenon. It is more efficient to limit the resolution to the frontal area instead of applying the required resolution unnecessarily to the entire domain.

Figure 14.8 shows the "dynamic" local grid resolution move through the domain. A calculation of dependent variable gradients identifies the location of the front and may be used to determine just how much resolution is required.

14.6.2   Multilevel Methods

Multilevel methods is the name given to a broad category of grid generation methods where several related grids are used in concert to achieve such objectives as enhanced rate of convergence and accuracy on adaptive grids, while

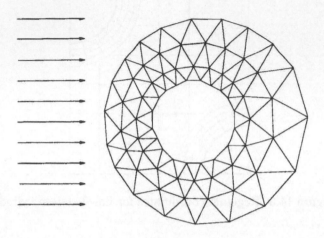

Figure 14.7: Irregular coordinates for flow around a cylinder.

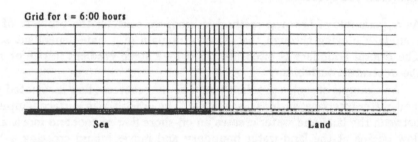

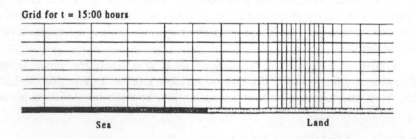

Figure 14.8: Solution adaptive grid for a moving seabreeze front.

avoiding some of the computational difficulties of nonuniform grids. Two some-what distinct methods will be introduced.

Multigrids

Multigrids are often used to enhance the rate of convergence. The general frame work for multigrid use is the construction of a sequence of grids $\{G_k\}$ where grid G_{k+1} is "finer" than grid G_k.

Figure 14.9(a) shows how multigrids would be used in the case of an outside corner in the computational domain. Note that grid G_2 includes all of the nodes of grid G_1 plus additional nodes for increased resolution, especially near the outside corner. The same is true for grids G_2 and G_3.

Composite Grids

A composite grid may be roughly defined as the union of uniform grids upon which the actual calculations are made. Because the component grids are uniform, convergence rates and accuracy improvements over nonuniform grids are attained. Yet, when the composite grid is considered, the solution is effectively found for the nonuniform, adaptive grid.

Figure 14.9(b) shows how the individual, uniform grid can be used to make up a composite grid that has increased resolution in one of the corners of the square domain.

The reader is referred two recent publications that provide very good information on multilevel, multigrid methods. The book by McCormick [129] gives a solid mathematical introduction to multilevel methods including fast adaptive composite grid methods. The monograph by Briggs [23] offers, as its title states, a multigrid tutorial.

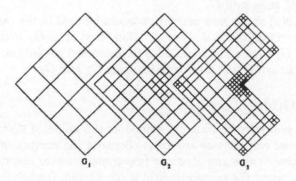

(a) Multigrid sequence for an outside corner.

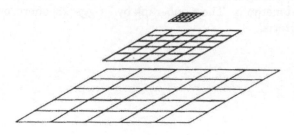

(b) Composite grid for a square domain.

Figure 14.9: Multigrid and composite grid examples.

Chapter 15

Elliptic Grid Generation

This chapter focuses on the use of the elliptic operator as a means of generating computational grids for irregular physical domains. Elliptic grid generation, as opposed to other differential methods such as hyperbolic and parabolic, has been widely used for several reasons. For bounded domains, the boundary functions mapping the physical boundaries to the logical boundaries result in boundary fitted coordinates. Specifying boundary conditions on all four sides makes the problem well-posed, including stability with respect to the boundary functions.

The chapter begins by presenting some of the properties of the Laplacian elliptic operator. Next, various methods for elliptic generation that are based on the Laplacian are introduced. The drawback to elliptic methods is the possibility of folded grids. Methods for overcoming this problem are outlined in subsequent sections.

The Poisson equation is introduced next. We will see how the source term, or terms, provide control over smoothness of the interior grid and packing of coordinate lines near boundaries.

15.1 Harmonic Functions

Perhaps the simplest elliptic planar grid generator is the Laplacian operator

$$L(f) = f_{xx} + f_{yy} = \nabla^2 f, \tag{15.1}$$

where f is a twice differentiable function in ξ and η. A function f that satisfies Laplace's equation is called a *harmonic* function. Note that the Laplacian and associated harmonic functions have wide-spread application to many fields of study including potential and viscous flow in fluids, electromagnetic field theory and atomic physics to name a few.

Two very important properties of harmonic functions are presented in this section that will be useful in understanding certain details in elliptic coordinate generation, such as control of grid smoothness and coordinate line spacing. Instead of developing the necessary theoretical background for providing formal

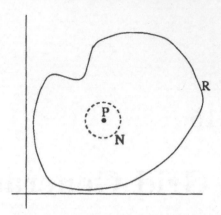

Figure 15.1: Interior point P and neighborhood N.

proofs for these properties, we offer hueristic arguments for why the properties are true. We refer the reader to texts such as that by Flanigan [68] for complete arguments concerning the following properties.

15.1.1 Average Value

The first property concerns how the value of the harmonic function f at a specific point P is related to values of f "near" P. Let R be a region on which $\nabla^2 f = 0$. Let N represent a "small" neighborhood about the point P, as shown in Figure 15.1. The point P is in the *interior* of R, which means we can find a small enough neighborhood N such that all points in N lie within R.

Property 15.1 *(Average Value) If f is harmonic on a region R, and N is a neighborhood of point P contained entirely in R, then*

$$\nabla^2 f\big|_P \ (\bar{f} - f_P),$$

where the Laplacian of f and f itself are evaluated at P, and $\bar{f}$ is the average value of f on the neighborhood N.

The following argument is meant as a loose justification of Property 15.1. If f is sufficiently differentiable, we may use a Taylor polynomial expansion to express f at points near P. That is,

$$
\begin{aligned}
f(x,y) \ = \ & f_P + f_x\big|_P\, x + f_y\big|_P\, y + \frac{1}{2}\left[f_{xx}\big|_P\, x^2 + f_{xy}\big|_P\, 2xy \right. \\
& \left. + f_{yy}\big|_P\, y^2 \right] + \cdots .
\end{aligned}
\tag{15.2}
$$

Now, calculate an average value for f, represented by $\bar{f}$, on the square $2h \times 2h$ region. The value for $\bar{f}$ is calculated using the following double integral

$$\bar{f} = \frac{1}{2h \times 2h} \int_{-h}^{h} \int_{-h}^{h} f\, dx\, dy .
\tag{15.3}$$

Substituting for f in Equation 15.3 using the Taylor expansion in Equation 15.2 gives

$$\bar{f} = f_p + \frac{h^2}{6} \nabla^2 f \bigg|_P + O(h^4).$$ (15.4)

Solving for the Laplacian of f evaluated at P results in

$$\nabla^2 f \big|_P \approx \frac{6}{h^2} \left[\bar{f} - f_P \right].$$ (15.5)

The last expression states that the Laplacian of f evaluated at point P is proportional to the difference of f at P and the average of f on the small domain about P. Therefore, if f is harmonic, then $\nabla^2 f = 0$ and the value of f at P is the same as the average value of f on a neighborhood around P.

The Poisson equation in two dimensions is $\nabla^2 f = Q$. Following the results of Property 15.1, if $Q > 0$, the Poisson equation implies the average value of f surrounding point P is greater than the value of f at P. Similarly, if $Q < 0$, then the average value of f around P is less than the value of f at P. It will be shown how the Poisson equation and the results of this section apply to grid generation techniques in Section 15.4.

15.1.2   Maximum and Minimum Principle

The next property concerns the location of extreme values (maximums or minimums) of a harmonic function f on an open region R. A region R is *open* if every point of R has a neighborhood (an open disc) contained in R. The function f has a *maximum value* at point P_0 if $f(P_0) \geq f(P)$ for all P in a small neighborhood of P_0. A similar statement holds for a *minimum value* except $f(P_0) \leq f(P)$.

Property 15.2 *(Maximum and Minumum Principle) Suppose f is a nonconstant harmonic function on an open region R. Then there is no point P_0 in R where f has a maximum or minimum.*

As with the first property, a loose justification for the validity of Property 15.2 will be offered. The justification begins by supposing there is a point P_0 in R at which f has a maximum value. That is, $f(P_0) \geq f(P)$ for all P in some neighborhood N. Since f is nonconstant, there exists a point P_m in N such that $f(P_m) < f(P_0)$. The continuity of f guarantees the existence of a neighborhood N_m of P_m, with finite area, on which the value of f is strictly less than $f(P_0)$. Consequently, the average value of f on N_m is less than $f(P_0)$. The average value of f on N not including N_m cannot be greater than $f(P_0)$ because of the assumption of the maximum. Taking the average value of f on both neighborhoods into consideration, it must be concluded that the average value of f on the whole of N is less than $f(P_0)$, which contradicts the fact that f is harmonic on R through Property 15.1.

15.2   Simple Elliptic Generator

The most straight-forward elliptic grid generator for the planar case, sometimes referred to as either the AH (Amsden-Hirt) or "length" generator, is given by the following set of PDEs

$$x_{\xi\xi} + x_{\eta\eta} = 0, \tag{15.6}$$
$$y_{\xi\xi} + y_{\eta\eta} = 0, \tag{15.7}$$

with boundary conditions specified as

$$x(\xi, 0) = x_b(\xi), \qquad x(\xi, 1) = x_t(\xi),$$
$$x(0, \eta) = x_l(\eta), \qquad x(1, \eta) = x_r(\eta),$$
$$y(\xi, 0) = y_b(\xi), \qquad y(\xi, 1) = y_t(\xi),$$
$$y(0, \eta) = y_l(\eta) \quad \text{and} \quad y(1, \eta) = y_r(\eta).$$

Note that boundary condition are specified for both x and y on all four boundaries of the unit square in the logical domain. The subscripts b, t, l and r refer the bottom, top, left and right boundaries, respectively, of the logical domain.

The Laplacians of x and y, given in Equations 15.7 and 15.7, respectively, may be solved using a simple explicit finite difference scheme. Alternatively, one may use more sophisticated methods, such as a FA solution, in conjunction with a semi-implicit scheme.

As an example, let us consider generating a grid for the section of the annulus pictured in Figure 15.2. The unit square in the logical space shows 21 ξ grid lines and 11 η grid lines (including the boundaries in both cases). These will be mapped to the physical space using the following boundary conditions on x and y.

$$x_b\left(\frac{i}{20}\right) = 1 + \frac{1}{2}cos\left(\pi - i\frac{\pi}{20}\right), \qquad x_t\left(\frac{i}{20}\right) = 1 + cos\left(\pi - i\frac{\pi}{20}\right),$$

$$x_l\left(\frac{j}{10}\right) = \frac{1}{2}\left(1 - \frac{j}{10}\right), \qquad x_r\left(\frac{j}{10}\right) = \frac{1}{2}\left(1 + \frac{j}{10}\right),$$

$$y_b\left(\frac{i}{20}\right) = 1 + \frac{1}{2}sin\left(\pi - i\frac{\pi}{20}\right), \qquad y_t\left(\frac{i}{20}\right) = 1 + sin\left(\pi - i\frac{\pi}{20}\right),$$

$$y_l\left(\frac{j}{10}\right) = 0 \quad \text{and} \quad y_r\left(\frac{j}{10}\right) = 0,$$

where the index i goes from 0 to 20, and the index j goes from 0 to 10. The results of simple elliptic map are shown in Figure 15.3(a). Note that there are 21 radial segments pictured, and 11 semicircular arcs.

Figure 15.3(b) shows the results of mapping a grid to the entire annulus. The boundary functions for this case are very similar to those for the upper annulus. The important point to make in this example is that the left and

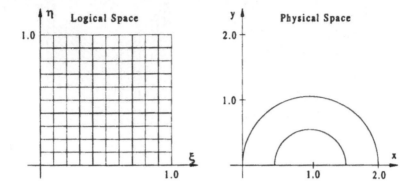

Figure 15.2: Logical and physical space for the upper annulus.

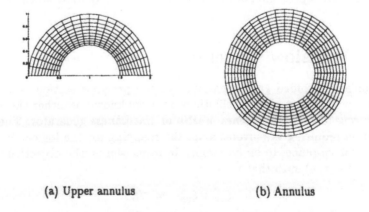

(a) Upper annulus   (b) Annulus

Figure 15.3: Simple elliptic grid transformations.

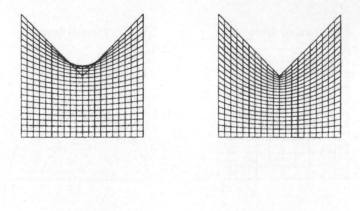

(a) Folded map   (b) Winslow transformation

Figure 15.4: Unfolding a transformation.

right boundaries of the logical space are both mapped to the segment given by $\frac{3}{2} \le x \le 2$ on the x axis.

The length elliptic generator is very simple to use. The primary drawback to the simple elliptic generator is the possibility of folded grids, especially on convex regions. As an example, Figure 15.4(a) shows to result of the mapping between the logical domain and the v-shaped domain. Note the slight folding that occurs at the top boundary near the valley of the v. The use of an inverse transformation is presented in the next section as a way to avoid the folded grid lines.

15.3   Winslow Generator

The problem of folded grids mentioned in the previous section is avoided by using one of the most popular elliptic generators known as either the *Winslow*, *homogeneous Thompson-Thames-Mastin* or *smoothness* generator. The method is based on requiring the inverse map, the mapping for the logical $\xi\eta$-plane to the physical xy-plane, to be harmonic. In other words, the objective is to find $\xi(x, y)$ and $\eta(x, y)$ such that

$$\xi_{xx} + \xi_{yy} = 0, \qquad (15.8)$$
$$\eta_{xx} + \eta_{yy} = 0, \qquad (15.9)$$

while satisfying the required boundary conditions for the well-posed problem. The difficulty in this "inverse" problem is knowing the required boundary conditions. That is, the value of $\xi(\vec{x})$ for an arbitrary point $\vec{x}$ on the boundary is unknown. However, the boundary conditions for x and y as functions of ξ and η are readily prescribed, so Equations 15.8 and 15.9 may be inverted to solve for x and y. The inverted system is

$$g_{22}x_{\xi\xi} - 2g_{12}x_{\xi\eta} + g_{11}x_{\eta\eta} = 0, \qquad (15.10)$$

$$g_{22}y_{\xi\xi} - 2g_{12}y_{\xi\eta} + g_{11}y_{\eta\eta} = 0, \qquad (15.11)$$

where the coefficients are defined as

$$g_{11} = x_\xi^2 + y_\xi^2,$$
$$g_{12} = x_\xi x_\eta + y_\xi y_\eta,$$
$$g_{22} = x_\eta^2 + y_\eta^2.$$

The details of the inverting process are not presented here because of length considerations. The interested reader is referred to the book by Knupp and Steinberg [106] for a full explanation.

Inverting the system of Laplacians results in a more complicated, yet solvable, system of PDEs. As opposed to the system of Laplacian equations, the inverted equations are coupled and nonlinear. They may still be solved using finite difference approximations, but convergence in an iterative scheme is slow with an explicit formulation.

The Winslow result to the v-shaped physical domain is provided in Figure 15.4(b) as evidence for the lack of folding when the inverse functions ξ and η are required to be harmonic.

15.4 Poisson Generator

Elliptic generation methods are very effective in creating smooth grids given continuous boundary conditions on the physical domain. However, the Laplace equation, or its inverse PDE, provide little control for the spacing, or packing, of grid lines in the interior of the physical domain. There are certain flow problems where more refinement in the grid is desirable near a boundary, for example, but the nature of the elliptic generator is such that grid spacing is essentially uniform throughout the physical domain.

This section introduces the Poisson equation methods for grid generation. The nonhomogeneous terms provide a mechanism for controlling grid resolution. We begin the section by providing a qualitative understanding of the effect of the nonhomogeneous term in the case of a one-dimensional grid. This understanding is useful in expanding a qualitative understanding for the Poisson equation in the two dimensional case. The nonhomogeneous term is sometimes referred to as a *control function* because of its effect on grid spacing. The section concludes with a presentation of various control functions.

15.4.1   Qualitative Analysis

One-dimensional Case

The development begins by considering the one-dimensional case. The physical space dimension will be represented by x and the logical space length will be given by ξ. For sake of an example, suppose the physical domain is the interval $0 \leq x \leq 2$, and the logical space is given by the unit interval $0 \leq \xi \leq 1$. The homogeneous differential equation with boundary conditions prescribed below

$$x_{\xi\xi} = 0,$$
$$x(0) = 0 \quad \text{and} \quad x(2) = 1,$$

results in $x(\xi) = 2\xi$, a linear distribution of x nodes as a function of ξ, as indicated in Figure 15.5(a).

Next, consider the following nonhomogeneous equation and boundary conditions

$$x_{\xi\xi} = P,$$
$$x(0) = 0 \quad \text{and} \quad x(2) = 1,$$

where P is a constant parameter. The solution to this boundary value problem is $x = \frac{P}{2}\xi^2 + \left(2 - \frac{P}{2}\right)\xi$. If $P > 0$, then the graph of $x(\xi)$ is concave up, which results in a packing of x nodes near $x = 0$, as shown in Figure 15.5(b) for $P = 4$. If P is negative, then the graph of $x(\xi)$ would be concave down and the transformation would result in a packing of nodes near $x = 2$.

As a final example, consider the case where the nonhomogeneous term is a function of ξ. For example, suppose the boundary value problem is given as

$$x_{\xi\xi} = P\left(\frac{1}{2} - \xi\right),$$
$$x(0) = 0 \quad \text{and} \quad x(2) = 1.$$

Note that when $P > 0$, the concavity of the graph of $x(\xi)$ is positive for $\xi < \frac{1}{2}$ and negative for $\xi > \frac{1}{2}$. The result is a grid on the x axis with packing of nodes near both boundaries ($x = 0$ and $x = 2$). Figure 15.5(c) shows the case for $P = 12$.

Two-dimensional Case

The Poisson equation method for the inverse problem in two dimensions has the following general frame work,

$$\xi_{xx} + \xi_{yy} = P(\xi, \eta), \tag{15.12}$$
$$\eta_{xx} + \eta_{yy} = Q(\xi, \eta), \tag{15.13}$$

where the *control functions* P and Q are, in general, functions of both ξ and η. Because Equations 15.13 and 15.13 are inverted to solve for x and y as functions

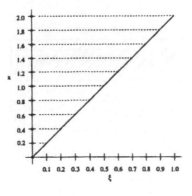

(a) Homogeneous case

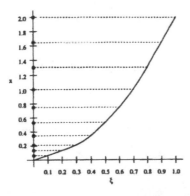

(b) Packing of point near the origin

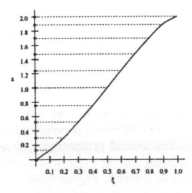

(c) Packing of point near the boundaries

Figure 15.5: One-dimensional examples for Poisson mapping.

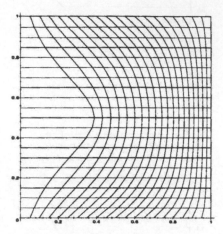

Figure 15.6: Nonhomogeneous Winslow: Packing of grid points near the $x = 1$ boundary.

of ξ and η, the boundary conditions are given in terms of x and y.

$$x(\xi, 0) = x_b(\xi), \qquad x(\xi, 1) = x_t(\xi),$$
$$x(0, \eta) = x_l(\eta), \qquad x(1, \eta) = x_r(\eta),$$
$$y(\xi, 0) = y_b(\xi), \qquad y(\xi, 1) = y_t(\xi),$$
$$y(0, \eta) = y_l(\eta), \quad \text{and} \quad y(1, \eta) = y_r(\eta).$$

Inverting Equations 15.13 and 15.13 results in the following system of PDEs:

$$g_{22}x_{\xi\xi} - 2g_{12}x_{\xi\eta} + g_{11}x_{\eta\eta} = -g(Px_\xi + Qx_\eta), \qquad (15.14)$$
$$g_{22}y_{\xi\xi} - 2g_{12}y_{\xi\eta} + g_{11}y_{\eta\eta} = -g(Py_\xi + Qy_\eta). \qquad (15.15)$$

As a simple initial example, consider a mapping from the unit square to the unit square, with $P = 4$ and $Q = 0$. A result analogous to the constant concavity case for the one-dimensional problem should be expected. However, since the inverse problem is being solved, the x node shift should be to the $x = 1$ boundary. This is the result as shown if Figure 15.6.

15.4.2   Control Functions

Several possible control functions that may be used to control grid spacing in the computational domain are presented in this section. As an example, consider the case of flow between offset cylinders, or rod bundles, as depicted in Figure 4.3(a).

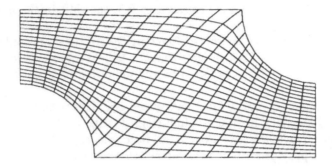

Figure 15.7: Grid pattern for offset tube array using the homogeneous Winslow generator.

If the Winslow generator is used with P and Q both identically zero, the resulting grid transformation is shown in Figure 15.7. Because the flow field pattern is assumed to be repeated from one row to the next and symmetric along the horizontal lines that cut the cylinders on their diameters, the grid is generated for only two consecutive cylinders. The boundaries of the physical domain are the lines of symmetry and the surfaces of the quarter-cylinders contained within a rectangle defined by the lines of symmetry.

Note that the image of $\eta = 0$ forms the "bottom" boundary of the physical domain, while the image of $\eta = 1$ forms the "top" boundary. Similarly, the left boundary is the image of $\xi = 0$, and the right boundary is that of $\xi = 1$. This is guaranteed by Property 15.2.

Perhaps the biggest weakness of the grid map shown in Figure 15.7 is the separation between the $\eta = 0$ coordinate and the next η coordinate near where the cylinder surface meets the horizontal boundary. Various control function may be used to improve the grid spacing in this region. One of the more popular forms for P and Q is an exponential form proposed by Thompson with

$$P(\xi) = -\sum_{i=1}^{M} a \frac{\xi - \xi_i}{|\xi - \xi_i|} e^{-b|\xi - \xi_i|}, \qquad (15.16)$$

$$Q(\eta) = -\sum_{i=1}^{N} a \frac{\eta - \eta_i}{|\eta - \eta_i|} e^{-b|\eta - \eta_i|}, \qquad (15.17)$$

where a and b are positive constant parameters that often must be determined by "experimentation". The constants M and N are the number of ξ and η grids lines, respectively. Figure 15.8(a) shows the resulting grid pattern for the case $P = 0$ and Q as defined in Equation 15.17 with $a = 1.0$ and $d = 0.75$.

As mention before, these values of a and d were found by trial and error. If the value of a is increased to 1.5, the η lines will shift too much toward the boundaries. If the value of a is decreased to 0.5, the shift in the η coordinates is negligible.

Figure 15.8(b) is a plot the control function $Q(\xi, \eta)$ that was used to generate the grid shown in Figure 15.8(a). Note that the Q field is symmetric with respect to the line $\xi = 0.5$, which is expected by the definition of Q given in Equation 15.17.

The previous set of control functions caused grid lines to be shifted relative to boundaries. There may be desire to induce grid refinement based on individual points as well. The following set of control functions expand upon those offered previously and provide for individual point refinement.

$$
\begin{aligned}
P(\xi, \eta) &= -\sum_{m=1}^{M} a_m \frac{\xi - \xi_i}{|\xi - \xi_m|} e^{-b_m|\xi - \xi_m|} \\
&\quad - \sum_{i=1}^{I} c_i \frac{\xi - \xi_i}{|\xi - \xi_i|} e^{-d_i[(\xi - \xi_i)^2 + (\eta - \eta_i)^2]^{\frac{1}{2}}} , \tag{15.18}
\end{aligned}
$$

$$
\begin{aligned}
Q(\xi, \eta) &= -\sum_{i=1}^{N} a_n \frac{\eta - \eta_m}{|\eta - \eta_m|} e^{-b_n|\eta - \eta_m|} \\
&\quad - \sum_{i=1}^{I} c_i \frac{\eta - \eta_i}{|\eta - \eta_i|} e^{-d_i[(\xi - \xi_i)^2 + (\eta - \eta_i)^2]^{\frac{1}{2}}} , \tag{15.19}
\end{aligned}
$$

where the parameters a_m, b_m, a_n, b_n, c_i and d_i are determined by experimentation. The points represented by (ξ_i, η_i) must be specified, and represent those to which the generated grid is attracted. The number M is the number of grid lines in either direction in the ξ direction, N is the number of grid lines in the η direction and the number I is the total number of points to which the grid is to be that attracted.

As an example, consider the staggered array of cylinders again. This time, an attempt will be made to attract the generated grid to the points where the cylinder surfaces intersect with the horizontal lines of symmetry. Figure 15.8(c) show the resulting grid for the case $a_m = b_m = c_i = d_i = 0$ in Equation 15.18, and $a_n = b_n = 0$, $c_i = 20$, and $d_i = 0.75$ in Equation 15.19. The grid points are indeed pulled in toward the two point mentioned above.

Figure 15.8(d) shows the distribution of the control function Q for this case. The field is quite similar to that in Figure 15.8(b) in terms of the symmetry. Note that the grid field resulting from this Q function does not cause as much separation in the η line in the center of the domain.

As one final example, the grid result in the case where the center node is the single node of attraction is presented. The resulting grid is shown in Figure

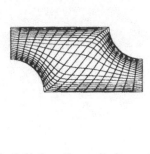

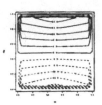

(a) Grid for offset cylinders using control functions of Equations 15.16 and 15.17

(b) Control function Q

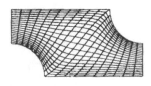

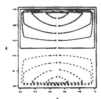

(c) Grid generated with $c_i = 20$ and $d_i = 0.75$ in Equation 15.19 and $c_i = d_i = 0$ in Equation 15.18

(d) Control function Q

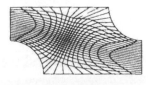

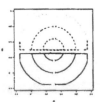

(e) Grid generated with $c_i = 5$ and $d_i = 0.75$ in Equation 15.19 and $c_i = 5$ and $d_i = 0.75$ in Equation 15.18

(f) Control function Q

Figure 15.8: Additional examples of grid generation using control functions.

15.8(e), with generating Q function plotted in Figure 15.8(f).

Chapter 16

Equations in ξ and η Coordinates

Solving fluid flow and heat transfer problems may require generation of a boundary fitted coordinate system on a complex physical domain. Coordinate transformation functions $x(\xi, \eta)$ and $y(\xi, \eta)$ are a result of this domain transformation. Likewise, the governing differential equations must be transformed to equivalent expressions in the new $\xi - \eta$ coordinate system. The transformation process is presented in the chapter. Constructing first and second derivative expressions for a general function is the initial step in this process. Once this formulation is in place, the two-dimensional continuity and transport equations are transformed to PDEs with derivatives with respect to ξ and η.

16.1 Derivative Transformations

The general form of the two-dimensional transport equation is show in Equation 16.1.

$$R(\phi_t + u\phi_x + v\phi_y) = \phi_{xx} + \phi_{yy} + Rf(x, y, t, \phi_j) \tag{16.1}$$

In order to introduce new independent variables ξ and η into the transport equation, general relationships between derivatives in the different coordinate systems are developed. With that objective in mind, the coordinate transformation functions $x(\xi, \eta)$ and $y(\xi, \eta)$ are assumed to be homeomorphic. Consequently, inverse mappings $\xi(x, y)$ and $\eta(x, y)$ exist.

The following expressions for the derivatives of ϕ are a result of applying the chain rule:

$$\phi_x = \phi_\xi \xi_x + \phi_\eta \eta_x , \tag{16.2}$$

$$\phi_y = \phi_\xi \xi_y + \phi_\eta \eta_y , \tag{16.3}$$

where the subscripts x, y, ξ and η denote the derivatives of ϕ with respective the given variable. The derivatives of ξ and η with respect to x and y are

geometric coefficients. Expression for these terms are found using the fact that
the coordinate transformation functions are inverses of each other. Therefore,

$$x(\xi(x,y),\eta(x,y)) = x, \tag{16.4}$$

$$y(\xi(x,y),\eta(x,y)) = y. \tag{16.5}$$

Differentiating both of these equations with respect to x and y gives

$$x_\xi \xi_x + x_\eta \eta_x = 1, \qquad x_\xi \xi_y + x_\eta \eta_y = 0, \tag{16.6}$$

$$y_\xi \xi_x + y_\eta \eta_x = 0, \qquad y_\xi \xi_y + y_\eta \eta_y = 1. \tag{16.7}$$

The matrix form of Equations 16.6 and 16.7 is

$$\begin{bmatrix} x_\xi & x_\eta \\ y_\xi & y_\eta \end{bmatrix} \begin{bmatrix} \xi_x & \xi_y \\ \eta_x & \eta_y \end{bmatrix} = \begin{bmatrix} 1 & 0 \\ 0 & 1 \end{bmatrix}. \tag{16.8}$$

The Jacobian matrix $\mathcal{J}$ for the transformation from the $x-y$ plane to the $\xi-\eta$
plane is

$$\mathcal{J} = \begin{bmatrix} x_\xi & x_\eta \\ y_\xi & y_\eta \end{bmatrix}. \tag{16.9}$$

Equation 16.8 implies that $\mathcal{J}$ is invertible, with its inverse $\mathcal{J}^{-1}$ given by

$$\mathcal{J}^{-1} = \begin{bmatrix} \xi_x & \xi_y \\ \eta_x & \eta_y \end{bmatrix}. \tag{16.10}$$

The Jacobian J of the coordinate transformation is given by the determinant
of the Jacobian matrix. That is,

$$J = det(\mathcal{J}) = x_\xi y_\eta - x_\eta y_\xi. \tag{16.11}$$

Now, the determinant of the inverse of a matrix is the multiplicative inverse of
the determinant. That is, $det(\mathcal{J}^{-1}) = \frac{1}{J}$. Using the formula for the inverse of
a 2×2 matrix gives the following alternative form for $\mathcal{J}^{-1}$:

$$\mathcal{J}^{-1} = \frac{1}{J} \begin{bmatrix} y_\eta & -x_\eta \\ -y_\xi & x_\xi \end{bmatrix}. \tag{16.12}$$

Comparing Equations 16.10 and 16.12 results in the following formulas for the
geometric coefficients:

$$\xi_x = \frac{y_\eta}{J}, \qquad\qquad \xi_y = \frac{-x_\eta}{J}, \tag{16.13}$$

$$\eta_x = \frac{-y_\xi}{J}, \qquad\qquad \eta_y = \frac{x_\xi}{J}. \tag{16.14}$$

The second derivatives ϕ_{xx} and ϕ_{yy} are obtained by differentiating Equations
16.2 and 16.3 by x and y, respectively. In order to complete the formulation, the

derivatives of Equations 16.2 and 16.3 with respect to ξ and η must be formed as well. The results may be neatly summarized using the matrix equation below:

$$
\begin{bmatrix} \phi_x \\ \phi_y \\ \phi_{xx} \\ \phi_{xy} \\ \phi_{yy} \end{bmatrix} = \begin{bmatrix} \xi_x & \eta_x & 0 & 0 & 0 \\ \xi_y & \eta_y & 0 & 0 & 0 \\ \xi_{xx} & \eta_{xx} & \xi_x^2 & 2\xi_x\eta_x & \eta_x^2 \\ \xi_{xy} & \eta_{xy} & \xi_x\xi_y(\xi_x\eta_y + \xi_y\eta_x) & \eta_x\eta_y & \eta_x\eta_y \\ \xi_{yy} & \eta_{yy} & \xi_y^2 & 2\xi_y\eta_y & \eta_y^2 \end{bmatrix} \begin{bmatrix} \phi_\xi \\ \phi_\eta \\ \phi_{\xi\xi} \\ \phi_{\xi\eta} \\ \phi_{\eta\eta} \end{bmatrix} . \quad (16.15)
$$

Expressions for the geometric coefficients of second order (ξ_{xx}, ξ_{xy}, ...) in Equation 16.15 are found by differentiating Equations 16.13 or 16.14 by the appropriate variable. The results are

$$
\begin{aligned}
\xi_{xx} &= \left[-J_\xi y_\eta^2 + J y_\eta y_{\xi\eta} + J_\eta y_\xi y_\eta - J y_\xi y_{\eta\eta} \right] / J^3, \\
\xi_{xy} &= \left[J_\xi x_\eta y_\eta - J y_\eta x_{\xi\eta} - J_\eta y_\xi x_\eta + J y_\xi x_{\eta\eta} \right] / J^3, \\
\xi_{yy} &= \left[-J_\xi x_\eta^2 + J x_\eta x_{\eta\xi} + J_\eta x_\xi x_\eta - J x_\xi x_{\eta\eta} \right] / J^3, \\
\eta_{xx} &= \left[-J_\eta y_\xi^2 + J y_\xi y_{\xi\eta} + J_\xi y_\xi y_\eta - J y_\eta y_{\xi\xi} \right] / J^3, \\
\eta_{xy} &= \left[J_\eta x_\xi y_\eta - J y_\xi x_{\xi\eta} - J_\xi y_\eta x_\xi + J y_\eta x_{\xi\xi} \right] / J^3, \quad \text{and} \\
\eta_{yy} &= \left[-J_\eta x_\xi^2 + J x_\xi x_{\xi\eta} + J_\xi x_\xi x_\eta - J x_\eta x_{\xi\xi} \right] / J^3. \quad (16.16)
\end{aligned}
$$

Expressions for the derivatives of the J are

$$
J_\xi = x_{\xi\xi} y_\eta + x_\xi y_{\xi\eta} - x_{\xi\eta} y_\xi - x_\eta y_{\xi\xi}, \quad (16.17)
$$

$$
J_\eta = x_{\xi\eta} y_\eta + x_\xi y_{\eta\eta} - x_{\eta\eta} y_\xi - x_\eta y_{\xi\eta}. \quad (16.18)
$$

The derivative relationships given in Equations 16.16 are used to transform the two-dimension continuity and transport equations to the logical $\xi - \eta$ plane in the next section.

16.2 Transformed Equations

16.2.1 The Continuity Equation

The continuity equation is an expression of zero divergence of mass. Its form in two-dimensions is

$$
u_x + v_y = 0. \quad (16.19)
$$

Using the derivative transformation given above gives

$$
\xi_x u_\xi + \eta_x u_\eta + \xi_y v_\xi + \eta_y v_\eta = 0. \quad (16.20)
$$

If $\vec{q}$ represents the velocity components in the curvilinear system, the two-dimensional case is

$$
q_1 = \frac{\partial \xi}{\partial t} = \xi_t
$$

and

$$
q_2 = \frac{\partial \eta}{\partial t} = \eta_t.
$$

The chain rule is used to express the time derivatives of ξ and η in terms x and y. That is,

$$
\begin{aligned}
\xi_t &= \xi_x x_t + \xi_y y_t \\
&= \xi_x u + \xi_y v,
\end{aligned}
$$

and

$$
\begin{aligned}
\eta_t &= \eta_x x_t + \eta_y y_t \\
&= \eta_x u + \eta_y v.
\end{aligned}
$$

Using these expressions in Equation 16.20 results in the following form for the continuity equation in $\xi - \eta$ coordinates

$$
[J(\xi_x u + \xi_y v)]_\xi + [J(\eta_x u + \eta_y v)]_\eta = 0, \tag{16.21}
$$

where the geometric coefficients ξ_x, ξ_y, η_x and η_y are given in Equation 16.13 and 16.14.

16.2.2 The Transport Equation

The linearized two-dimensional transport equation in two dimensions is

$$
R\phi_t + 2A\phi_x + 2B\phi_y = \phi_{xx} + \phi_{yy} + RF, \tag{16.22}
$$

where R is a flow parameter, and A, B and F are representative constants for the flow on a small finite analytic element.

In order to transform Equation 16.22 to one with derivatives of ϕ with respect to variables ξ and η, the first and second order derivative transformations given in Equation 16.15 are used. Making the appropriate substitutions gives

$$
\begin{aligned}
R\phi_t + 2A\left(\xi_x \phi_\xi + \eta_x \phi_\eta\right) + 2B\left(\xi_y \phi_\xi + \eta_y \phi_\eta\right) = \ & \xi_{xx}\phi_\xi + \eta_{xx}\phi_\eta + \xi_x^2 \phi_{\xi\xi} \\
& + 2\xi_x\eta_x \phi_{\xi\eta} + \eta_x^2 \phi_{\eta\eta} \\
& + \xi_{yy}\phi_\xi + \eta_{yy}\phi_\eta + \xi_y^2 \phi_{\xi\xi} \\
& + 2\xi_y\eta_y \phi_{\xi\eta} + \eta_y^2 \phi_{\eta\eta} \,,
\end{aligned}
$$

which, after combining like derivatives of ϕ, simplifies to

$$
R\phi_t + 2\tilde{A}\phi_\xi + 2\tilde{B}\phi_\eta = g^{22}\phi_{\xi\xi} + g^{11}\phi_{\eta\eta} + \tilde{F}, \tag{16.23}
$$

where

$$
\begin{aligned}
\tilde{A} &= A\xi_x + B\xi_y - \frac{\xi_{xx} + \xi_{yy}}{2}\,, \\
\tilde{B} &= A\eta_x + B\eta_y - \frac{\eta_{xx} + \eta_{yy}}{2}\,, \\
g^{22} &= \xi_x^2 + \xi_y^2\,, \\
g^{11} &= \eta_x^2 + \eta_y^2 \quad \text{and} \\
\tilde{F} &= RF + (2\xi_x\eta_x + 2\xi_y\eta_y)\,\phi_{\xi\eta}\,.
\end{aligned} \tag{16.24}
$$

The first- and second-order geometric coefficients in Equations 16.24 are given in Equation 16.15.

This completes the transformation of the Navier-Stokes equations in two dimensions into equivalent equations in $\xi - \eta$ coordinates. The process is similar for the energy equation.

Chapter 17

Diagonal Cartesian Method

A diagonal Cartesian method for generating computational grid for flows on irregular domains is presented in this chapter. The diagonal Cartesian method uses a structured Cartesian grid as a basis, and complex boundaries are approximated using both the Cartesian grid lines and diagonal segments between Cartesian grid nodes. The diagonal Cartesian method is relatively simple because of the structured grid, yet more accurate than the Cartesian approximation due to the use of the diagonal segments. Without the diagonal segments, complex boundaries are approximated with a jagged *saw-tooth* shape. The method has been successfully applied by Lin et al. [118, 119, 122, 121] to fluid flows on domains with irregular geometries.

Additionally, the diagonal Cartesian method is an attempt to achieve problem independence and automation of grid generation for problems with complex boundaries. At the same time, the emphasis is put on improving the accuracy of geometric approximations and the corresponding fluid solutions found using the simpler saw-tooth approximation. These goals are consistent with the urge made by Thompson [167] in his detailed 1990s review of grid generation methods. The study concluded that the major objectives in comprehensive grid codes must include automation and graphical interaction.

17.1 Current Methods

The boundary-fitted coordinate method proposed by Thompson [166] is a popular choice for modeling complex boundaries because the "irregular" boundary surface is fitted with a new coordinate line coincident with the body contour. Examples of its implementation include those by Takanashi [164], Balu and Unnikrishnan [12] and Jang et al. [93]. Undoubtedly, this method has resulted in great advances for flow simulation. However, it requires additional programming effort and computational resources in generating the new coordinates for each new domain. Additionally, the change of variables associated with the boundary fitted technique results in complex governing equations, as shown in

Chapter 16.

Boundary fitting techniques have difficulty in the cases of sharp boundaries or complex multibody systems. Automation is especially difficult [57]. Current algorithms for boundary-fitted coordinate generation are still problem-dependent and require much human-machine interaction. It is estimated that up to 80% of the total simulation effort is spent on grid generation for problems involving complex boundaries [167].

Overset, or composite, grids have great versatility and are especially attractive for multibody systems with bodies in relative motion [17, 16]. However, concerns have been continually raised about the accuracy of the interpolation necessary to transfer data between component grids, particularly the lack of conservation at the interfaces of the different grid blocks [17].

Unstructured meshes offer another alternative in numerical grid techniques [10]. They use both triangular and quadrilateral elements to approximate complex geometries by a series of linear or spline segments. Unstructured grids require information regarding nodal location and connectivity. As a result, the system of discretized equations is more difficult to solve than when structured grids are employed.

Advanced grid generation algorithms for which tetrahedrons are used for domain boundary approximation continue to be the focus of much research. However, further developments are needed before nonspecialists will be able to routinely and efficiently generate tetrahedral grids of high quality [4, 11]. Marviplis [128] had summarized that, for unstructured grids, viscous flow analysis using the Reynolds-averaged Navier-Stokes equations are severely hindered for large cases by seemingly insatiable CPU-time and memory requirements, even for steady-state calculations. In particular, most cases are limited by hardware memory costs, which have not decreased as rapidly as the cost of CPU-cycles over the last five years [128].

The Cartesian grid approach is attractive and gaining popularity because of its inherent simplicity and potential for automation. Many studies choose to approximate very complex geometries using only Cartesian coordinate grid lines. We will refer to this method as the *saw-tooth Cartesian method*, or simply the *saw-tooth* method, because of the jagged boundary approximations that sometimes develop (see Figure 17.1(a)).

Currently, this method is widely used for large scale problems such as in environmental and hydraulic engineering [138], as well as heat transfer in complex engineering devices [33]. A structured grid with fine grid spacing is generally utilized in this approach. Chai and Patankar [33] used this method to simulate radiative heat transfer on a complex geometry and obtained satisfactory results. The advantage of this approach is that it is much simpler, both in the governing equations and numerical procedures, than boundary fitted or unstructured grid methods mentioned above. One drawback to this approximation is that it forms a saw-tooth or stair-like boundary surface, and the boundary will remain rough with sharp 90° angles even if the grid is refined.

Another classical method used to treat complex geometries in Cartesian coordinates is to introduce additional nodes at the intersection of the complex

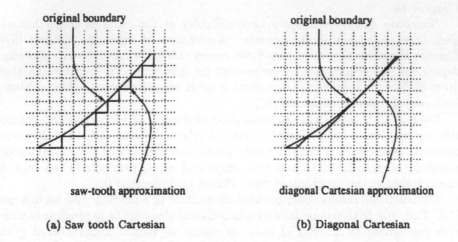

Figure 17.1: Saw tooth and diagonal Cartesian approximations.

geometry boundary and the Cartesian grid [109]. Although this approach provides a much smoother boundary approximation, it creates numerical instability in solving the system of discretized equations derived from the governing equations.

17.2   Diagonal Cartesian Method

The Diagonal Cartesian method approximates the contour of a complex geometry shape using both Cartesian grid lines and diagonal line segments as shown in Figure 17.1(b). The boundary approximation resulting from the diagonal Cartesian method is obviously closer to the original contour than the saw-tooth method as shown in Figure 17.1(a).

The proposed approximation method can be used for any complex geometry. However, for complex boundaries, the tedious approximation results in a daunting task if done manually when a large number of nodes are involved. Therefore, an automatic method would be beneficial.

17.2.1   Automatic Boundary Approximation

In order to describe the proposed "automatic" boundary approximation method, it is assumed a uniform Cartesian grid has been imposed on the computational domain. Suppose a complex two-dimensional boundary is given by the contour S, as shown in Figure 17.2. Further, suppose this continuous curve is described by a discrete set of data points $\{A, B, C, \ldots \}$ that may, coincidentally, correspond to the nodes of the Cartesian grid system. Suppose A and B are two given neighboring points on the original contour S. Let a be the Cartesian grid node that "approximates" A. To automate the selection of the node b on the

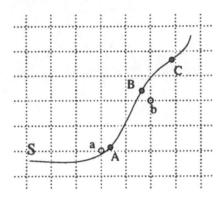

Figure 17.2: Approximation by the shortest distance method.

Cartesian grid to approximate the contour point B, the *minimum distance rule* is used. To that end, the distance between B and each neighboring grid node is calculated. According to the minimum distance rule, the Cartesian grid node having the shortest distance to B is considered as the approximate grid node of B, and is labeled "b" in Figure 17.2. This selection process is repeated for each of the contour points in the discrete set. Then, the union of the straight line segments connecting the grid nodes closest to the contour nodes is a polygonal approximation to the contour S. We will refer to this polygonal boundary approximation as the *grid boundary*.

Some considerations in how the discrete set of points $\{A, B, C, \ldots \}$ is determined are explored in the next section.

17.2.2 Boundary Point Selection

A "complex" curve or surface that defines a boundary may be specified either by an analytical function or, most likely, by a discrete set of points. In either case, when numerical solution methods are employed, the complex boundary is eventually specified by a discrete set of data points we will refer to as *boundary points*. With respect to the Cartesian grid imposed on the computational domain, one may ask how to determine the *optimal* set of boundary points to be used in the automatic grid boundary approximation. There are at least two important questions that must be considered in order to identify an optimal set of boundary points.

- What boundary points are required in the discrete set to adequately "describe" the original boundary?

- How many boundary points are appropriate for a given Cartesian grid?

These two questions are investigated, respectively, in the following subsections.

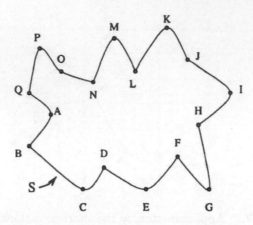

Figure 17.3: Principle of local monotonicity.

Local Monotonicity Principle

Suppose the curve labeled "S" in Figure 17.3 is the original complex contour, or boundary. In order to provide sufficient information about the nature of the contour with the least number of discretized data points, a *local monotonic principle* is used to select the required boundary points that belong to the discrete set. Local monotonicity is defined in the following manner: *For any two neighboring boundary points, such as A and B on contour S of Figure 17.3, the contour segment between A and B should vary monotonically with respect to both the x and y directions.* That is, either of the following must be true:

1. Given a segment of a body contour from A to B, there are no local maximum and minimum points with respect to the x or y coordinates.

2. If the boundary curve is sufficiently smooth, there is no point with zero first derivative on the segment between A and B, except if the segment is parallel to either x or y axis. Alternatively, there is no change in the sign of the second derivative along the segment between A and B on curve S.

Figure 17.3 shows the boundary points, labeled A, B, C, etc., that <u>must</u> be included so that the local monotonicity principle be satisfied. As an example, suppose the set did not include point B. Then, the segment of S between A and C would have a minimum in the x coordinate at point B, which violates the first statement of the local monotonicity principle.

Grid Considerations

Supposing the boundary points are properly selected by the local monotonicity principle, there may be a problem of mismatch between this discrete set of points and the Cartesian grid nodes. This problem is solved by comparing the spacing of the boundary points with that of the Cartesian grid.

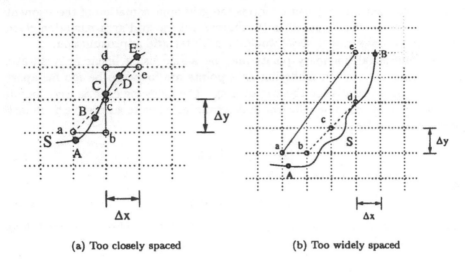

(a) Too closely spaced   (b) Too widely spaced

Figure 17.4: Improper spacing of boundary points.

The first example of a potential problem is illustrated in Figure 17.4(a). Suppose S is one segment of the original curve contour. The Cartesian grid has uniform spacing Δx and Δy in the x and y directions, respectively. Suppose the five boundary points A, B, C, D and E are specified to describe the segment S by the principle of local monotonicity. The corresponding grid approximation of S by the proposed shortest distance method is the segment $\overline{abcde}$, which turns out to be a saw-tooth like segments. However, the ideal approximation for the given segment S would be the segment $\overline{ace}$. This reveals that specifying too many boundary points for a given grid spacing may result in an undesirable jagged grid approximation.

In this case, removing "extra" boundary points will prevent the jagged grid approximation. Determining which boundary points are acceptable is done by comparing the spacing of the points to the spacing of the grid. If two boundary points are "too close" to each other relative to the grid spacing, at least one of the points must be eliminated. After some numerical testing, the following criteria were found suitable to prevent boundary points from being too close:

$$|x_1 - x_2| \le 0.70\Delta x \qquad (17.1)$$

and

$$|y_1 - y_2| \le 0.70\Delta y . \qquad (17.2)$$

That is, if two data points (x_1, y_1) and (x_2, y_2) meet the above two conditions, then one of these two neighboring points will be deleted.

Another problem arises if there are not enough boundary points in the description of the original contour, relative to the given grid spacing. Figure 17.4(b) shows that for the given segment S with only two given data points A

and B, a straight line segment $\overline{ae}$ forms the grid approximation of the segment S. With more boundary points in the discrete set the grid approximation given by the polygonal segment $\overline{abcde}$ would be a better grid approximation.

The additional boundary points may be added using linear or quadratic interpolation once two neighboring data points are found to be too far apart from each other, relative to the grid spacing. The following criteria were found, by experimentation, to be sufficient in determining when to successive boundary points were too far apart:

$$|x_1 - x_2| \geq 1.40\Delta x \qquad (17.3)$$

and

$$|y_1 - y_2| \geq 1.40\Delta y . \qquad (17.4)$$

That is, if two data points (x_1, y_1) and (x_2, y_2) meet the above two conditions at least one additional body data will be specified.

17.3   Accuracy Considerations

A method for automatically approximating an irregular boundary by a grid boundary using both diagonal segments and grid lines was outlined in the previous section. The accuracy of both the proposed diagonal Cartesian and the saw-tooth approximations will be examined in this section.

It is obvious that a good approximation of an irregular boundary should be as close as possible to the original curve or surface (if a three-dimensional application is considered). The method of least-square distance between the original contour and various approximating contours would likely provide an accurate determination of which contour approximation is "best." However, both the original contour and the approximating grid contour are, in general, not described by analytic functions. Instead, they are described by discrete sets of points. Consequently, the method of least-square distance between the original and approximating contours cannot be accurately calculated.

However, the concept of least-squares will be used to devise quantitative measures, referred to as $E1$ and $E2$, by which the relative accuracy of approximating contours may be judged.

17.3.1   Relative Length Error E1

Since both the original contour and the approximated one are described by a discrete set of points, the concept of relative length error $E1$ of total lengths between the original contour and the approximating may be defined and calculated as follows.

Consider solid curve, representing the original boundary, and the dashed curve, which gives the diagonal Cartesian grid boundary approximation, in Figure 17.5. They are described by two discrete sets of points. Uppercase letters A, B, C, ... identify the points on the original boundary, and lowercase letters

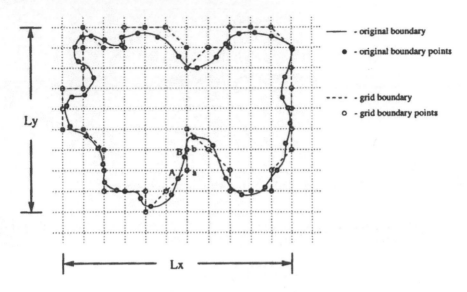

Figure 17.5: Original and diagonal Cartesian contours.

$a, b, c, \ldots$ denote the Cartesian grid points used in the grid boundary approximation, as seen in Figure 17.5. Let L_O be the length of the original contour and L_A the length of the approximating contour. They can be calculated in the following way:

$$L_O = \sum_{i=1}^{no} (\Delta l_o)_i \qquad (17.5)$$

and

$$L_A = \sum_{i=1}^{na} (\Delta l_a)_i , \qquad (17.6)$$

where

$$(\Delta l_o)_i = \sqrt{(\Delta x_o)_i^2 + (\Delta y_o)_i^2} , \qquad (17.7)$$

$$(\Delta l_a)_i = \sqrt{(\Delta x_a)_i^2 + (\Delta y_a)_i^2} , \qquad (17.8)$$

and no and na are the number of points on the original and approximated contours, respectively. The quantities Δx and Δy are the differences in x and y coordinates, respectively, for two neighboring nodes. The subscript "o" on A and B signifies the original contour, while the subscript "a" for a and b signifies the approximated contour in Figure 17.5.

After L_O and L_A are evaluated, the relative length error $E1$ is defined as

$$E1 = \frac{|L_O - L_A|}{L_O} \times 100\% .$$

A "better" approximation for the original contour should have a "smaller" relative length error $E1$. Therefore, the proposed diagonal Cartesian approximation

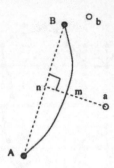

Figure 17.6: Calculating the normal distance.

should have a smaller $E1$ value than the saw-tooth approximation. This prediction is verified in subsequent sections of this chapter.

17.3.2 Average Normal Distance E2

Sharp edges on a boundary may lead to flow separations and vorticity development in the nearby regions. Therefore, large normal distances between the original and approximated boundaries may mean large errors will result in the numerical solutions of fluid flows around the jagged approximated boundaries. In order to ensure that an approximate contour is appropriate, a measure of the *separation* of the original and grid boundary contours is sought.

Consider, once again, the original and grid approximate contours shown in Figure 17.5. The normal distances from approximated grid nodes to the original contour, denoted as line segment $\overline{am}$ in Figure 17.6, should be small for a good approximation. Since the original contour is described by a discrete set of points, the line segment $\overline{an}$, rather than $\overline{am}$, is used to represent the normal distance from a to the original contour, where $\overline{an}$ is normal to $\overline{AB}$. Using the characteristic length of the computational domain LD as a reference length, a dimensionless average normal distance $E2$ can be defined by

$$E2 = \frac{AND}{LD} \, . \tag{17.9}$$

In the above equation, AND is the *average normal distance* and is evaluated by

$$AND = \frac{\sum_{i=1}^{na} (an)_i}{na} \, , \tag{17.10}$$

where na is the number of grid nodes on the approximated contour and an is the absolute normal distance from a to the original contour. The characteristic length is given by

$$LD = \frac{(Lx + Ly)}{2} \, , \tag{17.11}$$

where Lx and Ly are the characteristic lengths of the body domain shown in Figure 17.5. The dimensionless average normal distance $E2$ is another criterion to test the appropriateness of the approximation.

17.4 Two-dimensional Example

To illustrate the appropriateness of the diagonal Cartesian approximation method, a diagonal Cartesian approximation to a two-dimensional square cavity boundary, rotated $30°$ with respect to the x axis, is presented. The boundary of the rotated cavity presents a "complex" contour with respect to Cartesian coordinates.

The saw-tooth and diagonal Cartesian approximations are automatically generated by computer code once the original contour points and the grid size are specified. Error estimations are also automatically calculated by the code.

Figure 17.7(a) shows the Cartesian saw-tooth approximation to the rotated boundary. The grid size of 41×41 is used. The relative length error $E1$ in this approximation is 36.0%, and the average normal distance error $E2$ is calculated to be 0.865%.

The diagonal Cartesian approximation for the rotated boundary on the 41×41 grid is depicted in Figure 17.7(b). The improvement in the approximation is visually evident. The approximation errors $E1$ and $E2$ are calculated to be 14.9% and 0.644%, respectively. Confirming, numerically, the superior contour approximation offered by the diagonal Cartesian method.

17.5 Three-dimensional Approximations

Automatic grid boundary approximation of irregular or complex surfaces in three dimensions is extremely challenging. The diagonal Cartesian method described for the approximation of two-dimensional complex contours in the above sections can be extended to treat the three-dimensional complex geometries.

The 3D diagonal Cartesian approximation method begins with a series of equally spaced *calculation planes* running parallel to planes through $(x, y, 0)$, $(x, 0, z)$ and $(0, y, z)$. Any point of intersection of three mutually perpendicular planes constitutes a *foundation point* of the calculation domain. Complex geometrical boundaries of the problem domain will intersect with the calculation planes. The intersection of the boundary and a given calculation plane is a curve in the calculation plane. Each such curve in each calculation is approximated using the two-dimensional diagonal Cartesian method presented in section 17.2. The collection of approximation nodes in the (x, y, z) coordinates creates an approximation the the original complex surface.

The key issue of successful diagonal Cartesian approximation of irregular geometries is the specification of Cartesian grid positions representing the approximated three-dimensional geometries.

17.6 Three-dimensional Example

The grid generation for complex three-dimensional boundaries is very difficult with current methods, especially when automatic grid generation is desired. Figures 17.9(a) and 17.9(b) show the approximations of a sphere in a structured

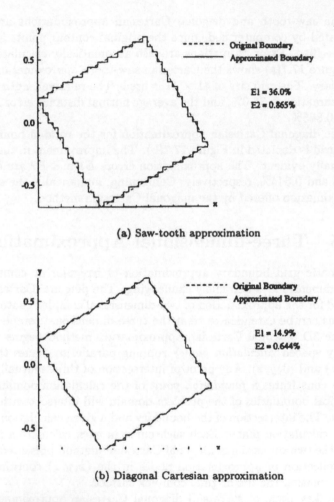

(a) Saw-tooth approximation

(b) Diagonal Cartesian approximation

Figure 17.7: Saw tooth and diagonal Cartesian approximation of a rotated two-dimensional cavity on a 41 × 41 grid.

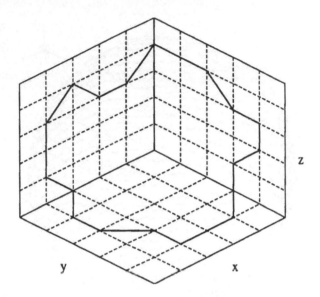

Figure 17.8: Approximation of three-dimensional complex boundaries.

Cartesian grid saw-tooth and diagonal Cartesian methods, respectively. A grid of $121 \times 121 \times 31$ is employed. The diagonal approximation is smoother than the saw-tooth approximation.

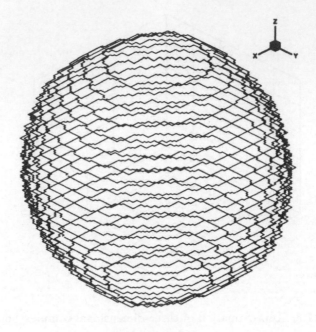

(a) Saw tooth

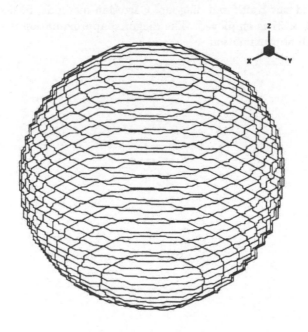

(b) Diagonal Cartesian

Figure 17.9: Approximations of a sphere.

Chapter 18

FA Method on Diagonal Cartesian Coordinates

The diagonal Cartesian method presented in the previous chapter uses both Cartesian grid lines and diagonal line segments to approximate complex geometry. When a diagonal segment is used, the computational element changes from a rectangle (square if the x and y spacing is uniform) to a triangle, as shown if Figure 18.1. The 9-point FA discretization (see Chapter 9) is used on the rectangular elements. However, the triangular elements include only five nodes, so the 9-point FA coefficient formulas do not apply. The present chapter is devoted to the development of the 5-point FA method for use on the triangular cells.

18.1   5-Point FA Scheme for Uniform Grids

The element on the left of Figure 18.2 shows a typical 5-point FA element as it would appear in a uniform grid, and the element on the right shows the same element rotated in terms of the $\xi - \eta$ coordinates which will be used to develop the 5-point FA method.

Using a simple backward difference to approximate the time derivative, and using a representative velocity within each element to remove the non-linear nature, the two-dimensional momentum equations can be written for a single element as

$$\phi_{xx} + \phi_{yy} = 2A\phi_x + 2B\phi_y + g, \qquad (18.1)$$

where A, B and g are constants for each element given by

$$
\begin{aligned}
2A &= RuP, \\
2B &= Rv_P \quad \text{and} \\
g &= R\frac{\phi^m - \phi_P^{m-1}}{\tau} + f_P^{m-1}.
\end{aligned}
\qquad (18.2)
$$

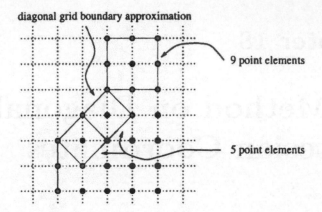

Figure 18.1: Triangular 5-point FA element for complex boundaries.

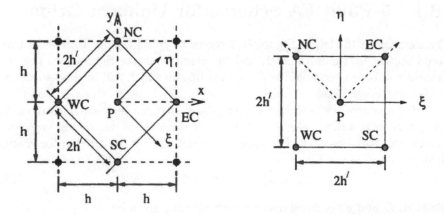

Figure 18.2: 5-point FA element for complex boundaries.

The subscript "P" denotes the quantity is calculated at the node P. The superscript denotes the time step at which the quantity is taken; the previous (known) time step is represented by $m - 1$.

To obtain a form of the momentum equation in the rotated $\xi - \eta$ coordinate frame, the relationship between the two coordinate systems

$$\xi = \frac{x - y}{\sqrt{2}} \tag{18.3}$$

and

$$\eta = \frac{x + y}{\sqrt{2}} \tag{18.4}$$

can be used to obtain the following derivatives:

$$\phi_x = \frac{\phi_\xi + \phi_\eta}{\sqrt{2}} ,$$

$$\phi_{xx} = \frac{\phi_{\xi\xi} + \phi_{\eta\eta}}{2} + \phi_{\xi\eta} ,$$

$$\phi_y = \frac{\phi_\eta - \phi_\xi}{\sqrt{2}} \quad \text{and}$$

$$\phi_{yy} = \frac{\phi_{\xi\xi} + \phi_{\eta\eta}}{2} - \phi_{\xi\eta} .$$

Substituting these derivatives into Equation (18.1) gives

$$\phi_{\xi\xi} + \phi_{\eta\eta} = 2A^{'} \phi_\xi + 2B^{'} \phi_\eta + g , \tag{18.5}$$

where

$$A^{'} = \frac{A - B}{\sqrt{2}} \quad \text{and} \quad B^{'} = \frac{A + B}{\sqrt{2}} .$$

The source term g can be absorbed by a change of variables. Defining $\bar{\phi}$ as

$$\bar{\phi} = \phi + g \frac{A^{'}\xi + B^{'}\eta}{2(A^{'2} + B^{'2})} , \tag{18.6}$$

and substituting for ϕ in Equation (18.5), the desired form of the momentum equation is obtained.

$$\bar{\phi}_{\xi\xi} + \bar{\phi}_{\eta\eta} = 2A^{'} \bar{\phi}_\xi + 2B^{'} \bar{\phi}_\eta \tag{18.7}$$

Equation (18.7) may be solved uniquely if the boundary conditions for the top, bottom, left and right boundaries of the rotated 5-point element in Figure 18.2 are specified. There are two nodes on each 5-point element boundary. To approximate ϕ on each boundary in terms of these two points, functions with a constant and exponential terms are used to specify the boundary profiles. For example, the top boundary condition is approximated by

$$\bar{\phi}_T = a_T e^{2A^{'}\xi} + c_T , \tag{18.8}$$

where a_T and b_T are determined using $\tilde{\phi}_T(h') = \tilde{\phi}_{EC}$ and $\tilde{\phi}_T(-h') = \tilde{\phi}_{NC}$. As a result,

$$a_T = \frac{\tilde{\phi}_{EC} - \tilde{\phi}_{NC}}{e^{2A'h'} - e^{-2A'h'}} \quad \text{and}$$

$$c_T = \frac{\tilde{\phi}_{NC}e^{2A'h'} - \tilde{\phi}_{EC}e^{-2A'h'}}{e^{2A'h'} - e^{-2A'h'}}.$$

The three remaining boundary profiles, $\tilde{\phi}_B$, $\tilde{\phi}_L$ and $\tilde{\phi}_R$, are approximated by

$$\tilde{\phi}_B(\xi) = a_B e^{2A'\xi} + c_B , \tag{18.9}$$

$$\tilde{\phi}_L(\eta) = a_L e^{2B'\eta} + c_L \quad \text{and} \tag{18.10}$$

$$\tilde{\phi}_R(\eta) = a_R e^{2B'\eta} + c_R . \tag{18.11}$$

In a manner analogous to the development of the 9-point FA method, the method of separation of variables is used to solve Equation (18.7) with boundary conditions given in Equations (18.8) - (18.11). The solution takes the form

$$\tilde{\phi}_P = \left[\tilde{\phi}_{EC}e^{-A'h'-B'h'} + \tilde{\phi}_{NC}e^{A'h'-B'h'} \right.$$
$$\left. + \tilde{\phi}_{SC}e^{-A'h'+B'h'} + \tilde{\phi}_{WC}e^{A'h'+B'h'} \right] E , \tag{18.12}$$

where

$$E = \sum_{m=1}^{\infty} \frac{-(-1)^m \lambda_m}{2h' \cosh\mu_m h'} \left[\frac{1}{A'^2 + \lambda_m^2} + \frac{1}{B'^2 + \lambda_m^2} \right] ,$$

$$\lambda_m = \frac{m\pi}{2h'}$$

and

$$\mu_m = \sqrt{A'^2 + B'^2 + \lambda_m^2} .$$

Substituting $\tilde{\phi} = 1$ (a particular solution to equation (18.7)) into Equation (18.12), the following analytic expression can be found for E,

$$E = \frac{1}{4\cosh(A'h')\cosh(B'h')} . \tag{18.13}$$

Using the analytic expression for E in Equation (18.13), computation of an infinite series is avoided. Converting the dependent variable from $\tilde{\phi}$ back to ϕ, and evaluating this expression at point $P = (0,0)$, the 5-point FA formula for Equation (18.1) with uniform grid spacing becomes

$$\phi_p = \alpha_p \sum_{j=1}^{4} (C_{nb}\phi_j) + (1 - \alpha_p)\phi_p^0 - \alpha_p Q_p , \tag{18.14}$$

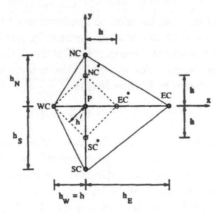

Figure 18.3: 5-point FA element for complex boundaries.

where

$$\alpha_p = \frac{1}{(1 + (C_p Re/\tau)}\ ,$$

$$Q_p^u = C_p Re \frac{\partial P}{\partial x}\ ,$$

$$Q_p^v = C_p Re \frac{\partial P}{\partial y}$$

(18.15)

and ϕ_p^0 is the value from the previous time step. The FA coefficients C_{nb} and C_P are given by

$$C_{EC} = E \cdot e^{-A'h'-B'h'}\ ,$$

$$C_{NC} = E \cdot e^{A'h'-B'h'}\ ,$$

$$C_{SC} = E \cdot e^{-A'h'+B'h'}\ ,$$

(18.16)

$$C_{WC} = E \cdot e^{A'h'+B'h'} \quad \text{and}$$

$$C_P = \frac{-h'}{2(A'^2 + B'^2)} \Big[C_{EC}(A' + B') + C_{NC}(-A' + B')$$

$$+ C_{SC}(A' - B') + C_{WC}(-A' - B') \Big]\ .$$

18.2 5-Point FA Scheme for Nonuniform Grids

For non-uniform grid spacing, interpolation can be used to obtain an applicable FA formula from Equation (18.14). Consider the case shown in Figure 18.3, where $h_W < h_E$, h_N and h_S. A smaller square element with side length $2h'$ ($h' = h/\sqrt{2} = h_W/\sqrt{2}$) and with interior point P located at the center is added as shown.

Interpolation functions can be used to approximate the unknown nodal values ϕ_{NC^*}, ϕ_{SC^*} and ϕ_{EC^*} of the smaller element in terms of the known nodal values of the larger non-uniform element. The interpolation functions used have linear and exponential terms. For example, the interpolation function on the x-axis of the element shown in Figure 18.3 is

$$\phi(x) = a(e^{2Ax} - 1) + bx + c, \tag{18.17}$$

where A is defined in Equation (18.2). The values of a, b and c can be determined using $\phi(x = 0) = \phi_P$, $\phi(x = -h_W) = \phi_{WC}$ and $\phi(x = h_E) = \phi_{EC}$. Evaluating Equation (18.17) at $x = h$ yields an expression for ϕ_{EC^*} in terms of the known nodal values. The other unknown nodal values can be found in a similar manner.

Applying the 5-point FA method to the smaller element and substituting the expressions for the unknown nodal values, the 5-point FA formula for an element with non-uniform grid spacing becomes

$$\phi_p = \beta_p \sum_{j=1}^{4}(b_j\phi_j) + (1 - \beta)\phi_p^0 - \beta_p Q_p, \tag{18.18}$$

where

$$\beta_p = \frac{1}{(G + (b_p Re/\tau))},$$

$$Q_p^u = b_p Re \frac{\partial P}{\partial x},$$

$$Q_p^v = b_p Re \frac{\partial P}{\partial y},$$

$$G = 1 - b_1 C_{EC} - b_2 C_{NC} - b_3 C_{SC},$$

$$b_{EC} = a_1 C_{EC},$$

$$b_{NC} = a_2 C_{NC} + a_3 C_{SC},$$

$$b_{WC} = C_{WC} + c_1 C_{EC},$$

$$b_{SC} = c_2 C_{NC} + c_3 C_{SC}$$

and

$$b_P = C_P.$$

The FA coefficients C_{nb} in the above equations are defined in Equations (18.17), and a_i, b_i and c_i, $i = 1, 2, 3$ are given by following equations.

$$a_1 = \frac{h_W(e^{2A'}h_W - 1) + h_W(e^{2A'}h_W - 1)}{h_E(e^{-2A'}h_W - 1) + h_W(e^{2A'}h_W - 1)},$$

$$b_1 = \frac{-(h_E + h_W)(e^{2A'}h_W - 1) + h_W(e^{2A'}h_E - 2A'h_W)}{h_E(e^{-2A'}h_W - 1) + h_W(e^{2A'}h_W - 1)} + 1,$$

$$c_1 = \frac{h_W(e^{2A'}h_W - 1) - h_W(e^{2A'}h_E - 1)}{h_E(e^{-2A'}h_W - 1) + h_W(e^{2A'}h_E - 1)},$$

$$a_2 = \frac{h_S(e^{2B'}h_W - 1) + h_W(e^{-2B'}h_S - 1)}{h_N(e^{-2B'}h_S - 1) + h_S(e^{2B'}h_N - 1)},$$

$$b_2 = \frac{-(h_N + h_S)(e^{2B'}h_W - 1) + h_W(e^{2B'}h_N - e^{-2B'}h_S)}{h_N(e^{-2B'}h_S - 1) + h_S(e^{2B'}h_N - 1)},$$

$$c_2 = \frac{h_N(e^{2B'}h_W - 1) - h_W(e^{2B'}h_N - 1)}{h_N(e^{-2B'}h_S - 1) + h_S(e^{2B'}h_N - 1)},$$

$$a_3 = \frac{h_S(e^{-2B'}h_W - 1) - h_W(e^{-2B'}h_S - 1)}{h_N(e^{-2B'}h_S - 1) + h_S(e^{2B'}h_N - 1)},$$

$$b_3 = \frac{-(h_N + h_S)(e^{-2B'}h_W - 1) - h_W(e^{2B'}h_N - e^{-2B'}h_S)}{h_N(e^{-2B'}h_S - 1) + h_S(e^{2B'}h_N - 1)} + 1$$

and

$$c_3 = \frac{h_N(e^{-2B'}h_W - 1) - h_W(e^{2B'}h_N - 1)}{h_N(e^{-2B'}h_S - 1) + h_S(e^{2B'}h_N - 1)}.$$

18.3 Three-Dimensional Case

The three-dimensional transport equation (Equation (10.1) must be discretized on each of the resulting diagonal Cartesian cells used in approximating a complex three-dimensional domain. For an arbitrary cell, the general form of the discretized transport equation is

$$\phi_p = \sum_{i=1}^{n_x} C_i^x \phi_i^x + \sum_{j=1}^{n_y} C_j^y \phi_j^y + \sum_{k=1}^{n_z} C_k^z \phi_k^z + C_p g_p , \qquad (18.19)$$

where C_i^x, C_j^y, C_k^z and C_p are the FA coefficients that must be determined. The dependent variable (one of the velocity components) is represented by ϕ, while g_p contains a finite difference representation for the time derivative as well as the source term (pressure gradient). The reader may want to refer to chapter 10 for more detail. For sake of reference, the superscript variable represents the dimension held constant on the plane. That is, ϕ_i^x represent ϕ values on the yz plane.

Recall from section 17.5 that each three-dimensional diagonal Cartesian cell is defined in terms of three mutually perpendicular, intersecting planes. The point of intersection is the foundation point of the cell. A cell near the surface boundary of the complex three-dimensional domain may have one or more planes that intersect with the boundary. If so, the resulting intersection is a curve in the given plane, and the curve is approximated using the Cartesian segments or the diagonal. Depending on how many planes of a given cell intersect with the boundary, the four possible configurations for the diagonal Cartesian cell are shown in Figure 18.4

If the complex boundary does not (see Figure 18.4(a)) intersect with any of the three planes (which would be the case for the majority of the cells), then a

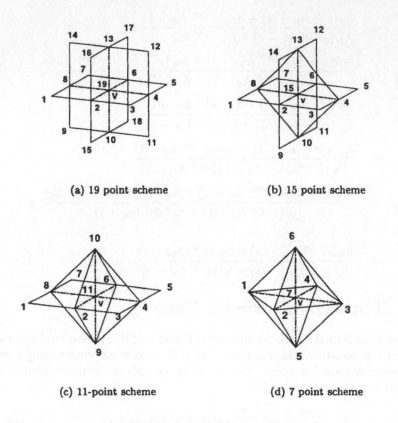

(a) 19 point scheme   (b) 15 point scheme

(c) 11-point scheme   (d) 7 point scheme

Figure 18.4: Four possible three-dimensional diagonal Cartesian cell configurations.

nine-point FA method (chapter 9) for each plane is used to calculate each set of coefficients in Equation (18.19. Consequently, $n_x = n_y = n_z = 8$. The resulting three-dimensional FA discretization is the 19-point formula presented in section 10.4.

If one of the three planes intersects with the boundary, then a cell configuration as shown in Figure 18.4(b) would result. In this case, the 9-point FA method would be used on the two planes free from an intersection with the boundary, while a 5-point FA scheme (see section 18.1)is used for the intersecting plane. If the 5-point plane is the yz plane, then $n_x = 4$, while $n_y = n_z = 8$. The result is a 15-point FA formula (the node numbering is shown in Figure 18.4(b)).

When two planes intersect with the complex boundary, 5-point formulas are used on each, while the standard 9-point formula is used on the third. The result is an 11-point FA formula. The geometric shape and node numbering is shown in Figure 18.4(c). The fourth case is depicted in Figure 18.4(d), where all three planes intersect with the boundary. 5-point FA formulas are used for each, resulting in a 7-point overall discretization formula.

In any of the four possible cases, the following FA algebraic equation results.

$$\phi_p = \alpha_p \sum_{i=1}^{n} (C_i \phi_i) + (1 - \alpha_p) \phi_p^0 - \alpha_p Q_p^\beta . \qquad (18.20)$$

The superscript "0" denotes a value taken from the previous time step. The upper limit "n" on the sum is either 18, 14, 10 or 6 depending on which cell configuration is used. For sake of reference, the superscript variable represents the dimension held constant on the plane. That is, ϕ_i^x represent ϕ values on the yz plane (where x is constant). The coefficients α_p and Q_p are defined as

$$\alpha_p = \frac{1}{\left(1 + \frac{C_p Re}{\tau}\right)} \quad \text{and}$$

$$Q_p^\beta = C_p Re \left(\frac{\partial p}{\partial \beta}\right)\bigg|_p ,$$

where β is either x for $\phi = u$, y for $\phi = v$ or z for $\phi = w$.

18.4   Exercises for Part III

1. (a) What is the meaning of the Laplace operator $\nabla^2 f$ on the differentiable function $f(x, y)$?

   (b) Show that

   $$\nabla^2 f\big|_p \doteq \frac{6}{h^2} \left(\bar{f} - f_p \right) ,$$

   where $\quad \bar{f} = \dfrac{1}{4h^2} \displaystyle\int_{-h}^{h} \int_{-h}^{h} f\, dx\, dy \quad$ and $\quad f_p = f(x_p, y_p) .$

   (c) Use the result of part B to explain the behavior of the solution f of

   $$\nabla^2 f = Q$$

   when Q is (i) zero, (ii) positive and (iii) negative.

2. Derive the expression for ξ_{xy} given in Equation (16.15).

3. Provide the details for deriving the expression for η_{xx} shown in Equation (16.16) from Equations (16.13) and (16.14).

4. Fill in the details for the derivation of Equation (16.23) from Equation (16.22).

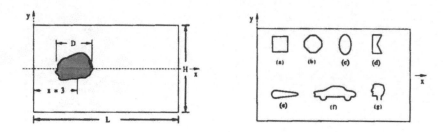

(a) Computational domain.   (b) Possible object shapes.

Figure 18.5: Boundary fitted coordinates exercise.

5. (Computer Exercise) Use available computer code (FORTRAN code is available at the web site **www.finiteanalytic.com**), or write your own, to construct at least two sets of boundary fitted coordinates for the assigned body shape shown in Figure 18.5(b). The body shaped used should be enclosed in the computational domain depicted in Figure 18.5(a). Generate the first set of coordinates using (a) $\nabla^2 \xi = 0$ and $\nabla^2 \eta = 0$. The

second set should be generated using (b) $\nabla^2 \xi = P$ and $\nabla^2 \eta = Q$. Use your choice for functions P and Q. The grid resolution should be at least 20×60. The characteristic length D of your object should be 1. Your report should include:

(a) an explanation of your solution procedure,

(b) a listing or reference of the computer code used,

(c) a plot of the resulting coordinates for case (a), and at least one plot for case (b) (specifying formulas for P and Q), and

(d) a discussion of your results and conclusion.

6. (Computer Exercise) Follow the directions outlined in the previous computer exercise but for a computational domain depicted in Figure 18.6. The outer boundary is a circle of radius 10. The characteristic length D of the object is 2.

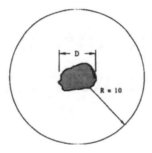

Figure 18.6: Circular computational domain.

Part IV

Computational Considerations

Part IV of the text is devoted to implementing the numerical finite analytic method introduced in part II of the text.

The momentum and continuity equations are "coupled" through the presence of velocity components in each. The equations are solved, numerically, one at a time. In order to find the correct velocity and pressure values that satisfy all equations "simultaneously," a solution algorithm will include an iterative feature wherein the dependent variables of velocity and pressure are recalculated for each equation until one solution set of dependent variables (two or three velocity components and pressure) solves all equations.

Algorithms designed for solving the coupled momentum and continuity equations may be categorized, roughly, into two groups: those using staggered grids and those using non-staggered grids. Various algorithms for solving the coupled momentum and continuity equations on staggered grids are introduced in Chapter 19. Alternative nonstaggered grid methods are, in many cases, based on the staggered grid algorithms adjusted to the nonstaggered grids. The methods are introduced in Chapter 20.

Various types of boundary conditions for pressure and velocity are presented in Chapter 21 of part IV. The iterative nature of solution algorithms for the coupled momentum and continuity equations require some consideration of convergence. Convergence criteria, as well as means for accelerating convergence, are examined in the third chapter as well.

Chapter 19

Velocity, Pressure and Staggered Grids

Most applications involving the Navier-Stokes equations require velocity calculation. That is, the velocity components (u, v and w) are unknown, variable quantities except on the boundaries. Therefore, the momentum equation is nonlinear in the convection terms. The nonlinearity in the momentum equation is handled by an iterative solution process.

The source term in the momentum equation includes a pressure gradient. The more challenging aspect is calculating the pressure field when it is not prescribed. There is no explicit equation for pressure. Instead, pressure is calculated using the continuity equation, which also involves the velocity components, thus exposing the coupling of the momentum and continuity equations.

Various algorithms for solving the coupled momentum and continuity equations on staggered grids are introduced in this chapter. Staggered grid systems are used to alleviate the "checkerboard" problem, which is described in the first section.

19.1   The Checkerboard Problem

Numerical methods result in dependent variables, such as velocity components and pressure, being calculated at discrete nodal locations. Referring to Figure 19.1, suppose both velocity components and pressure are calculated at the same nodal locations. The following scenario is possible. Let both velocity components be zero at every node, while pressure takes on a "checkerboard" pattern where all pressure values have the same absolute value, but alternate between positive and negative in each direction. The continuity equation is trivially satisfied by the identically zero velocity fields. For the momentum equations, the pressure gradient term would be zero if calculated using a centered difference. In the absence of any other forces, the zero velocity components satisfy the momentum equations as well.

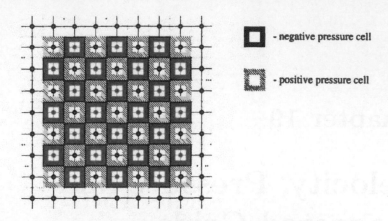

□ - negative pressure cell

▢ - positive pressure cell

Figure 19.1: Checkerboard pressure pattern.

However, we would expect some motion in the fluid because of the pressure gradients between adjacent nodes. This describes what is know as the *checkerboard problem* or paradox [145] in the numerical solution to the Navier-Stokes equations.

Harlow and Welch [81] introduced the use of staggered grids, as shown in Figure 19.2, in an attempt to solve the checkerboard problem. Three sets of nodes, one for pressure P, one for the horizontal velocity u, and one for the vertical velocity v, are used in a staggered grid to construct the discrete analog versions of the continuity and momentum equations for two-dimensional problems. The algebraic analog of the continuity equation uses the difference of adjacent velocity components. Moreover, the pressure gradient term in algebraic analog of the momentum equations is calculated using the difference in adjacent pressure nodes, removing the possibility of the checkerboard paradox.

An algorithm for calculating velocity and pressure on staggered grids is presented in each of the following sections. The semi-implicit method for pressure linked equations (SIMPLE) is presented first. A rather detailed description of the method is given even though it is not the most popular algorithm. The detail is warranted, however, because two of the more popular methods presented in subsequent sections are modifications of the SIMPLE algorithm.

19.2   The SIMPLE Algorithm

The SIMPLE algorithm was introduced by Patankar and Spalding [147] in 1972. It is presented in the context of the finite analytic algebraic equations for momentum. It begins by considering the FA equations for velocity components u and v on the staggered grid as pictured in Figure 19.2.

For the sake of notation simplicity, uniform grid spacing of h in both the x and y directions is assumed. The development is similar for nonuniform grids.

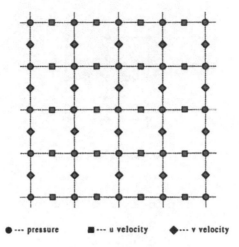

● ··· pressure   ■ ··· u velocity   ◆ ··· v velocity

Figure 19.2: Staggered grid pattern.

The FA control volume for pressure P is shown in Figure 19.3(a).

In the interest of calculating the pressure at node P, the four cell face velocities u_e, u_w, v_n, and v_s need to be determined. The equation for the u_e, the horizontal velocity component on the east face of the cell, is given in Equation (19.1).

$$u_e = \alpha_e \sum_{j=NE}^{NC} C_j^e u_{e_j} - \alpha_e S_P^e \left(\frac{P_E - P_P}{h} \right) + (1 - \alpha_e)u_e^0 \qquad (19.1)$$

The superscript "0" on the u_e term on the right hand side of Equation (19.1) denotes the value from the previous time step, while all other variables without a superscript denote values from the current time step. The FA coefficients are denoted by C_j^e ($j = NE, EC, \ldots , NC$) and C_P^e, and the coefficients α_e and S_P^e are defined as

$$\alpha_e = \left(1 + \frac{C_P^e R}{\tau} \right) \quad \text{and} \quad S_P^e = C_P^e R \,,$$

respectively, with grid spacing h, time step τ and Reynolds number R. The FA element for the east face velocity for the pressure control volume in question is shown in Figure 19.3(b). Note that the pressure control volume is also indicated in this figure in order to show its spatial relationship with the FA element for u_e.

A similar equation for the north-face vertical velocity, v_n, of the cell is shown in Equation (19.2). The FA element for the north face velocity is depicted in Figure 19.3(c). Note that the elements for u_e and v_n are defined by different nodal locations, which means the FA coefficients are, in general, different for the different elements. This difference is denoted by the superscripts e and n in

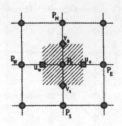

(a) Pressure control volume

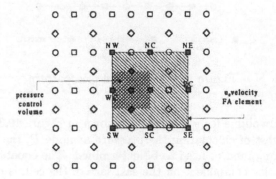

(b) FA element for u_e (locations denoted by □)

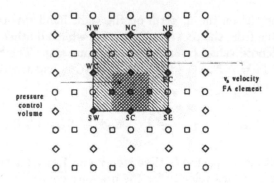

(c) FA element for v_n (locations denoted by o)

Figure 19.3: Pressure and velocity elements for the FA method on a staggered grid.

Equations (19.1) and (19.2).

$$v_n = \alpha_n \sum_{j=NE}^{NC} C_j^n v_{n_j} - \alpha_n S_P^n \left(\frac{P_N - P_P}{h} \right) + (1 - \alpha_n)v_n^0 \qquad (19.2)$$

Because pressure values in Equations (19.1) and (19.2) are from the current time step (therefore, initially unknown), the momentum equations are initially solved using "guessed" pressure values, P^*. These pressure values are taken from the previous time step for the initial iteration. Guessed pressure values result in guessed velocity values as well. Writing Equations (19.1) and (19.2) in terms of the guessed values gives,

$$u_e^* = \alpha_e \sum_{j=NE}^{NC} C_j^e u_{e_j}^* - \alpha_e S_P^e \left(\frac{P_E^* - P_P^*}{h} \right) + (1 - \alpha_e)u_e^0 \qquad (19.3)$$

and

$$v_n^* = \alpha_n \sum_{j=NE}^{NC} C_j^n v_{n_j}^* - \alpha_n S_P^n \left(\frac{P_N^* - P_P^*}{h} \right) + (1 - \alpha_n)v_n^0 . \qquad (19.4)$$

Subtracting, respectively, Equations (19.3) and (19.4) from Equations (19.1) and (19.2) gives

$$u_e' = \alpha_e \sum_{j=NE}^{NC} C_j^e u_{e_j}' - \alpha_e S_P^e \left(\frac{P_E' - P_P'}{h} \right) \qquad (19.5)$$

and

$$v_n' = \alpha_n \sum_{j=NE}^{NC} C_j^n v_{n_j}' - \alpha_n S_P^n \left(\frac{P_N' - P_P'}{h} \right) , \qquad (19.6)$$

where

$$u_e' = u_e - u_e^*, \qquad v_n' = v_n - v_n^*, \qquad P_e' = P_e - P_e^* \qquad \text{and} \qquad P_w' = P_w - P_w^*$$

represent the difference, or "correction," between the guessed values and the correct values.

In the SIMPLE method, the summation terms in Equations (19.5) and (19.6) are dropped in order to more easily determine the velocity correction. That is,

$$u_e' = -c_e \left(P_E' - P_P' \right) \qquad \text{and} \qquad v_n' = -c_n \left(P_N' - P_P' \right), \qquad (19.7)$$

where

$$c_e = \frac{\alpha_e S_P^e}{h} \qquad \text{and} \qquad c_n = \frac{\alpha_n S_P^e}{h}. \qquad (19.8)$$

The discretized version of the continuity equation is used to calculate the pressure correction. For the correct velocity components, u_e, u_w, v_n and v_s, the algebraic version of the continuity equation is

$$\frac{u_e - u_w}{h} + \frac{v_n - v_s}{h} = 0. \qquad (19.9)$$

Multiplying both sides by h, and expressing the velocity components in terms of the their guessed and corrected parts results in

$$u_e^* + u_e^{'} - (u_w^* + u_w^{'}) + (v_n^* + v_n^{'}) - (v_s^* + v_s^{'}) = 0. \tag{19.10}$$

Substituting for the velocity corrections in terms of the corrected pressure and rearranging gives

$$c_e(P_P^{'} - P_e^{'}) - c_w(P_w^{'} - P_P^{'}) + c_n(P_P^{'} - P_n^{'}) - c_s(P_s^{'} - P_P^{'}) = -D^*, \tag{19.11}$$

where

$$D^* = u_e^* - u_w^* + v_n^* - v_s^*$$

is a measure of the dilation in the guessed velocity field. If D^* is zero, the guessed velocity and pressure quantities are, in fact, correct.

Solving Equation (19.11) for $P_P^{'}$ results in a pressure correction formula for node P, i.e.,

$$c_P P_P^{'} = c_e P_e^{'} + c_s P_s^{'} + c_w P_w^{'} + c_n P_n^{'} - D^*, \tag{19.12}$$

where

$$c_P = c_e + c_s + c_w + c_n.$$

Once the pressure correction is determined, it is used to correct the guessed velocities u_e^*, u_w^*, v_n^* and v_s^*. The coupling of the velocity and pressure variables in the momentum equation implies that if either change, then the new values may not solve the equation. Therefore, an iterative method is used until the corrected quantities converge to values that solve the momentum equations.

The SIMPLE algorithm is outlined below:

1. Select a guessed pressure field P^* (usually that from a previous time step or iteration).

2. Use the guessed pressure field in the FA algebraic versions of the momentum equations [Equations (19.3) and (19.4] to calculate u^* and v^*.

3. Solve the pressure correction equation [Equation (19.12)] for $P^{'}$, and correct the pressure using

$$P = P^* + \lambda_P P^{'},$$

   where λ_P is an under-relaxation factor.

4. Correct velocities u^* and v^* using the velocity correction equations [Equations (19.7)].

5. Repeat steps 2 through 4 until the pressure and velocities satisfy a prescribed convergence criteria.

The pressure correction equation [Equation (19.12)] and momentum equations [Equations (19.3) and (19.4)] are usually solved in an semi-implicit manner by such methods as the line-by-line Thomas algorithm. Repeated horizontal or vertical sweeps through the computational domain are made until the differences between pressure values in consecutive sweeps fall below some prescribed tolerance. If an equation is nonlinear, such as the momentum equations, it is not numerically economical to require strict tolerance for a given variable since some coefficients in the algebraic equations are functions of the dependent variables. It is a common practice to perform only several sweeps through the computational domain, update the variables and coefficients, and then recalculate the quantities.

The SIMPLE algorithm is not the most popular method for solving the coupled momentum and continuity equations because of poor convergence characteristics. However, a detailed description of the methods is presented because two of the more popular methods are improvements on the original SIMPLE algorithm. The first of these improvements is presented in the next section.

19.3 The SIMPLEC Algorithm

The SIMPLE Consistent (SIMPLEC) algorithm, introduced by van Doormaal and Raithby [63, 153], follows the same steps as the SIMPLE method except the term

$$\alpha_e u_e' \sum_{j=NE}^{NC} C_j^e$$

is subtracted from both sides of Equation (19.5) giving

$$u_e' - \alpha_e u_e' \sum_{j=NE}^{NC} C_j^e = \alpha_e \sum_{j=NE}^{NC} C_j^e u_{e_j}' - \alpha_e u_e' \sum_{j=NE}^{NC} C_j^e - \alpha_e S_P^e \left(\frac{P_E' - P_P'}{h} \right).$$
(19.13)

Using the fact that $\sum_{j=NE}^{NC} C_j^e = 1$ (see Property 11.2), and the approximation $u_{e_j} \approx u_e$ for $j = NE, EC, \ldots, NC$ results in the following expression for the velocity correction:

$$u_e' = -d_e(P_E' - P_P'),$$
(19.14)

where

$$d_e = \frac{\alpha_e S_P^e}{h(1 - \alpha_e)}.$$

Similar velocity correction expressions may be derived for the other three cell face velocities.

These new velocity correction expressions replace the old in the continuity equation [Equation 19.10] expressed in terms of the guessed and corrected velocities. The resulting pressure correction equation is

$$d_P P_P' = d_e P_e' + d_s P_s' + d_w P_w' + d_n P_n' - D^*,$$
(19.15)

where

$$d_P = d_e + d_s + d_w + d_n,$$

and D^* is the dilations as defined for the SIMPLE algorithm.

The SIMPLEC algorithm is almost identical to the SIMPLE algorithm except for the velocity correction [Equation (19.14)]. Recall that in SIMPLE algorithm the approximation $\sum_{j=NE}^{NC} C_j^e u_{e_j}^* = 0$ is made, while in the SIMPLEC algorithm the velocity correction is derived using the approximation

$$\sum_{j=NE}^{NC} C_j^e (u_e^* - u_{e_j}^*) = 0.$$

Certainly the latter approximation is better because the difference of the nodal velocity and the surrounding velocities is more likely to be closer to zero. Because of the overall similarity between the two algorithms, an outline for the SIMPLEC method will not be listed.

19.4   The SIMPLER Algorithm

The SIMPLER (SIMPLE Revised) algorithm was devised by Patankar [146] because it was felt that the approximations

$$\sum_{j=NE}^{NC} C_{e_j} u_{e_j} = \sum_{j=NE}^{NC} C_{n_j} v_{n_j} = 0$$

were responsible for the slow convergence of the pressure field. The method begins by calculating *pseudo-velocities* $\hat{u}_e$ and $\hat{v}_n$ using the momentum equations without the pressure gradient terms, as shown in Equations 19.16 and 19.17.

$$\hat{u}_e = \alpha_e \sum_{j=NE}^{NC} C_j^e u_{e_j} + (1 - \alpha_e) u_e^0 \qquad (19.16)$$

and

$$\hat{v}_n = \alpha_n \sum_{j=NE}^{NC} C_j^n v_{n_j} + (1 - \alpha_n) v_n^0, \qquad (19.17)$$

The velocity terms are written in component form as

$$u_e = \hat{u}_e - c_e(P_E - P_P) \qquad \text{and} \qquad v_n = \hat{v}_n - c_n(P_N - P_P), \quad (19.18)$$

where c_e and c_n are defined as in Equations (19.8). Substituting the component forms of the velocity expressions into Equation (19.9) results in the pressure equation

$$c_P P_P = c_e P_e + c_s P_s + c_w P_w + c_n P_n - \hat{D}, \qquad (19.19)$$

where c_P and $\hat{D}$ are defined as

$$c_P = c_e + c_s + c_w + c_n \qquad \text{and} \qquad \hat{D} = \hat{u}_e - \hat{u}_w + \hat{v}_n - \hat{v}_s,$$

respectively. This newly determined pressure field takes the place of the guessed pressure field in the SIMPLE algorithm. The resulting guessed velocity components u_e^* and v_n^* are used to determine a pressure correction as in Equation (19.12). The pressure correction is used to correct the velocities u_e^*, u_w^*, v_n^*, and v_s^*. The SIMPLER algorithm is outlined below:

1. Use the current velocity field to determine the pseudo-velocities $\hat{u}_e$, $\hat{v}_s$, $\hat{u}_w$, and $\hat{v}_n$ using equations Equations (19.16) and (19.17).

2. Calculate an updated pressure field using Equation (19.19).

3. Calculate u_e^*, v_s^*, u_w^* and v_n^* using Equations (19.3) and (19.4).

4. Solve the pressure correction equation [Equation (19.12)].

5. Correct the guessed velocities using Equations(19.7).

6. Repeat steps 1 through 5 until suitable convergence is achieved.

19.5   The PISO Algorithm

The pressure-implicit with splitting of operators (PISO) method proposed by Issa [87] is a time-marching procedure in which, during each time step, there is a *predictor* step and one or more *corrector* steps. In the description that follows, variables marked with the superscript "0" are quantities known from the previous time step, those marked with a single star ($*$) are predictor quantities, those with two stars ($**$) are the first corrected values, and those with three stars ($* * *$) represent variables corrected for the second time.

The method begins by calculating the predicted velocities using the pressure quantities from the previous time step. The equation for u_e^* is

$$u_e^* = \alpha_e \sum_{j=NE}^{NC} C_j^e u_{e_j}^* - \alpha_e S_P^e \frac{(P_E^0 - P_P^0)}{h} + (1 - \alpha_e)u_e^0. \qquad (19.20)$$

Similar equations are used to predict values for u_w, v_n, and v_s. Corrected values for velocity are expressed using predicted values for pressure, as shown in Equation (19.21).

$$u_e^{**} = \alpha_e \sum_{j=NE}^{NC} C_j^e u_{e_j}^* - \alpha_e S_P^e \frac{(P_E^* - P_P^*)}{h} + (1 - \alpha_e)u_e^0 \qquad (19.21)$$

Subtracting Equation (19.20) from Equation (19.21) gives another form for correcting the u_e velocity

$$u_e^{**} = \hat{u}_e - d_e(P_E^* - P_P^*), \qquad (19.22)$$

where

$$\hat{u}_e = u_e^* + d_e(P_E^0 - P_P^0) \qquad (19.23)$$

and

$$d_e = \frac{\alpha_e S_P^e}{h}.$$

Equation (19.22) is the actual formula for correcting the velocity for the first time. Similar equations for u_w^{**}, v_n^{**} and v_s^{**} can be constructed. These expressions for the corrected cell face velocities are substituted into the discretized continuity equation [Equation (19.9)] to determine the predictor pressures in Equation (19.22). The following pressure equation results

$$d_P P_P^* = d_e P_E^* + d_w P_W^* + d_s P_S^* + d_n P_N^* - \hat{D}, \qquad (19.24)$$

where

$$\hat{D} = \hat{u}_e - \hat{u}_w + \hat{u}_n - \hat{u}_s$$

and

$$d_P = d_e + d_s + d_w + d_n.$$

The second set of corrector equations is derived in a similar way. Begin by writing the expression for u_e^{***} in terms of the P^{**} field,

$$u_e^{***} = \alpha_e \sum_{j=NE}^{NC} C_j^e u_{e_j}^{**} - \alpha_e S_P^e \frac{(P_E^{**} - P_P^{**})}{h} + (1 - \alpha_e) u_e^0. \qquad (19.25)$$

Subtract Equation (19.21) from Equation (19.25) to get the second corrector equation for the east-face velocity,

$$u_e^{***} = \hat{\hat{u}}_e - d_e(P_E^{**} - P_P^{**}), \qquad (19.26)$$

where

$$\hat{\hat{u}}_e = u_e^{**} + \alpha_e \sum_{j=NE}^{NC} C_j^e \left(u_{e_j}^{**} - u_{e_j}^* \right) - d_e \left(P_P^* - P_E^* \right). \qquad (19.27)$$

The calculation of the second correction for pressure is done as above. Equations for u_w^{***}, v_n^{***} and v_s^{***}, like that of Equation (19.27), are used in the discretized continuity equation. The result is

$$d_P P_P^{**} - d_e P_E^{**} - d_s P_S^{**} - d_w P_W^{**} - d_n P_N^{**} = -\hat{\hat{D}}, \qquad (19.28)$$

where

$$\hat{\hat{D}} = \hat{\hat{u}}_e - \hat{\hat{u}}_w + \hat{\hat{v}}_n - \hat{\hat{v}}_s.$$

The outline of the PISO algorithm follows.

1. Use dependent variable values u^n, v^n and P^n from the previous time step to calculate predictor quantities u_e^* and v_n^* using equations of the form of Equation (19.20).

2. Use the predicted velocities to calculate the predicted pressure P^* using Equation (19.24).

3. Use Equation (19.22) with the pressures calculated in step 2 to correct the velocities.

4. Using the corrected velocities found in step 3, solve the pressure correction equation [Equation (19.28)].

5. Correct the velocities for the second time by using the corrected pressure quantities in Equation (19.26).

6. March on to the next time step.

19.6 Convergence Criteria

Except for the PISO algorithm, the methods presented in this chapter for solving the coupled momentum and pressure equations involve an iterative component. This iteration process continues until the change, or *residual*, in each of the variables between successive iterations falls below some prescribed tolerance.

These residuals may represent the maximum change of the respective variables between iterations. That is,

$$\Delta P_{max} = Max\left|P_{i,j}^n - P_{i,j}^{n-1}\right|, i = 1\ldots i_{max}, \ j = 1\ldots j_{max} \quad (19.29)$$

$$\Delta u_{max} = Max\left|u_{i,j}^n - u_{i,j}^{n-1}\right|, i = 1\ldots i_{max}, \ j = 1\ldots j_{max} \quad (19.30)$$

$$\Delta v_{max} = Max\left|v_{i,j}^n - v_{i,j}^{n-1}\right|, i = 1\ldots i_{max}, \ j = 1\ldots j_{max} \quad (19.31)$$

where i and j are grid point indices ranging from 1 to i_{max} and 1 to j_{max}, respectively. Superscripts n and $n-1$ represent the present and previous iterations, respectively. Thus, the residual for each variable is simply the maximum difference, over the entire computational domain, between the values of that variable at the present and previous time steps.

Convergence for a given time step is attained when ΔP_{max}, Δu_{max} and Δv_{max} are all less than 10^{-4}.

An alternative way to define the residuals is

$$\Delta P_{max} = \frac{\sum_{i,j=1}^{i_{max},j_{max}} \frac{\left|P_{i,j}^n - P_{i,j}^{n-1}\right|}{\left|P_{i,j}^{n-1}\right|}}{N_{total}}, \quad (19.32)$$

$$\Delta u_{max} = \frac{\sum_{i,j=1}^{i_{max},j_{max}} \frac{\left|u_{i,j}^n - u_{i,j}^{n-1}\right|}{\left|u_{i,j}^{n-1}\right|}}{N_{total}} \quad \text{and} \quad (19.33)$$

$$\Delta v_{max} = \frac{\sum_{i,j=1}^{i_{max},j_{max}} \frac{\left|v_{i,j}^n - v_{i,j}^{n-1}\right|}{\left|v_{i,j}^{n-1}\right|}}{N_{total}}, \quad (19.34)$$

where N_{total} is the total numbers of fluid nodes in the computational domain. Note that the evaluations in Equations (19.32) to (19.34) are problematic when zero values of pressure or velocity appear.

Experience indicates there is little difference in solution convergence when using the different convergence criteria outlined above. Since the dimensionless governing equations are used in the simulation, the convergence criteria in Equations (19.29) to (19.31) are considered acceptable for most engineering problems.

19.7   Performance Summary

Coupling between the momentum and mass conservation equations for incompressible flows is often the major cause of the slow convergence of iterative solution techniques [153]. The SIMPLE, or variations such as SIMPLEC and SIMPLER, have been used in resolving the pressure-velocity coupling in incompressible flow problems. The main disadvantage of the SIMPLE algorithm is the slow convergence of the pressure field [146]. The SIMPLER algorithm has better overall convergence characteristics but requires more CPU time per iteration than the SIMPLE algorithm. The SIMPLEC algorithm is another variation on the SIMPLE method, and appears to exhibit more robust behavior than the SIMPLER method, especially in time-marching formulations [92].

The PISO algorithm was originally intended for obtaining time accurate solutions for the incompressible flows. The method is suitable for both steady and unsteady flows. It demonstrates robust convergence behavior and requires less computational effort than the SIMPLER and SIMPLEC methods for problems in which the momentum equations are not coupled with a scalar quantity [87, 88, 92]. However, when a scalar quantity is strongly coupled with momentum, such as temperature appearing in the source term, the iterative SIMPLER and SIMPLEC methods tend to perform better than the two-stage PISO method [92].

Chapter 20

Nonstaggered Grid Methods

Staggered grids are used to avoid the checkerboard problem. However, the various staggered grid algorithms presented in Chapter 19 have some drawbacks. Using different nodal locations for the different variables means different coefficient values in the algebraic equations used to calculate the dependent variables. Consequently, staggered grids may require a large amount of CPU time and memory to calculate and store the corresponding sets of coefficients, especially in the case for three-dimensional flows. The same may be true for problems with complex geometries, as pointed out by Tien [169] and experienced by Lin and Chen [116, 118, 119, 117, 122, 121, 120].

Initial attempts to use *nonstaggered*, or *regular* grids, incorporated simple averaging as a way of calculating pressure and velocity in "desirable" locations. However, methods for nonstaggered grids most be more sophisticated then simple averaging. Zang and Date [182, 61] report a loss of accuracy when using certain averaging techniques.

Perhaps the first attempt to use nonstaggered grids is that by Rhie and Chow [155] in 1983. They successfully implemented the SIMPLE algorithm on a regular grid arrangement. Peric [150] used a similar approach in conjunction with the finite volume method to model three-dimensional flow in complex ducts.

Three different algorithms for nonstaggered grid implementations are presented in this chapter. The first is an adaptation of the SIMPLEC to nonstaggered grids.

20.1   Pressure Weighted Interpolation Method

One of the early successful attempts to use nonstaggered grids was accomplished by Miller and Schmidt [131] using their pressure weighted interpolation method (PWIM). The algorithm is based on the SIMPLEC method for staggered grids,

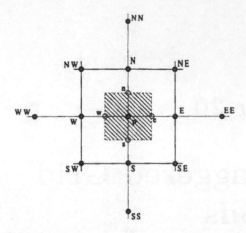

Figure 20.1: FA element in a nonstaggered grid.

and was originally used in a finite difference scheme. An outline of this method adapted for use with the FA method is presented in this section.

Figure 20.1 shows a section of the nonstaggered grid. In this grid arrangement, both velocity components and pressure are calculated at the nodal locations marked by the solid circles, and denoted by the capital letters NE, E, SE, ... , and P.

The method begins by calculating velocity components at the nodal locations using the FA algebraic formulations as shown in Equations (20.1) and (20.2).

$$u_P = \alpha_P \sum_{j=NE}^{NC} C_j u_j - \alpha_P S_P \left(\left. \frac{\partial P}{\partial x} \right|_P \right) + (1 - \alpha_P)u_P^0 \qquad (20.1)$$

$$v_P = \alpha_P \sum_{j=NE}^{NC} C_j v_j - \alpha_P S_P \left(\left. \frac{\partial P}{\partial y} \right|_P \right) + (1 - \alpha_P)v_P^0 \qquad (20.2)$$

The coefficients C_j in Equations (20.1) and (20.2) are those determined using the two-dimensional FA formulation presented in Chapter 9, with α_P and S_P given by

$$\alpha_P = \frac{1}{1 + \frac{C_P Re}{\tau}} \qquad \text{and} \qquad S_P = C_P Re.$$

The terms with the superscript "O" are taken from the previous time step, and Re is the associated Reynolds number.

The pressure gradient term in the momentum equations represents that determined using a current pressure field. Since the current field is unknown, a "guessed" field represented by P^*, is used in its place. As a result, the two velocity components are also "guessed" quantities as well, as indicated by the

"*" superscript in Equations (20.3) and (20.4).

$$u_P^* = \alpha_P \sum_{j=NE}^{NC} C_j u_j^* - \alpha_P S_P \left(\left. \frac{\partial P^*}{\partial x} \right|_P \right) + (1 - \alpha_P) u_P^0 \qquad (20.3)$$

$$v_P^* = \alpha_P \sum_{j=NE}^{NC} C_j v_j^* - \alpha_P S_P \left(\left. \frac{\partial P^*}{\partial y} \right|_P \right) + (1 - \alpha_P) v_P^0 \qquad (20.4)$$

The velocity components determined from the guessed pressure field will, most likely, not satisfy the continuity equation. The velocity components and pressure need to be adjusted, or corrected, to ensure they solve both the momentum and continuity equations simultaneously. The correction for each variable is the difference between the guessed and true values. That is,

$$P' = P - P^*, \qquad u' = u - u^* \qquad \text{and} \qquad v' = v - v^*.$$

The velocity correction is expressed by subtracting the guessed values, given by Equations (20.3) and (20.4), from the "true" values, given in Equations (20.1) and (20.2). The following equations for the velocity corrections result.

$$u_P' = \alpha_P \sum_{j=NE}^{NC} C_j u_j' - \alpha_P S_P \left(\left. \frac{\partial P'}{\partial x} \right|_P \right) \qquad (20.5)$$

$$v_P' = \alpha_P \sum_{j=NE}^{NC} C_j v_j' - \alpha_P S_P \left(\left. \frac{\partial P'}{\partial y} \right|_P \right) \qquad (20.6)$$

Following the SIMPLEC method, the neighboring velocities u_j' and v_j' are represented by u_P' and v_P', respectively. With the sum of the FA coefficients adding to unity, we have the following expressions for the velocity corrections

$$u_P' = -\tau \left(\left. \frac{\partial P'}{\partial x} \right|_P \right) \qquad (20.7)$$

and

$$v_P' = -\tau \left(\left. \frac{\partial P'}{\partial y} \right|_P \right). \qquad (20.8)$$

An expression for the pressure correction P' will be derived using the continuity equation. However, to avoid the troublesome checkerboard pressure pattern mentioned in the Section 19.1, the discretized continuity equation will be solved using the velocities at the cell faces, represented by the open circles and labels e, s, w and n in Figure 20.1. These cell face velocities will be calculated using the same guessed pressure field, and must be corrected in a similar way. Thus, the following correction equations for the cell face velocities result,

$$u_e = u_e^* - \tau \left(\left. \frac{\partial P'}{\partial x} \right|_P \right) \qquad (20.9)$$

and

$$v_n = v_n^* - \tau \left(\left. \frac{\partial P'}{\partial y} \right|_P \right).$$ (20.10)

Equations (20.9) and (20.10), and similar expressions for u_w and v_n are substituted into the discretized continuity equation. The following pressure correction equation results

$$4aP_P' = a\left(P_E' + P_S' + P_W' + P_N'\right) - D^*,$$ (20.11)

where

$$a = \frac{\tau}{h^2} \quad \text{and} \quad D^* = \frac{u_e^* - u_w^*}{h} + \frac{v_n^* - v_s^*}{h}.$$

The cell face velocities in the dilation term, D^*, are calculated using the algebraic form of the momentum equation, as shown in Equations (20.12) and (20.13).

$$u_e^* = \alpha_e \sum_{j=NE}^{NC} C_j u_{e_j}^* - \alpha_e S_e \left(\left. \frac{\partial P^*}{\partial x} \right|_e \right) + (1 - \alpha_e) u_e^0$$ (20.12)

$$v_n^* = \alpha_n \sum_{j=NE}^{NC} C_j v_{n_j}^* - \alpha_n S_n \left(\left. \frac{\partial P^*}{\partial y} \right|_n \right) + (1 - \alpha_n) v_P^0$$ (20.13)

The coefficients α_e, α_n, C_{e_j}, C_{n_j}, S_e and S_n are not known because the velocities are those at the cell faces. However, simple averaging using known quantities will give values for α_i and $\alpha_i S_i$ coefficients. That is,

$$\alpha_e = \frac{\alpha_P + \alpha_E}{2}, \qquad \alpha_n = \frac{\alpha_P + \alpha_N}{2},$$ (20.14)

$$\alpha_e S_e = \frac{\alpha_P S_P^u + \alpha_E S_E^u}{2} \quad \text{and} \quad \alpha_n S_n = \frac{\alpha_P S_P + \alpha_N S_N}{2}.$$ (20.15)

Concentrating on the east face for the time being, a similar average is executed for the summation term

$$\alpha_e \sum_{j=NE}^{NC} C_j^e u_{e_j}^* = \frac{\left(\alpha_P \sum_{j=NE}^{NC} C_j^P u_{P_j}^* + \sum_{j=NE}^{NC} C_j^E u_{E_j}^* \right)}{2}.$$ (20.16)

Using Equation (20.3), Equation (20.16) may be written as

$$\alpha_e \sum_{j=NE}^{NC} C_j^e u_{e_j}^* = \frac{u_P^* + u_E^*}{2} + \frac{\alpha_P S_P^u (\partial P^*/\partial x)_P + \alpha_E S_E^u (\partial P^*/\partial x)_E}{2}$$

$$- \frac{(1 - \alpha_P) u_P^0 + (1 - \alpha_E) u_E^0}{2}.$$ (20.17)

The relationship given in Equation (20.17) makes it possible to complete the derivation of an explicit formula for the east face velocity,

$$
u_e^* = \frac{u_P^* + u_E^*}{2} + \frac{\alpha_P S_P^u (\partial P^*/\partial x)_P + \alpha_E S_E^u (\partial P^*/\partial x)_E}{2}
$$
$$
- \frac{(1 - \alpha_P)u_P^0 + (1 - \alpha_E)u_E^0}{2}
$$
$$
+ [1 - 0.5(\alpha_P + \alpha_E)] - 0.5(\alpha_P S_P^u + \alpha_E S_E^u) \left(\left. \frac{\partial P^*}{\partial x} \right|_e \right). \quad (20.18)
$$

The formula for the velocity at the north cell face is

$$
v_n^* = \frac{v_P^* + v_N^*}{2} + \frac{\alpha_P S_P^v (\partial P^*/\partial y)_P + \alpha_N S_N^u (\partial P^*/\partial y)_N}{2}
$$
$$
- \frac{(1 - \alpha_P)v_P^0 + (1 - \alpha_N)v_N^0}{2}
$$
$$
+ [1 - 0.5(\alpha_P + \alpha_N)] - 0.5(\alpha_P S_P^v + \alpha_N S_N^v) \left(\left. \frac{\partial P^*}{\partial y} \right|_n \right), \quad (20.19)
$$

and similar expressions are used to determine the guessed velocities at the southern and western faces as well.

Note that Equation (20.18) requires pressure values at nodes P, E and EE. Likewise, the calculation of the north cell face velocity in Equation (20.19) requires pressure values at nodes P, N and NN.

Once the cell face velocities are calculated, they are used to determine the dilation D^* in Equation (20.11) from which the pressure correction P' is calculate. Next, the velocities are corrected at both the nodal and facial locations. The PWIM algorithm is outlined below.

1. Use a "guessed" pressure field to calculate u^* and v^* with Equations (20.3) and (20.4). Use the pressure field from the previous time step for the first iteration.

2. Calculate the cell face velocities using Equations (20.18) and (20.19).

3. Calculate the dilation using the cell face velocities and solve the pressure correction equation [Equation (20.11)].

4. Correct the nodal velocities using Equations (20.7) - (20.8).

5. Correct the cell face velocities using Equations (20.9) and (20.10).

6. Return to step 1 until some convergence criteria is met for all variables.

The PWIM algorithm requires values for pressure on each of the computational domain boundaries. Therefore, after the pressure correction equation has been solved, the pressure correction values on the boundaries are set using

$$
\frac{\partial P'}{\partial n} = 0,
$$

where n is the coordinate normal to the boundary under consideration. Research by Aksoy and Chen [7] showed that a first order approximation to the derivative provides the same rate of convergence as a second order approximation. Once the correction for the boundary pressures is determined, the actual pressure boundary conditions are set using

$$P = P^* + P'.$$

The pressure correction P' is automatically zero wherever the boundary pressure is known.

20.2   Poisson Pressure Equation Method

The Poisson pressure equation method (PPEM) for nonstaggered grids was first suggested by Abdallah [3, 2] and used in research by Mansour and Hamed [124]. The method, as the name suggests, uses a Poisson-type equation to solve for pressure.

The method begins by differentiating the u and v momentum equations by x and y, respectively, as shown in Equations (20.20) and (20.21).

$$\frac{\partial}{\partial x}\left[\frac{\partial u}{\partial t} + u\frac{\partial u}{\partial x} + v\frac{\partial u}{\partial y}\right] = -\frac{\partial P}{\partial x} + \frac{1}{Re}\left(\frac{\partial^2 u}{\partial x^2} + \frac{\partial^2 u}{\partial y^2}\right)\right] \qquad (20.20)$$

$$\frac{\partial}{\partial y}\left[\frac{\partial v}{\partial t} + u\frac{\partial v}{\partial x} + v\frac{\partial v}{\partial y}\right] = -\frac{\partial P}{\partial y} + \frac{1}{Re}\left(\frac{\partial^2 v}{\partial x^2} + \frac{\partial^2 v}{\partial y^2}\right)\right] \qquad (20.21)$$

Adding the results, and moving the pressure terms to the left hand side gives

$$\frac{\partial^2 P}{\partial x^2} + \frac{\partial^2 P}{\partial y^2} = \sigma - \frac{\partial D}{\partial t} + \frac{1}{R}\left(\frac{\partial^2 D}{\partial x^2} + \frac{\partial^2 D}{\partial y^2}\right), \qquad (20.22)$$

where

$$\sigma = -(uu_x + vu_y)_x - (uv_x + vv_y)_y, \qquad (20.23)$$

and the dilation D is given as

$$D = \frac{\partial u}{\partial x} + \frac{\partial v}{\partial y}. \qquad (20.24)$$

Terms with subscripts in Equation (20.23) denote derivatives of that variable with respect to the subscript.

Dropping the viscous terms from the source term in Equation (20.22) and approximating the unsteady term as

$$\frac{\partial D}{\partial t} = \frac{D^{k+1} - D^k}{\tau} = -\frac{D^k}{\tau}, \qquad (20.25)$$

one has

$$\frac{\partial^2 P}{\partial x^2} + \frac{\partial^2 P}{\partial y^2} = \sigma + \frac{D^k}{\tau}. \qquad (20.26)$$

The reason D^{k+1} is set to zero is to force the dilation to vanish as the iterations proceed. Following the suggestions of Harlow and Welch [81], the unsteady term is carried in the source term to overcome the nonlinear instabilities.

The FA algebraic solution for the pressure equation [Equation (20.26)] for location P shown in Figure 20.1 is

$$P_P = \sum_{j=NE}^{NC} C_j P_j - C_P\left(\sigma_P + \frac{D_P^k}{\tau}\right), \qquad (20.27)$$

where σ_P is evaluated using consistent finite differencing following the suggestions of Abdallah [3, 2]. Referring to Figure 20.1 and using central differences with uniform grids ($\Delta x = \Delta y = h$), $h\sigma_P$ is expressed as

$$h\sigma_P = (uu_x + vu_y)_w - (uu_x + vu_y)_e + (uv_x + vv_y)_s - (uv_x + vv_y)_n, \quad (20.28)$$

where the subscripts e, w, n and s represent the cell face locations of east, west, north and south, respectively. The velocities and velocity derivatives are obtained using averages and central differences. For example, the expressions for the east cell face are given by

$$(uu_x)_e = \frac{(u_E + u_P)(u_E - u_P)}{2h}, \qquad (20.29)$$

$$(vu_y)_e = \frac{(v_E + v_P)(u_{NE} + u_N - u_{SE} - u_S)}{8h} \qquad (20.30)$$

and

$$(u)_e = \frac{u_E + u_P}{2}. \qquad (20.31)$$

The checkerboard problem described in Section 19.1 may result if central differences only were used to calculate σ_P of the source term. The choice of locations for discretizing the derivatives in the source term is important in eliminating oscillations in the pressure. When the velocity terms are average halfway between nodal points for the terms in Equations (20.29) - (20.31), the pressure at adjacent nodes is linked removing the nonphysical oscillations characterizing the checkerboard problem.

20.3 Momentum Weighted Interpolation Method

The momentum weighted interpolation method (MWIM) was introduced by Aksoy and Chen [6] in 1989. The algorithm is similar to that for the PWIM, except for determining the cell face velocities for the calculation of the dilation in the pressure correction equation. The formula for the east face velocity is

$$u_e^* = \alpha_e \sum_{j=NE}^{NC} C_j^e u_{e_j}^* - \alpha_e S_e^u\left(\left.\frac{\partial P^*}{\partial x}\right|_e\right) + (1 - \alpha_e)u_e^0. \qquad (20.32)$$

Recall that the coefficients α_e, S_e^u and C_j^e are not known for the cell face sites, only for the nodal locations. Therefore, these coefficients and cell face velocities are interpolated from the nodal quantities by

$$\alpha_e = \frac{\alpha_P + \alpha_E}{2}, \qquad S_e^u = \frac{S_P^u + S_E^u}{2}$$

$$C_j^e = \frac{C_j^P + C_j^E}{2} \quad \text{and} \quad u_{e_j} = \frac{u_{P_j} + u_{E_j}}{2}, \tag{20.33}$$

$$\text{where} \qquad j = NE, E, \ldots, NC.$$

Similar interpolation is used to calculate coefficients for discrete FA formulations for velocities at the other facial locations.

Equation (20.32) is solved explicitly for each cell face velocity. The actual velocity components at the nodal locations are determined using Equations (20.1) and (20.2) in an implicit scheme. Non-physical pressure oscillations are avoided in the MWIM scheme because adjacent pressures are used in the pressure gradient term of Equation (20.32).

Once the cell face velocities have been determined, the dilation is calculated and the pressure correction equation [Equation (20.11)] is used to determine the pressure correction P'. The pressure is corrected by

$$P = P^* + P',$$

and the cell-centered velocities are corrected using Equations (20.7) and (20.8). Because of the similarity to the PWIM algorithm, the MWIM algorithm will not be outlined here. Instead, the reader should use the PWIM steps replacing step 2 with the MWIM method for calculating the cell face velocities.

The primary advantage of MWIM over the PWIM method is not requiring the use of pressure quantities at nodes NN, EE, SS and WW.

20.4   Method Comparisons

Aksoy and Chen [6] used the lid driven flow in a two-dimensional cavity to compare the performances of the SIMPLER, SIMPLEC, PWIM, PPEM and MWIM algorithms. Flow in the cavity is generated by the lower wall moving at a constant horizontal speed (please refer to Section 9.5). Recall that the first two methods are designed for staggered grids. They compared the computational effort, the number of iterations, and the maximum dilation as the computations achieve a specified level of accuracy. The number of iterations refers to how often the algorithm marches in time before the steady state solution is obtained. In each case, the algebraic equations were solved using a line-by-line Gauss-Seidel method with the application of the tridiagonal matrix algorithm (TDMA). Grid sizes of 21 x 21 and 41 x 41 nodes were used for two Reynolds number flows of 100 and 1000. The dimensionless time step was varied from 0.05 to 2.2 so that the effort, number of iterations and maximum dilation could be compared as a function of the time step.

They summarized their conclusion into three main points.

1. In general, the performance of all algorithms deteriorates as grid size is refined. this behavior is more dramatic with the SIMPLER algorithm because its optimum time step is very close to the time step at which it is unstable. However, if the optimum time step is used, performance is very good and best satisfies the continuity equation.

2. The nonstaggered PPEM algorithm has excellent convergence and stability properties. Unfortunately, the continuity equation is poorly satisfied. High values of dilation are observed close to the singularity points of the cavity flow, where the flow strikes the wall, and where the flow completes its circulation at the lower left corner. Since these high values of dilation tend to appear near the boundaries, grid refinement or nonuniform grids should be considered.

3. The nonstaggered MWIM and PWIM algorithms exhibit more or less the same performance in all cases. The range of time steps at which they are stable is generally larger than that of SIMPLER and SIMPLEC. In general, SIMPLEC requires fewer iterations than MWIM and PWIM. However, since SIMPLEC is a staggered grid algorithm, it requires the computation of two sets of coefficients and, therefore, more CPU time. In the case of a three-dimensional problem, three different sets of coefficients would need to be computed with both the SIMPLER and SIMPLEC algorithms resulting in even higher demand for CPU time.

The study concludes with the general remark that nonstaggered grids, on the whole, possess no real disadvantages relative to staggered grids. Non-staggered grids provide easier implementation, which would be especially advantageous with the combined use of multigrid and multilevel techniques.

Some researchers believe, because of the large similarities between the staggered and nonstaggered grid methods, the term "nonstaggered" is not correct in describing this class of algorithms. Indeed, methods such as PWIM and MWIM actually calculate velocities at cell face locations that are *staggered* from the nodal locations. This *must* be so in order to avoid the checkerboard problem. Consequently, one must ask if calculation of velocity at the nodal locations is at all necessary. If velocity at the nodal locations is needed, it may be calculated by averaging the staggered (cell face) velocities.

The actual difference between the methods has more to do with the number of different sets of coefficients that must be calculated to solve for all of the variables. For staggered grids, coefficients for u, v and any other variable (such as temperature) calculated at the nodal location are all different. This may not be significant in simple discretization schemes such as upwind differencing, but more elaborate schemes such as the FA method may be better suited for nonstaggered implementations. This is especially true for 3D applications on complex domains.

Chapter 21

Boundary Conditions

The importance of implementing boundary conditions properly in numerical methods cannot be overstated. In many cases they are straight forward; such as a velocity component taking on prescribed constant value (known as a *Dirichlet* boundary condition) on a stationary boundary. More delicate situations arise when a derivative, normal to the boundary surface, of a particular variable is prescribed, or when a boundary is not stationary. The former case is referred to as a *Neumann* boundary condition.

For many applications the first tricky task is the specification of pressure on a given boundary. The divergence-free velocity condition is the most important requirement in the simulation of incompressible fluid flow [79, 75]. There is no explicit pressure equation to solve for pressure for incompressible applications because the equation of state is no longer available. As seen in Chapters 19 and 20, the pressure field is commonly calculated indirectly using the continuity equation. In many cases, proper pressure boundary conditions are necessary to guarantee the velocity fields satisfy the continuity equation. This is especially true for regions close to the boundaries. For these reasons, a great majority of this chapter is devoted to proper pressure boundary conditions.

21.1 Staggered Grids

Staggered grids (see Chapter 19) are used to avoid the checkerboard pressure oscillation problem. The classic approach of Harlow and Welch [81] proposed a staggered grid on which the boundary can be arranged to pass through the velocity points. No pressure boundary conditions are required with this method because they are inferred from information of boundary velocities [145, 63].

For SIMPLE-like pressure schemes, Doormaal and Raithby [63] recommend that the boundary conditions for the pressure correction equation be inferred from the Dirichlet velocity boundaries conditions. As an example, consider the case for a horizontal boundary, where the normal velocity v is zero. The

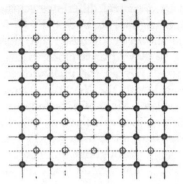

● cell-vertex arrangement

○ cell-center arrangement

Figure 21.1: Node placement on staggered grids.

momentum equation for two-dimensional, laminar flow is

$$\frac{\partial v}{\partial t} + u\frac{\partial v}{\partial x} + v\frac{\partial v}{\partial y} = -\frac{\partial p}{\partial y} + \frac{1}{R}\left[\frac{\partial^2 v}{\partial x^2} + \frac{\partial^2 v}{\partial y^2}\right]. \tag{21.1}$$

Because v is constant, and zero, along the boundary, Equation 21.1 simplifies to

$$\frac{\partial p}{\partial y} = \frac{1}{R}\left[\frac{\partial^2 v}{\partial y^2}\right]. \tag{21.2}$$

The partial derivatives are approximated on the discrete domain using forward (bottom boundary) or backward (top boundary) derivatives.

However, this treatment leads to a redundant pressure equation if a Neumann velocity boundary condition, rather than the velocity itself, is given at the boundary [63].

21.2 Nonstaggered Grids

All variables are computed at the same grid nodes on nonstaggered grids. There are two different arrangements of the computational nodes on the grid. Referring to Figure 21.1, cells defined by the grid are denoted by the solid lines. The *cell-vertex* arrangement places the computational nodes at the intersection of the solid lines. The *cell-centered* arrangement places them at the intersections of the dotted lines.

21.2.1 Cell-Vertex Nodes

The pressure calculation (correction) equation is derived from the discretized continuity and momentum equations (see Chapter 20). In order to eliminate the checkerboard phenomenon, the continuity equation is discretized on the smaller control volume *wsen*, as shown in Figure 21.2(a). Mass conservation is enforced on this small control volume. The pressure gradients in the momentum equations are evaluated by a central difference scheme, coupling the pressure values at node P with the neighboring nodes W, E, N and S.

The grid arrangement shown in Figure 21.2(a) does not guarantee the conservation of mass in the shadowed area close to the boundaries. The continuity equation is discretized only on the unshadowed area. Consequently, the cell-vertex grid arrangement with the Neumann pressure boundary conditions results in an ill-posed problem. Mass and momentum are not necessarily conserved near the boundaries. Since calculations begin with known quantities at the boundaries and work towards the interior, the errors in the near-boundary region are likely to propagate throughout the computational domain. This is one reason that pressure boundary conditions remain a difficult issue in the numerical simulation of incompressible fluid flows [148, 7].

Abdallah [3, 2] proposed a consistent and compatible pressure boundary condition for the cell-vertex grid. Using Green's theorem, the following Poisson pressure equation

$$P_{xx} + P_{yy} = \sigma \tag{21.3}$$

can be expressed as

$$\int_{y=0}^{1} \int_{x=0}^{1} \sigma \, dx dy = \int P_n dS \,, \tag{21.4}$$

where σ is the source term of the Poisson pressure equation as defined in Equation 20.23. Abdallah considered the above equation a compatibility pressure condition and stated that on a nonstaggered grid, the Neumann boundary condition for pressure, P_n, must be applied one half grid space from the boundary in order to achieve a compatible and consistent solution for incompressible flows. However, the equivalence between the conservation of mass at the boundary and this compatibility pressure boundary condition was not illustrated. Also, his proof of the compatibility condition was based only on a uniform grid. Therefore, it is not certain that the compatibility condition is satisfied and that mass conservation is preserved at the boundaries for a cell-vertex arrangement on a nonstaggered grid using Abdallah's method. Further, it is quite tricky to apply the Neumann pressure boundary condition one half grid space from the boundaries, since this involves the discretization of the second order derivative diffusion terms. This is especially true when a nonuniform grid or complex geometries are involved.

Another way to generate a consistent pressure boundary condition is through the direct discretization of the continuity equation at the boundaries, such as control volume P_1esw shown in Figure 21.2(b). In order to include the pressure equation at boundary node P_1, v_n needs to be evaluated using the momentum equations. However, the evaluation of cell surface velocity v_n with the momentum equations is very difficult since it is close to the boundary. Based on the authors' experience, a simple linear interpolation for v_n is not sufficient because large numerical errors often preclude convergence.

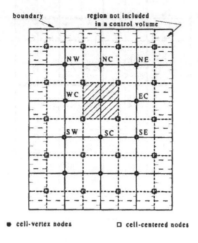

(a) Boundary space not covered by a control volume

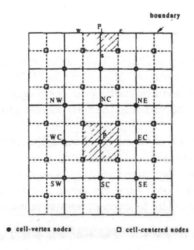

(b) Interior and boundary cell control volumes

(b) Interior and boundary cell control volumes

Figure 21.2: Cell-vertex method on a nonstaggered grid.

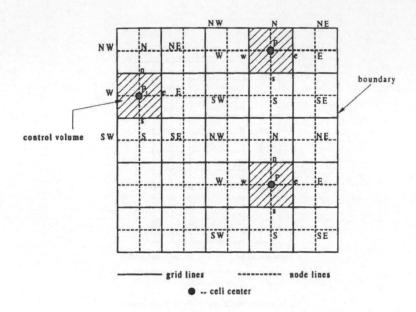

Figure 21.3: Cell-centered method on a nonstaggered grid.

21.2.2   Cell-Centered Nodes

Using cell-centered nodes (see Figure 21.3) on a nonstaggered grid, all variables are computed at the cell center P. Rhie and Choi [155] used this node arrangement to prevent the pressure oscillation on nonstaggered grids. The pressure correction equation was solved using the zero normal gradient of pressure at the boundaries.

For a cell-centered node adjacent to a boundary, such as P_1 shown in Figure 21.3, the pressure equation is obtained from mass conservation over control volume $Wsen$ using the discretized continuity equation. Value u_W is known from the Dirichlet velocity boundary condition, and need not be evaluated from the momentum equation. The following pressure equation results:

$$a^P P_P = a^E P_E + a^W P_W + a^N P_N + a^S P_S - D^* , \qquad (21.5)$$

where

$$D^* = \frac{U_e^* - U_W}{x_e - x_w} + \frac{V_n^* - V_s^*}{y_n - y_s}$$

and

$$a^W = 0.$$

No pressure value is required at boundary node W in spite of the appearance of pressure derivatives in the momentum equations, and the conservation of mass and momentum are preserved at both interior nodes P and boundary nodes P_1.

This is consistent, since the physics of the problem does not provide a boundary condition for pressure as it does for the velocities.

21.2.3 PPEM

A pressure Poisson equation may be derived from the momentum and continuity equations (see Section 20.2). This scalar pressure equation has the vector momentum equations as its implicit pressure boundary condition. It remains a challenging problem to numerically implement the proper pressure boundary conditions. Several different approaches to this problem have been proposed [77, 76, 75, 25].

Orszag and Israeli [142] have pointed out that either the normal or the tangential component of the momentum equations is permissible as a boundary condition for the pressure Poisson equation. The former leads to a Neumann boundary condition and the latter to a Dirichlet boundary condition. Moin and Kim [133] found that the Neumann and Dirichlet boundary conditions generally may not provide the same solutions, contradicting the study of Orszag and Israeli. Gresho [77] gave a thorough review of pressure boundary conditions for the incompressible Navier-Stokes equations. He concluded that for internal flows with a Dirichlet velocity boundary condition, only a Neumann boundary condition is always appropriate for the pressure Poisson equation.

Recall the term $\frac{\partial D}{\partial t}$, where D is the dilatation, is dropped in the formulation of the PPEM equation. This may cause some difficulties in the numerical implementation of the pressure boundary conditions. It has been pointed out by Abdallah [3, 2] and Sotiropoulos, et al. [160], that if the Neumann pressure boundary condition is employed, the unsteady dilatation term should be kept in the discretized Poisson pressure equation (approximated by $-\frac{D}{\Delta t}$) in order to obtain divergence free solutions. Gresho [77] also showed that the Poisson pressure equation approach, being of high order in the spatial derivatives, requires additional smoothness (differentiability) for both pressure and velocity variables. Thus, it is difficult to implement the Neumann pressure boundary condition because of the need to approximate second-order velocity derivatives at the boundaries.

It is common for researchers to use both the normal and tangential momentum equations as pressure boundary conditions [148]. Some use linear extrapolation to obtain the pressure values at the boundary from the interior nodes. This treatment is also used for quasiparabolic problems in high Reynolds number flows. Others try to avoid this difficult problem of specifying pressure boundary conditions by introducing pressure correction fields and setting the normal gradient of these fields to zero [155, 7].

Abdallah mentions that the viscous terms in the momentum equations do not appear in the source term of the Poisson pressure equation, but do appear in the Neumann boundary conditions for pressure. To satisfy the compatibility condition (see Abdallah [3]), which is a consequence of Green's Theorem, the integral of the viscous terms over the boundary contour should vanish. This is achieved by writing the viscous terms in the pressure boundary conditions as the curl of the vorticity vector. Therefore, the Neumann boundary conditions are written as

$$\frac{\partial P}{\partial x} = -u\frac{\partial u}{\partial x} - v\frac{\partial u}{\partial y} - \frac{1}{R}\frac{\partial \omega}{\partial y} \qquad (21.6)$$

and

$$\frac{\partial P}{\partial y} = -u\frac{\partial v}{\partial x} - v\frac{\partial v}{\partial y} + \frac{1}{R}\frac{\partial \omega}{\partial x}. \qquad (21.7)$$

These boundary conditions are discretized at half-steps from the boundary so as to be consistent with the discretization of the Poisson pressure equation. The vorticity ω on the boundary is evaluated using second-order forward and backward finite differences. Abdallah [3, 2] shows that the compatibility condition is exactly satisfied in discrete form on nonstaggered grids when both the foregoing boundary conditions and the consistent differencing in Equations (20.29) - (20.31) are used to solve the Poisson pressure equation.

21.2.4   Ghost Boundary Nodes

Some numerical simulations of fluid flows include conjugate heat transfer, where heat conduction in the solid walls is coupled with convective heat transfer in the fluid. The computational domain for conjugate heat transfer is divided into fluid and solid domains. An example of a possible grid arrangement is shown in Figure 21.4. Conservation of mass, momentum and energy must be satisfied in the fluid domain, while only heat conduction needs to be considered in the solid domain. [5].

An extra computational boundary needed in this case. In addition to the outer boundary, there is an interface boundary (inner boundary) where the fluid domain meets the solid domain. The final solution of fluid flows with conjugate heat transfer relies on both outer and inner boundary conditions. Referring to Figure 21.4, note that when cell-centered nodes are employed on a nonstaggered grid, the inner boundary does not pass through any nodes.

In order to enforce the inner boundary conditions, the *ghost boundary node method* is introduced. The inner boundary is described by a series of ghost boundary nodes located between existing grid nodes, such as nodes SW, W and NW (marked with empty circles) shown in Figure21.4. The corresponding velocity boundary conditions are specified on these ghost boundary nodes. Therefore, the velocity values at ghost nodes SW, W and NW can be used for the conservation of both momentum and energy on volume $WSEN$. Mass conservation can be enforced at the inner boundary on the smaller control volume $Wsen$. No pressure boundary condition is required for the pressure equation [Equation (21.5)] with this arrangement. The ghost boundary node method is conceptually simple, but requires considerable effort to implement.

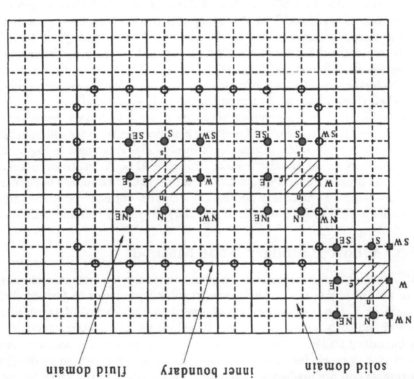

● - regular node for u, v and p

■ - outer boundary node   ○ - ghost boundary node

Figure 21.4: Ghost boundary nodes for fluid flows with conjugate heat transfer.

21.3   2D Diagonal Cartesian Method

The diagonal Cartesian method may be used for problems with complex boundaries (see Chapter 18). Complex boundaries are approximated using Cartesian coordinate grid lines and diagonal segments, as shown in Figure 21.5. In the case of conjugate heat transfer, there is an interface boundary (or inner boundary) where the fluid domain meets the solid domain as shown in Figure 21.5. In order to properly treat inner boundary conditions, ghost boundary nodes and diagonal surface nodes, which consist of the inner boundary and are marked with open circles and rectangular, are employed. The mass conservation and the pressure boundary conditions at the ghost boundary nodes and diagonal surface nodes, with cell-centered nodes, are discussed in following sections.

When cell-centered nodes on a nonstaggered grid are used for regular boundaries, the velocities are specified at the boundary nodes and ghost boundary nodes to avoid the difficulty of specifying pressure boundary conditions. Mass conservation is enforced by discretizing the continuity equation on selected control volumes close to the boundaries so that the false pressure oscillation (checkerboard phenomenon) in the computation is eliminated.

For problems involving complex boundaries, some boundary nodes may be coincident with the Cartesian grid nodes. These boundary nodes will be referred to as *diagonal surface nodes*. Examples of these nodes are shown in Figure 21.5. Ghost boundary nodes are also evident in Figure 21.5. A treatment similar to that used for pressure boundary conditions at the regular boundary nodes can be applied for the ghost boundary nodes. The diagonal surface nodes are handled using either a local control volume or an enlarged control volume method.

21.3.1   Ghost Boundary Node Method

Ghost boundary nodes methods at complex boundaries are to those used in simulating the conjugate heat transfer over regular boundaries in Section 21.2.4. In Figure 21.5, the inner boundary is described by a series of ghost boundary nodes and diagonal surface nodes. Based on the cell-centered method, the ghost boundary nodes are located between existing grid nodes and are marked denoted with open circles. The diagonal surface nodes are located at the diagonal segments and are marked with the open rectangles. On the ghost boundary nodes, such as SW, S and SE for the node of P_3 in Figure 21.5, the corresponding velocity boundary conditions are specified. Therefore, the velocity values at these ghost nodes can be used for the conservation of momentum on volume "$WSEN$" and the mass conservation on the smaller control volume "$Senw$." With this arrangement, no pressure boundary condition is required for the pressure equation [Equation (21.5)].

21.3.2   Diagonal Surface Nodes

In the simulation of fluid flows over complex geometries, what is of the utmost concern is the conservation of mass and momentum close to the diagonal

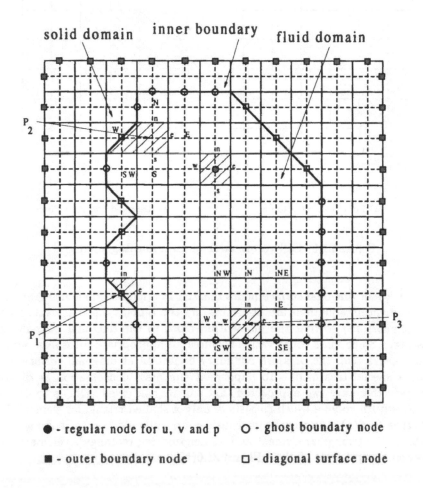

Figure 21.5: Ghost boundary nodes and diagonal surface nodes for complex boundaries.

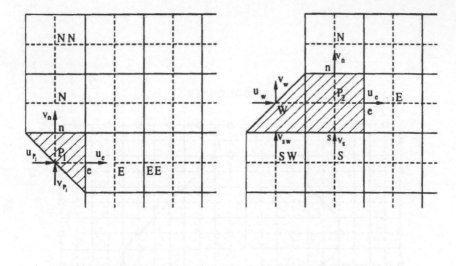

(a) Local control volume   (b) Enlarged control volume

Figure 21.6: Cell-vertex method on a nonstaggered grid.

boundaries, or diagonal surface nodes marked in open rectangle in Figure 21.5. The momentum equations are discretized by the 5-point finite analytic scheme (see Chapter 18) so that the momentum close to the diagonal boundaries is conserved. Since mass conservation requires an appropriate pressure boundary condition [79, 75], the pressure boundary condition at diagonal surface nodes needs a special treatment. To implement the mass conservation at diagonal boundary there are two approaches that can be conceived. One is to consider a local control volume which consists of only a shaded triangular element P_1en in Figure 21.6(a). Another is to consider an enlarged control volume which combines the triangular element and its neighboring rectangular element, such as $Wsen$ around node of P_2 in Figure 21.6(b).

Mass Conservation on a Local Control Volume

For the local control volume approach, the establishment of the pressure equation for the diagonal surface node P_1, as shown in Figures 21.5 and 21.6(a), is made by enforcing the mass conservation at a small shadowed triangle area. The discretized continuity equation is

$$\frac{u_e - u_{P_1}}{x_e - x_{P_1}} + \frac{v_n - v_{P_1}}{y_n - y_{P_1}} = 0 , \tag{21.8}$$

where u_{P_1} and v_{P_1} are known as $u_j^{P_1}$ and $v_j^{P_1}$. Using the definitions of pseudo-velocities in Equations (19.16) and (19.17),

$$u_e = \hat{u}_e - \alpha_e S_e^u \left(\frac{\partial p}{\partial x}\right)_e \qquad (21.9)$$

and

$$v_n = \hat{v}_n - \alpha_n S_n^v \left(\frac{\partial p}{\partial y}\right)_n \qquad (21.10)$$

Equation (21.8) becomes

$$\left[\frac{\hat{u}_e - u_j^{P_1}}{x_e - x_{P_1}} - \frac{\alpha_e S_e^u (\frac{\partial p}{\partial x})_e}{x_e - x_{P_1}}\right] + \left[\frac{\hat{v}_n - v_j^{P_1}}{y_n - y_{P_1}} - \frac{\alpha_n S_n^v (\frac{\partial p}{\partial y})_n}{y_n - y_{P_1}}\right] = 0. \qquad (21.11)$$

The pressure gradients in the above equation are evaluated by a central difference scheme. Therefore the following pressure equation is obtained

$$a^P p_{P_1} = a^E p_E + a^W p_W + a^N p_N + a^S p_S - \hat{D}, \qquad (21.12)$$

where

$$\hat{D} = \frac{\hat{u}_e - u_j^{P_1}}{x_e - x_{P_1}} + \frac{\hat{v}_n - v_j^{P_1}}{y_n - y_{P_1}}, \qquad (21.13)$$

and

$$a^W = 0, \qquad a^S = 0, \qquad (21.14)$$

$$a^E = \frac{\alpha_e S_e^u}{(x_e - x_{P_1})(x_E - x_{P_1})}, \qquad (21.15)$$

$$a^N = \frac{\alpha_n S_n^v}{(y_n - y_{P_1})(y_N - y_{P_1})}, \qquad (21.16)$$

$$a^P = a^E + a^W + a^N + a^S. \qquad (21.17)$$

The pseudo-velocities $\hat{u}_e$ and $\hat{v}_n$ in Equation (21.13) cannot be evaluated using the FA algebraic momentum equation because the finite analytic coefficients C_j^e and C_j^n at the diagonal surface node P_1 and the corresponding nodes at the solid domain are not defined. Therefore the following expressions are used for $\hat{u}_e$ and $\hat{v}_n$.

$$\hat{u}_e = \frac{\hat{u}_{P_1} + \hat{u}_E}{2} \qquad (21.18)$$

$$\hat{v}_n = \frac{\hat{v}_{P_1} + \hat{v}_N}{2} \qquad (21.19)$$

Pseudo-velocities $\hat{u}_{P_1}$ and $\hat{v}_{P_1}$ can be calculated by

$$\hat{u}_{P_1} = u_{P_1} + \alpha_{P_1} S_{P_1}^u \left(\frac{\partial p}{\partial x}\right)_{P_1} \qquad (21.20)$$

and

$$\hat{v}_{P_1} = v_{P_1} + \alpha_{P_1} S^v_{P_1} \left(\frac{\partial p}{\partial y}\right)_{P_1}. \tag{21.21}$$

Pressure gradients $(\frac{\partial p}{\partial x})_{P_1}$ and $(\frac{\partial p}{\partial y})_{P_1}$ are evaluated by

$$\left(\frac{\partial p}{\partial x}\right)_{P_1} = \frac{p_E^{(n-1)} - p_{P_1}^{(n-1)}}{x_E - x_{P_1}}$$

and

$$\left(\frac{\partial p}{\partial y}\right)_{P_1} = \frac{p_N^{(n-1)} - p_{P_1}^{(n-1)}}{y_N - y_{P_1}}.$$

The terms with superscript "$(n-1)$" are taken from the previous iteration.

The terms α_e, α_n, S^u_e and S^v_n in Equations (21.15) and (21.16) can be evaluated by the following equations

$$\alpha_e = \frac{\alpha_E + \alpha_{P_1}}{2}, \qquad \alpha_n = \frac{\alpha_N + \alpha_{P_1}}{2}, \tag{21.22}$$

$$S_e = \frac{S_E + S_{P_1}}{2} \quad \text{and} \quad S_n = \frac{S_N + S_{P_1}}{2}, \tag{21.23}$$

while α_{P_1} and S_{p_1} in Equations (21.20) to (21.23) are defined as

$$\alpha_{P_1} = \frac{1}{(1 + (C_{P_1} R/\tau)}$$

and

$$S^u_{P_1} = S^v_{P_1} = C_{P_1} Re.$$

The undefined FA coefficient C_{P_1} can be found from the neighboring FA coefficients, such as C_E, C_{EE}, C_N and C_{NN} in Figure 21.5(b), by linear extrapolation. However, linear extrapolation for the FA coefficient C_{P_1} may bring some numerical errors in the final solutions, especially when a coarse grid is used or the Reynolds number is high.

Mass Conservation on an Enlarged Control Volume

For an enlarged control volume, the mass conservation is enforced on an enlarged control volume for a diagonal surface node, as shown in Figures 21.5 and 21.6(a). Similar to the treatments of pressure boundary conditions for the regular boundary nodes in Section 21.2.2, and ghost boundary nodes in Section 21.3.1, velocity information at diagonal surface nodes is used for the pressure boundary conditions. Using the principle of mass conservation on the enlarged control volume with interior node P_2, the following equation is obtained

$$(u_e \Delta y_j - u_W \Delta y_j) + (v_n \Delta x_i - v_s \Delta x_i) + v_W \Delta x_{i-1} - v_{sw} \Delta x_{i-1} = 0,$$

or

$$\frac{u_e - u_W}{\Delta x_i} + \frac{v_n - v_s}{\Delta y_j} + \frac{v_W - v_{sw}}{\Delta x_i \Delta y_j} \Delta x_{i-1} = 0. \tag{21.24}$$

By the definitions of pseudo-velocities in Equations (19.16) and (19.17)

$$u_e = \hat{u}_e - \alpha_e S_e^u \left(\frac{\partial p}{\partial x}\right)_e,$$

$$v_n = \hat{v}_n - \alpha_n S_n^v \left(\frac{\partial p}{\partial y}\right)_n$$

and

$$v_s = \hat{v}_s - \alpha_s S_s^v \left(\frac{\partial p}{\partial y}\right)_s.$$

The pressure gradients $\frac{\partial p}{\partial x}$ and $\frac{\partial p}{\partial y}$ are now discretized by a central difference scheme, then the following pressure equation at the regular interior node P_2 is obtained:

$$a^P p_{P_2} = a^E p_E + a^W p_W + a^N p_N + a^S p_S - (\hat{D} + \hat{D}_W), \tag{21.25}$$

where

$$a^E = \frac{\alpha_e S_e^u}{\Delta x_i (x_E - x_P)},$$

$$a^W = 0,$$

$$a^N = \frac{\alpha_n S_n^v}{\Delta y_j (y_N - y_P)},$$

$$a^S = \frac{\alpha_s S_s^v}{\Delta y_j (y_P - y_S)},$$

$$a^P = a^E + a^W + a^N + a^S,$$

$$\hat{D} = \frac{\hat{u}_e - u_W}{\Delta x_i} + \frac{\hat{v}_n - \hat{v}_s}{\Delta y_j} \quad \text{and}$$

$$\hat{D}_W = \frac{v_W - v_{sw}}{\Delta x_i \Delta y_j} \Delta x_{i-1}.$$

The interface velocity v_{sw} can be evaluated by

$$v_{sw} = \frac{v_W + v_{SW}}{2}, \tag{21.26}$$

where v_W and v_{SW} are the velocities at the nodes W and SW in Figure 21.5(a). Therefore, no pressure boundary condition is required at the diagonal surface node that is adjacent to P_2, while the mass is conserved close to the diagonal boundary.

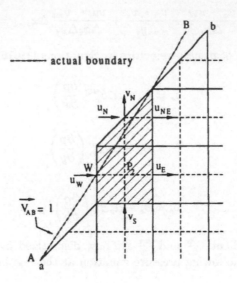

Figure 21.7: Dirichlet boundary condition on diagonal Cartesian coordinates.

21.3.3   Velocity Boundary Conditions

In the diagonal Cartesian method, the numerical simulation of incompressible fluid flow around an original boundary is actually approximated by the simulation of flow around an approximated boundary. The problem of specifying the appropriate velocity boundary conditions on the approximated boundaries is considered here.

Figure 21.7 depicts an original boundary contour with the Dirichlet velocity boundary condition $\vec{v}_{AB}=1$ in the direction parallel to the boundary $\overline{AB}$. Figure 21.7 shows the approximated boundary on a coarse Cartesian grid. For nodes along boundary ab, such as W in Figure 21.7, a normal velocity boundary condition may be utilized, as shown. Care must be used with this kind of velocity boundary condition or failure of mass conservation in the adjacent control volume may result. Note that the actual boundary passes through the control volume, so that the mass entering at node W must be balanced by an equal amount leaving at some other node in order to prevent mass form accumulating in the cell causing a physically unrealistic response at an interior control volume boundary.

21.4 Exercises for Part IV

1. Briefly summarize the similarities and differences of the SIMPLE, SIMPLEC and SIMPLER methods.

2. What is the continuity based pressure equation?

3. (a) What is *pressure correction*?

   (b) Explain why it is needed.

4. Provide the details for deriving the Poisson pressure equation given in Section 20.2.

5. (a) (Computer Exercise) Alter available FA code for 2D laminar code (at the web site **www.finiteanalytic.com**) so that the pressure is calculated using the PPEM method on a non-staggered grids.

   (b) Use the revised code to solve lid-driven cavity flow for a Reynolds number of 100. Use a uniform grid resolution of $\Delta x = \Delta y = 0.05$.

   (c) Comment on the comparison of your PPEM solution and a similar solution using either staggered grids, or some other non-staggered grid method. If possible, use a similar convergence criteria so that comparisons of efficiency may be made.

Part V

Applications of the FA Method

Selected applications of the FA method in solving flow and heat transfer problems are presented in the final part of the book. Flow and heat transfer problems may be distinguished, roughly, as laminar or turbulent, by the spatial dimension (2D or 3D), by a simple or complex domain, and the heat transfer as being simple or conjugate. Examples in Part V are selected to show the versatility of the FA method, perhaps in conjunction with the diagonal Cartesian method, in solving a range of flow and heat transfer problems. Because of length considerations, an example from each category of flow and heat transfer will not be given. However, for some types application, reference may be made to appropriate publication where the FA method was used.

Chapter 22

Turbulent Flows

Turbulence may be generally described as an unsteady, rotational, stochastic, three-dimensional phenomenon. Due to these characteristics, accurate numerical modeling of turbulent flows and heat transfer remains one of the most challenging areas of computational fluid dynamics and heat transfer. In Chapter 2, turbulence modeling for incompressible flows was introduced by taking the ensemble average of the instantaneous Navier-Stokes and energy equations, Equations (2.7) - (2.9). This resulted in the mean flow and energy equations, Equations (2.21) - (2.23), which include unknown quantities such as $\overline{u_i u_j}$ (Reynolds stresses) and $\overline{u_i \theta}$ (Reynolds heat fluxes). The modeling of these unknowns is termed the second order turbulence closure problem, and the reader may want to refer to section 2.4.2 for a discussion on various methods of closure.

The FA method has been used in conjunction with turbulence models to solve a variety of turbulent flow applications. Chen and Chang [36] used both the $k - \epsilon - A$ and $k - \epsilon - E$ turbulence models described in Section 2.5.1 to solve turbulent flow in a rectangular cavities. Their numerical results agree quite well with experimental data reported by Mills [132], Girard and Curlet [72], Grand [74] and Normandin [141].

Although some turbulent flow problems can be accurately modeled in a two-dimensional or axisymmetric manner, other problems are truly three-dimensional in nature, and must be modeled as such. An example of this is the study by Chen and Cheng [52], which examines three-dimensional turbulent flow past an inclined cylinder. This is a fundamental flow problem for ship propelling systems, because it simulates the propeller shaft under the ship hull as it is subjected to approaching flow at an angle of attack.

The FA study presented in this chapter concerns turbulent flow through disc-type valves. However, the next section is devoted to references to laminar flow and heat transfer numerical simulations integrated using the FA method.

22.1   Laminar Applications

The FA method has been used to successfully integrate the Navier-Stokes and
energy equations for laminar flows and heat transfer applications. Such results
will not be presented in order to shorten part V. Readers interested in such
applications are encouraged to consider FA studies such as the one by Ho and
Chen [83], who used a vorticity-streamfunction formulation to model laminar
flow over a backward-facing step. Their results were in close agreement with
experimental finding of Denham and Patrick [62].

Chen and Chen [37] simulated vortex shedding behind a rectangular block
using the FA method and a vorticity-streamfunction mathematical model. Numerically predicted Strouhal numbers of 0.156 and 0.125 for Reynolds numbers
of 100 and 500, respectively, are similar to those measured by Blevins [20]

Laminar flow past axisymmetric disc valves is modeled by Chen, et al. [50].
Flow rate of the re-circulation in the separation zone increases with increasing
Reynolds number, a phenomena confirmed by experiments of Mueller et al.
[136].

A short summary of laminar heat transfer FA applications is presented in
Chapter 23.

22.2   Fluid Dynamics of Disc Valves

A detailed study of turbulent flow past valves is of basic importance to fluid
mechanics, but it also has practical value in the design of heart valve prostheses
for replacement of diseased natural valves. Hemolysis, subhemolytic damage,
damage to the endothelial cells in the aortic wall and thromboembolic compli-
cations after prosthetic valve implantation have been thought to be caused by
large magnitudes of turbulent stress and wall shear stress, as well as regions of
flow separation and relative stasis. Thus, detailed analysis of flow characteristics
around valves may result in improvements in valve design.

22.3   Mathematical Model

The problem domain is shown in Figure 22.1. This problem considers steady,
axisymmetric turbulent flow through a constant-diameter pipe with a valve orifice ring $(\overline{B'BCC'})$ and a disc valve $(\overline{DD'E'E})$, where fluid enters from the
left and exits to the right. This disc-valve system can be thought to simulate
a caged disc heart valve, similar to the Kay-Shiley valve in the fully opened
position. All the dimensions in Figure 22.1 are normalized by the diameter D
of the pipe.

22.3.1   Governing Equations

The modified Jones and Launder low-Reynolds number $k - \epsilon$ turbulence model
[100] was adapted to close the turbulence problem. This model employs the

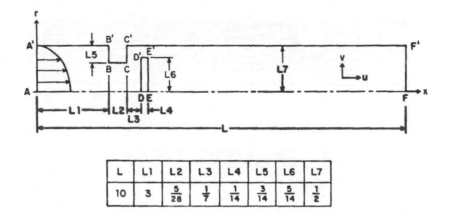

L	L1	L2	L3	L4	L5	L6	L7
10	3	$\frac{5}{28}$	$\frac{1}{7}$	$\frac{1}{14}$	$\frac{3}{14}$	$\frac{5}{14}$	$\frac{1}{2}$

Figure 22.1: Geometry of the axisymmetric disc valve model.

turbulent eddy viscosity concept to relate the Reynolds stresses to the mean velocity gradients. In order to reduce the computational time, the low-Reynolds number $k - \epsilon$ model was used in conjunction with the wall function method discussed in Section 2.5.3. The fluid was assumed to be Newtonian with constant properties. A primitive variable formulation was used. All variables were non-dimensionalized with the pipe diameter D and the mean inlet velocity U_m. The continuity, momentum and turbulence equations that govern the problem under these assumptions and models are given in [49] and [48].

22.3.2 Boundary Conditions

The inlet and exit velocity functions were assumed to be fully-developed turbulent velocity profiles, given by

$$u_i(r) = u_o(r) = (2r)^{1/n} \frac{(n+1)(2n+1)}{2n^2} , \qquad (22.1)$$

where $n = 6.6$ for $Re = \frac{U_m D}{\nu} = 10000$, and $n = 7.0$ for $Re = 200{,}000$. The inlet and outlet boundary conditions for k and ϵ were specified as

$$k_i = 0.001, \qquad \epsilon_i = \frac{0.1 k_i^{3/2}}{0.435(0.5 - r)} ,$$

$$\frac{\partial k_o}{\partial x} = 0 \quad \text{and} \quad \frac{\partial \epsilon_o}{\partial x} = 0.$$

22.4   Results and Discussion

The problem is solved using the 9-point FA method discussed in Chapter 9, on a staggered grid for the velocity components as explained in Chapter 19. A 59×25

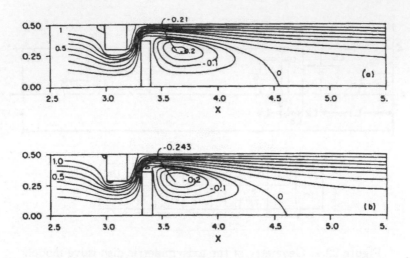

Figure 22.2: Stream function contours for $Re = 10,000$ and 200,000.

non-uniform grid is used, with most of the computational nodes concentrated in the region near the orifice ring and disc valve. The pressure-velocity coupling was treated using a modified SIMPLER algorithm.

Figure 22.2 shows stream function contours for $Re = 10,000$ and 200,000. In both cases, the flow first converges in the orifice, and is then forced to turn around the disc valve. The flow is accelerated to a very high velocity in the gap between the ring and the disc valve. The maximum velocity in the gap is about five times the mean velocity at the inlet, U_m. The turbulent flow patterns shown in Figure 22.2 look similar to those in laminar flow [50]. The flows shown in the figure have both primary and secondary separation zones. The two primary separation zones occur downstream of both the orifice ring and the disc valve. The three secondary separation zones common to both flows are where the upstream side of the orifice ring meets the wall, on the top surface of the disc, and on the upstream side of the disc at the centerline. For $Re = 200,000$, there is an additional secondary separation zone on the upstream side of the horizontal surface of the orifice ring. For the flows shown in Figure 22.2, the recirculation lengths are $1.15D$ for $Re = 10,000$ and $1.25D$ for $Re = 200,000$. This indicates that for turbulent flow, the separation length behind the disc is relatively insensitive to the Reynolds number of the flow. This is contrary to laminar flow, where the separation length varies significantly with Reynolds number.

Profiles for both axial and radial velocities at several cross-sections of interest are shown in Figure 22.3 for $Re = 10,000$ and 200,000. At about $0.8D$ upstream of the orifice ring (where again D is the tube diameter), both flows are fully developed. At the middle of the orifice ring, the axial velocity profile for $Re = 10,000$ shows an overshoot phenomenon. At the corresponding location for

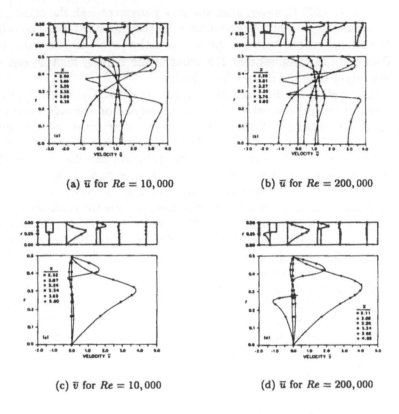

(a) $\bar{u}$ for $Re = 10,000$ (b) $\bar{u}$ for $Re = 200,000$

(c) $\bar{v}$ for $Re = 10,000$ (d) $\bar{u}$ for $Re = 200,000$

Figure 22.3: Axial $\bar{u}$ and radial $\bar{v}$ velocity plots at various cross sections in valve geometry.

$Re = 200,000$, the axial velocity profile shows a similar overshoot, but near the surface of the orifice ring the velocity is negative. Physically, this implies the existence of a small separation zone. Between the orifice and the disc, the axial velocity is significantly reduced from that at the orifice ring. The separation region behind the orifice ring is observed here as well. The fluid in this region is strongly accelerated in the radial direction. In the middle of the gap, the maximum radial velocity is about $4U_m$. At the top of the disc, the fluid is again accelerated in the axial direction, and reaches a maximum value of more than $3U_m$. Furthermore, the axial velocity near the disc wall is negative, indicating that separation occurs on the top disc surface. Downstream of the disc valve, the axial velocity is negative for a long portion of the cross-section, corresponding to the large primary separation zone there. Finally, far downstream, the velocity profiles show a gradual recovery to fully developed turbulent flow.

Contours of turbulent kinetic energy k and rate of turbulent kinetic energy dissipation ϵ for $Re = 10,000$ and $200,000$ are shown in Figure 22.4. k is normalized by U_m^2, and ϵ is normalized by U_m^3/D. The inlet turbulent kinetic energy was assumed to be 0.01. However, after the flow passes through the orifice, the value of k increases significantly to about four. In the region between the orifice ring and the disc, as well as between the disc and the tube wall, the value of k is larger than one. This implies that the order of the velocity fluctuations can become even larger than the mean inlet velocity.

Figure 22.5 displays the dimensionless wall shear stress, $C_f = \tau/\rho U_m^2$, along the tube wall, orifice and disc valve. Along the tube and orifice wall, two peaks of shear stress are observed. One is located at the rear corner of the orifice ring, and the other is located at the wall just downstream of the disc valve. The wall shear stress at the front stagnation point on the disc is zero, but increases very rapidly along the wall in the radial direction. Due to the flow separation that occurs at the upstream corner of the disc valve, the velocity inside the separation zone is much lower than that outside the separation zone. Consequently, the wall shear stress along the separation region is much lower than it is on the upstream side of the disc valve. For both Reynolds numbers shown, the maximum wall shear stress is seen to occur on the disc rather than on the orifice ring or the tube wall.

Numerical details and additional steady-state results for this problem can be found in [48] and [49]. Additionally, pulsatile turbulent flow results for the geometry considered here are given in [48]. Finally, a parametric study of the size and position of the disc valve for steady turbulent flow is performed in [181].

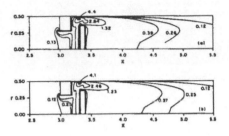

(a) TKE k for $Re = 10,000$ and 200,000

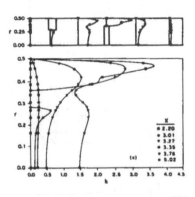

(b) TKE profiles for $Re = 200,000$

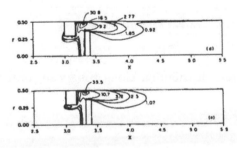

(c) Dissipation rate ϵ for $Re = 10,000$ and 200,000

Figure 22.4: Turbulent kinetic energy (TKE) and dissipation of TKE rates.

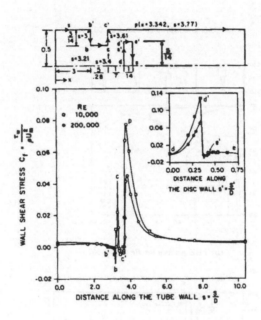

Figure 22.5: Shear stress distribution along tube wall, orifice ring, and disc valve.

Chapter 23

Turbulent Heat Transfer

The previous chapter focused on issues concerning the numerical modeling of incompressible flows that involved turbulence. This chapter extends these ideas to include the energy equation for turbulent flows. The example of such a problem offered in this chapter is the a two-dimensional, turbulent seabreeze integrated using the FA method on a regular (nonstaggered) grid. This atmospheric boundary layer flow is powered by *natural* convection. The results reported here are similar to those presented in originally published in [18] and [19]. As an additional application of the FA method in turbulent heat transfer, the reader is referred to the work by Chen and Yoon [47], who modeled turbulent *forced* convection in an incompressible flow past a cylindrical cavity. Before presenting the results for the selected example, a summary of laminar heat transfer applications based on the FA method is presented in the next section.

23.1 Laminar Applications

Laminar heat transfer studies include a paper by Chen and Obasih [41] that considered convective heat transfer associated with flow past a two-dimensional square cavity. Reynolds number flows of 10, 100, 300 and 1000 were simulated. They found the convective effect is quite evident as the Reynolds number exceeds 300.

Flow and heat transfer through tube bundles were studied by Chen and Wung [46]. They used boundary fitted coordinates to model flow around in-line and staggered arrays of circular cross-section tubes. In the case of in-line tubes, they found the temperature distribution heavily dependent on thermal diffusion at low Peclet numbers. convective diffusion play an increasingly important role for higher Peclet numbers. Their results show, in general, the averaged heat transfer coefficient for the staggered array pattern is higher than the in-line array. Correlations of Nusselt number with Reynolds and Prandtl numbers were slightly lower than the empirical findings of Zukauskas et al. [185].

Natural convection was the focus of the work of Talaie and Chen [165].

They simulated steady and transient circulations within various rectangular enclosures. They found that heat transfer between two adiabatic walls tends to be more effective at low aspect ratio. In terms of transient flows, they simulated the flow in a cavity of aspect ratio 10 where the bottom wall is heated and the top wall is cooled. Their results indicate that the initial motion is by the warm fluid away from the insulated walls. This forces the cool fluid down along the side walls. As a result, the circulation cells start at the ends and propogate to the center.

A study by Bravo et al. [22] considered the combined effect of convection and radiation heat transfer in an array of electronic chips. The 19-point FA discretization was used to solve the three dimensional model. The calculations show that radiation heat transfer can account for as much as 33% of the total heat transfer in a given electronic chip configuration.

Chen et al. [39] simulated the convective heat transfer of a heated rectangular block in a two-dimensional channel flow. Computations were made for Reynolds numbers from 10 to 500, and Prandtl numbers from 0.01 to 1.0. Their results indicate that the total heat transfer between the block surface and the fluid is relatively insensitive to the vortex shedding occurring in the wake region behind the block. The heat transfer from the front of the block is high, with the peak transfer occurring at the front corners where the boundary layer thickness is the smallest.

Chen and Bravo [34] use the FA method to solve the governing equations for fluid flow and heat transfer in two-dimensional staggered thin rectangular blocks in a channel flow heat exchanger. Their calculations show that a substantial increase in heat transfer can be obtained for a given block area when the block lengths and arrangement are properly chosen. The most effective heat transfer rate occurs when the separation distance is twice the block length and the block length and width are equal. However, the use of shorter block length increases the pressure drop.

23.2 Two-Dimensional Sea Breeze

23.2.1 Introduction

The sea breeze phenomena is a turbulent atmospheric boundary layer (ABL) circulation driven by contrasting surface temperatures across the cooler water and warmer land masses. The circulation cell, often accompanied by a sea breeze front, and the thermal internal boundary layer affect pollutant transport in metropolitan and industrial areas bounded by large bodies of water. In rural and agricultural areas, the sea breeze can effect the microclimates on which the production of certain crops depends.

The importance in studying the phenomena is evident in early observational studies made by Kimble [104] and Wexler [174]. It was not until the second half of this century that detailed comprehensive observational studies, including over-water and upper level measurements, could be made. Sea coast experimental studies were made by Fisher [67] on the coast of Rhode Island, Frizzola and Fisher [70] in the New York Harbor, Hsu [84] on the upper Texas coast, Simpson et al. [99] on the south coast of England, and Yoshikado and Hiroaki [180] on the south east shore of Japan. The two latter studies focused attention on the associated sea breeze front. Studies by Moroz [135] on the eastern shore of Lake Michigan and Lyons [125] on the southwest shore of Lake Michigan provide experimental data for the lake breeze events and comparison with the sea breeze circulations.

Estoque [65] used a two-layer approach to numerically simulate turbulent heat and momentum flux in a two-dimensional model that captured many sea breeze circulation characteristics, including the frontal event. Neumann and Mahrer [139] improved on Estoque's model with a proper treatment of mass conservation and the use of the full conservation equation for vertical momentum instead of the simpler hydrostatic version. Pielke [152] developed an elaborate hydrostatic, three-dimensional model for studying the effects of sea breeze circulations on thunderstorm activity over south Florida. Physick [151] used a sigma vertical coordinate (based on pressure) in a model designed to simulate sea breeze flow over variable surfaces. A heat balance relationship was used to calculate the land surface temperature, thus improving the model's prediction of inland frontal penetration.

The dynamic behavior of the sea breeze front and internal thermal boundary layer were predicted and analyzed by Kitada [105] in a sea breeze simulation incorporating the popular $k - \epsilon$ turbulence model. Papageorgiou [144] used a three-dimensional, hydrostatic, Boussinesq model to study the sea breeze-related pollutant dispersion over varied terrain in coastal areas. Turbulent aspects were modeled using a first-order closure scheme, based on a mixing length formula determined from similarity theory, and the universal function proposed by Dyer [64]. Finally, Sha et al. [159] used a two-dimensional, nonhydrostatic, compressible model to study the Kelvin-Helmholtz instability in the foremost part of the sea breeze head.

The sea breeze circulation offers a challenging numerical modeling problem various reasons, including:

- it is strongly convective in nature, especially near the shoreline where wind speeds may be as high as 15 m/s

- the vortex associated with the sea breeze circulation means model winds will be skew to the computational grid, and

- it is a turbulent flow requiring a dynamic turbulence model in order to capture the evolution of its turbulent character.

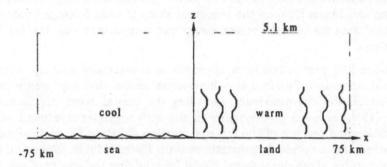

Figure 23.1: Seabreeze problem domain.

23.3 Mathematical Model

23.3.1 Model Equations

The seabreeze flow within the ABL is considered two-dimensional with a domain 5.1 km in height (z direction) and 150 km in width (x direction). Variation of the dependent variables in the y direction (parallel to the coast) is assumed to be zero. One half of the domain is over water and the other half is over land, as indicated in Figure 23.1.

The following system of partial differential equations models the transient flow of the sea breeze for a full 24-hour cycle. The Boussinesq approximation for density changes is made to account for buoyant effects:

$$U_x + W_z = 0 \tag{23.1}$$

$$U_t + U \cdot U_x + W \cdot U_z = -\frac{1}{\rho_0}P_x + \nu\,(U_{xx} + U_{zz}) - (\overline{uu})_x \\ -(\overline{wu})_z + f \cdot V , \tag{23.2}$$

$$V_t + U \cdot V_x + W \cdot V_z = \nu\,(V_{xx} + V_{zz}) - (\overline{uv})_x - (\overline{wv})_z \\ -f \cdot U , \tag{23.3}$$

$$W_t + U \cdot W_x + W \cdot W_z = -\frac{1}{\rho_0}P_z + \nu\,(W_{xx} + W_{zz}) - (\overline{uw})_x \\ -(\overline{ww})_z - \beta\theta g , \tag{23.4}$$

$$\Theta_t + U \cdot \Theta_x + W \cdot \Theta_z = \alpha\,(\Theta_{xx} + \Theta_{zz}) - (\overline{u\theta})_x - (\overline{w\theta})_z . \tag{23.5}$$

The dependent variables in Equations (23.1) - (23.5) are the mean (time averaged) flow velocities U, V, and W, in x, y, and z directions, respectively, the mean pressure P, and the mean departure Θ from a reference temperature T_0. The parametric quantities include the molecular kinematic viscosity ν (0.14×10^{-4} m^{2}s^{-1}), the molecular thermal diffusivity α (0.19×10^{-4} m^{2}s^{-1}), the acceleration of gravity g (9.8 ms^{-2}), the Coriolis parameter f (0.834×10^{-4} at latitude 35° N), the volumetric expansion rate of air β ($3.53x10^{-3}$K), and the mean air density ρ_0 (0.0125 kgm^{-3}). The lower case terms u, v, w, and θ represent time dependent deviations from their respective mean values, and their products give the turbulent Reynolds stresses and heat flux. They axe modeled by a generalized $k - \epsilon$ model (see Section 2.5.1).

23.3.2 Initial and Boundary Conditions

Initially, all dependent mean variables, with the exception of Θ, are zero over the entire computational domain to approximate the early morning hours. The initial vertical profile for Θ is that of a slightly stable atmosphere, common at early morning hours, where the temperature decreases at a rate of roughly $7^{\circ}K$ per km. Turbulent variables k, ϵ and ν_t are initially set to a small value (10^{-6}).

All dependent variables are invariant in time on the top boundary. On the lateral boundaries, the normal derivative of each dependent variable is zero. Steep vertical gradients of dependent variables near the earth's surface would require high resolution in the computational grid. The number of required grid points can be greatly reduced by incorporating a wall function technique (see Section 2.5.3) on the lower boundary. Values for the dependent variables at the first interior node are determined from the following:

$$U_1 = \frac{U^*}{\kappa} ln \left(\frac{z}{z_0} \right) , \qquad \Theta_1 = T^* \frac{Pr_t}{\kappa} ln \left(\frac{1}{Pr} + \frac{\kappa z^+}{Pr} \right) ,$$

$$k_1 = \frac{(U^*)^2}{\sqrt{C_k}} , \quad \text{and} \quad \epsilon_1 = \frac{(U^*)^3}{\kappa z^+} . \tag{23.6}$$

E ($=9.0$) and κ ($=0.41$) are the log-law and von Kármán constants, respectively. The friction velocity, U^* ($= \sqrt{\tau_w}/\rho$) is used to define a dimensionless distance from the boundary z^+ ($=U^*z/\nu$), and $T^*(= \alpha q_w/K)$ is the friction temperature. In the preceding expressions, τ_w and q_w are the wall shear stress and heat flux, respectively, and Pr is the Prandtl number. The surface roughness parameter z_0 is included in the formula for friction velocity, and is taken to be 10^{-2} m for land, and 10^{-4} m for sea [9]. The formulas given in Equations (23.6) are valid for $z^+ > 10$. For $z^+ \leq 10$, $U_1 = z^+U^*$, $\theta_1 = PrT^*z^+$, $k_1 = 0.1(z^+)^2$, and $\epsilon_1 = 0.2$. W_1 is zero in either case.

The thermal boundary condition for the earth's surface is

$$\Theta(x,0,t) = A(x)10sin(\pi t/43200) ,$$

where

$$A(x) = \begin{cases} 0, & \text{for} \quad -75 \text{ km} \leq x \leq 0 \text{ km,} \\ 1, & \text{for} \quad 0 \text{ km} \leq x \leq 75 \text{ km.} \end{cases}$$

This thermal boundary condition simulates the 24 hour cycle of heating by the sun on a land mass located for $x = 0$ km to $x = 75$ km. The sea surface temperature, on the interval 0 km $\leq x \leq 75$ km, is considered stationary so that the maximum temperature difference between the land and sea surfaces is $10°K$.

23.4   Results and Discussion

The governing equations outline in section 23.3.1 were solved using the finite analytic method on a nonstaggered grid using the momentum weighted interpolation method (see Chapter 20). The grid spacing is uniform in the horizontal direction. In the vertical, the height of the first computational node is 5 meters off the earth's surface. The second node is placed at a height of 20 meters, and the third node at 100 meters. The interval from 100 meters to 5.1 km is partitioned uniformly.

A time step corresponding to a real time of 12 seconds is used throughout the simulation. Typically, four to eight iterations, in which each dependent variable is calculated, are necessary to achieve convergence for a given time step. The model is integrated through a 24-hour heating cycle. The initial time is taken as 07:00 local standard time (LST), and corresponds to one hour after sunrise, allowing for a hour time lag in the surface heating.

Isotachs for selected hours are shown in Figures 23.2(a) - 23.2(c). An onshore flow clearly develops within two hours of the onset of surface heating. A slight return flow to the sea is also evident at this time. The depth of the inflow is approximately 1 kilometer. The circulation penetrates to about 7 kilometers on shore. However, by 10:00 LST the onshore flow has retreated slightly, probably due to the development of additional circulation, albeit weaker, over the heated land. This oscillation of the onshore flow in the initial stages of the circulation is consistent with observations made by Yoshikado and Hiroaki [180]. A slight seaward flow combines with the onshore flow to create a convergence zone between near the 8 km location, similar to observations of Lyons [125]. The onshore flow has reached a maximum of over 3.0 m/sec by this time, and the depth on the seabreeze has exceeded 1.0 km.

The heating reaches a maximum at 13:00 LST, and the inflow has increased to more than 4 m/sec, and it has begun its push further inland. In fact, as the cooling portion of the cycle continues, the momentum of the onshore flow carries it well inland. By 20:00 LST, the "front" has pushed as far inland as 60 km. Vertical velocity speeds during this time are typically half, or less, the magnitude of the horizontal velocity values. Isotherms are plotted in Figure 23.3, and they suggest the circulation is not strong enough to significantly alter the isotherm pattern.

Turbulent quantities for 13:00 LST are plot in Figure 23.4(a). The turbulent kinetic energy values grow to as much as 1.8 m^2s^{-1}. The combined effect of mechanical and thermal production is evident in the larger values near the shore where gradients in velocity and temperature are larger than elsewhere.

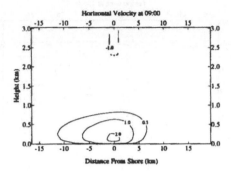

(a) 09:00 LST

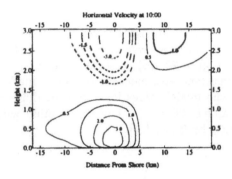

(b) 10:00 LST

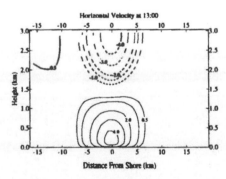

(c) 13:00 LST

Figure 23.2: Isotachs for horizontal velocity (ms^{-1}).

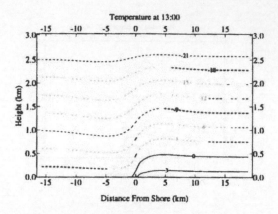

Figure 23.3: Isotherms (degrees Kelvin) for 13:00 LST.

The higher values for turbulent viscosity over land help diffuse the secondary circulations that try to develop over land. The rate of turbulent energy dissipation, plotted in Figure 23.4(b), reaches a maximum value in excess of 7 cm^2s^{-3}, which is consistent with measurements of ϵ made by Weill et al. [173] and Gryning [78]. The model values for turbulent diffusivity (Figure 23.4(c)) reach a maximum value of 80 m^2s^{-1}, which are similar with those estimated in an observational study by Bauer [14].

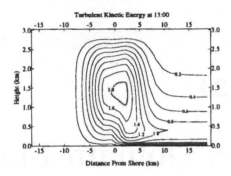

(a) Turbulent kinetic energy (m^2s^{-2})

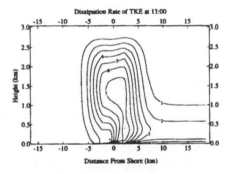

(b) Rate of dissipation of TKE (cm^2s^{-3})

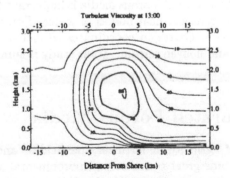

(c) Turbulent viscosity (m^2s^{-1})

Figure 23.4: Turbulence quantities for 13:00 LST.

Chapter 24

Complex Domain Flows

The applications presented thus far have involved only rather simple geometries, where the body surfaces involved lie along one or more coordinate lines. Typically, realistic applications involve more complex geometries, and thus more care is needed to accurately represent the flow boundaries.

The FA method has been used to solve for the flow field on several complex domains. Kim et al. [103] used a standard Cartesian grid with a fine mesh to model the blood flow the an artificial heart. Their results are in reasonable agreement with previously reported experimental results. Chen and Chen [54] completed a numerical study of supersonic flow in two-dimensional channels of varying geometries. They used the FA method based on characteristics (see Chapter 12), and boundary fitted coordinates to solve the governing partial differential equations on the irregular domain. In some cases considered, an exact solution is know, to which their numerical results agreed quite well. Choi and Chen [55] used the FA method in conjunction with zonal modeling to solve for turbulent flow past an axisymmetric body. Extension of these concepts to solution of flow around complete axisymmetric ship bodies can be found in [56].

The modeling of flow through porous media is important to numerical studies of chemical pollutant transport near bodies of water. Tsai et al. [171] used the FA method and diagonal Cartesian coordinates to model pollutant migration associated with lake and uneven terrain. Their study is summarized in the next section.

24.1 Unsaturated Porous Media

Recently, the use of agricultural fertilizer, pesticides, and the treatment of wastewater have become great concerns to environmental and agricultural engineers. Chemical substances are frequently the source of soil and water pollutants. These pollutants are transported with water flow through the soil by several processes, including advection and diffusion. To estimate the magnitude of the hazard posed by these pollutants, it is important to investigate the processes that control the movement of water from the ground surface to the water table. The transport of water in unsaturated porous media plays an important role in environmental safety and agricultural applications. Accurate prediction of the transient water flow in unsaturated porous media is essential for the optimum management and control of groundwater contamination. Unfortunately, the Richards equation, which is the governing equation for water flow in unsaturated porous media, is highly nonlinear. Because of this, exact analytical solutions cannot be obtained, and numerical solutions must be sought.

Application of the FA method to groundwater flow was first considered by Hwang et al. [86] for a two-dimensional groundwater solute transport problem. Li et al. [115] improved the FA method for the transient problem. They incorporated the Laplace transform into the solution of the linearized advection-diffusion-reaction equation. Tsai and Chen [170] proposed an optimal time-weighting factor for the FA method, to improve accuracy in the prediction of one- and two-dimensional groundwater solute transport. The diagonal Cartesian method described in Chapter 17 is used to extend the previous works to more complex domains.

In this numerical simulation, the optimal/variable time-weighting factor developed by Tsai and Chen [170] is incorporated into the FA method to improve the transient solution. The performance of the optimal FA solution is demonstrated via examples of infiltration into uniform clay loam, a layered-system soil with regular boundaries, and unsaturated flow of leachate from a landfill with complex boundaries.

24.2 Mathematical Model

The Richards equation [15], which describes transient, two-dimensional flow of water in unsaturated porous media, is given by

$$C(H)\frac{\partial H}{\partial t} - \frac{\partial}{\partial x}\left[K(H)\frac{\partial H}{\partial x}\right] - \frac{\partial}{\partial y}\left[K(H)\frac{\partial H}{\partial y}\right] - \frac{\partial K(H)}{\partial y} = 0, \qquad (24.1)$$

where H is the *pressure head* in the porous media; $C(H)$ is the *specific moisture capacity*, which in general is a nonlinear function of H and defined as $\partial\theta(H)/\partial H$, in which θ is the *volumetric water content*; K (hydraulic conductivity) and θ are nonlinear functions of H and generally determined from laboratory experiments; x and y are the horizontal and vertical space coordinates, respectively; and t is time. This equation assumes that there is no sink or source in the flow domain of the porous media, that the porous media are isotropic, and that the porosity and water density of the porous media remain constant.

Equation (24.1) is a second-order elliptic equation in space, but first-order in time, and thus has parabolic behavior. Therefore, to obtain a unique solution it is necessary to specify one initial condition and boundary conditions of H or the gradient of H on the spatial boundaries. The water flux can be defined as $\vec{q} = q_x\mathbf{i} + q_y\mathbf{j}$, in which q_x and q_y are components of the water flux in the x and y directions, and are defined by $q_x = -K(\partial H\partial x)$ and $q_y = -K(\partial H/\partial y + 1)$, respectively. The symbols i and j represent unit vectors in the x and y directions, respectively.

To begin, a linearized transport equation that incorporates an optimal/variable time-weighting factor is developed for the Richards equation. The time-weighting factor associated with the unsteady term, $\partial H/\partial t$, within one time step can be expressed as

$$\frac{H^n - H^{n-1}}{\Delta t} = \omega \left(\frac{\partial H}{\partial t}\right)^{n-1} + (1 - \omega)\left(\frac{\partial H}{\partial t}\right)^n , \qquad (24.2)$$

in which Δt is the time interval used in the numerical computation; n denotes the time step; and ω is a time-weighting factor. Based on investigation of the analytic solutions of a one-dimensional advection-diffusion equation, Tsai and Chen [170] proposed the following optimal value of ω:

$$\omega = 0.5 \left[\frac{\left(\frac{\partial H}{\partial t}\right)^{n-1}}{\left(\frac{H^n - H^{n-1}}{\Delta t}\right)}\right]^{-0.28} . \qquad (24.3)$$

Incorporating the optimal time-weighting factor in the FA solution improves the accuracy of the solution for the transient transport equation by avoiding numerical diffusion and oscillation [170]. This scheme, for simplicity, will be referred to as the *optimal finite analytic method*.

Equation (24.1) can be locally linearized and expressed as

$$\frac{\partial^2 H^n}{\partial x^2} + \frac{\partial^2 H^n}{\partial y^2} = 2A\frac{\partial H^n}{\partial x} + 2B\frac{\partial H^n}{\partial y} + g , \qquad (24.4)$$

where A, B and g are taken to be constants within each FA element, having explicit values from the central node P of the element at the $(n-1)$st time step. A, B and g can then be expressed as

$$A = -\frac{1}{2K_p^{n-1}}\left(\frac{\partial K}{\partial x}\right)^{n-1}_p , \qquad (24.5)$$

$$B = -\frac{1}{2K_p^{n-1}}\left(\frac{\partial K}{\partial y}\right)^{n-1}_p , \qquad (24.6)$$

$$g = R\frac{H_p^n - H_p^{n-1}}{(1 - \omega_p)\Delta t} + f_p, \qquad (24.7)$$

with

$$R = \frac{1}{K_p^{n-1}}C(H_p^{n-1}) \qquad (24.8)$$

and

$$f_p = 2B - R\frac{\omega_p}{1 - \omega_p}\left(\frac{\partial H}{\partial t}\right)^{n-1}_p . \qquad (24.9)$$

The subscript p in Equations (24.5) - (24.9) indicates that the value is evaluated at the central node of the local FA element.

Equation (24.4) is equivalent to the advection-dispersion equation in solute transport with advective velocities $2A$ and $2B$, and with dispersion coefficients both valued at one in the x and y directions, respectively. The A and B terms are obtained from the nonuniform distribution of the nonlinear hydraulic conductivity, which creates the equivalent of advection for water flow in the x and y directions, respectively. For homogeneous media, $\partial K/\partial x = \partial K/\partial y = 0$, and $A = B = 0$ accordingly. The term g is a source function. The spatial derivatives of K in Equations (24.5) and (24.6) may be calculated by a central difference approximation. Note that Equation (24.4) has the form of the generalized two-dimensional transport equation, and can thus be discretized using the 9-point and 5-point FA methods. The set of algebraic equations resulting from this can then be solved with the given boundary and initial conditions to yield the FA numerical solution.

24.3 Results and Discussion

The first problem considered here is a two-dimensional simulation of the infiltration of water in a uniform clay loam with a constant-flux strip source. The geometry of the problem is shown in Figure 24.1(a). The y coordinate is positive upward. A constant downward flux of water, $q_y = -0.0278$ cm/min $[q_y = -K(\partial H/\partial y + 1)]$, is imposed at the 5 cm wide strip source located at the upper left corner of the 75 x 85 cm domain. The rest of the top boundary is impermeable ($q_y = 0$). Also, no water is allowed to drain through the left boundary ($q_x = 0$). Only the right and bottom boundaries are allowed to drain (constant pressure head with $H = $ -250 cm). The initial water content is assumed to be uniformly distributed with the porous media in a very dry condition, $\theta = 2.3 \times 10^{-5}$, which in terms of pressure head H is -250 cm.

The exponential forms of the $K(H)$ and $\theta(H)$ relationships are used in this simulation, for which an analytic solution is available [172]. $K(H)$ and $\theta(H)$ are given as

$$K(H) = K_o e^{\alpha H} \qquad (24.10)$$

and

$$\theta(H) = (K_o/A_o)e^{\alpha H} , \qquad (24.11)$$

in which K_o, α and A_o are empirical constants for the soil. Using Panoche clay loam, these values are $K_o = 0.0694 cm/min$, $\alpha = 0.04$ and $A_o = 0.1388$ cm/min [32].

Four-minute time intervals are used for the numerical simulation. A nonuniform 11×13 grid is used in both the x and y directions, packing nodes near the upper left corner where the strip source is located. Δx ranges from 1 to 10 cm, and Δy ranges from 0.2 to 10 cm. At each time step, the solution is obtained with a line-by-line iteration from the bottom boundary to the top

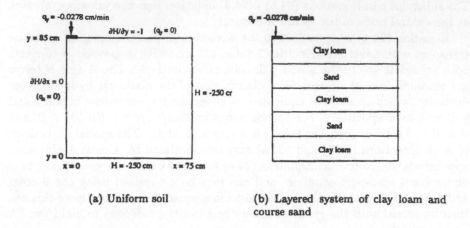

(a) Uniform soil   (b) Layered system of clay loam and
                                 course sand

Figure 24.1: Computational domain and boundary conditions for two-dimensional infiltration.

boundary until the difference in H between successive iterations is less than 0.001 cm.

In this problem, the strip source of constant flux along the top boundary produces a transient, two-dimensional infiltration of water. The optimal FA numerical solution is compared with solutions obtained using the fully implicit FA method ($\omega = 0$) and the two-dimensional Galerkin finite element computer code FEMWATER [179, 178]. The dependent variable of the flow equation in FEMWATER is pressure head. This computer code was designed to solve flows through saturated-unsaturated media with consideration of water and media compressibility.

Figure 24.2 presents the comparison of the pressure head contours obtained from the optimal (variable ω) and implicit ($\omega = 0$) FA methods, the FE method, and the analytic solution for a time of 72 minutes. The optimal FA method produces results very close to the analytic solution, and improves on the fully implicit FA method. The optimal FA solution is also more accurate than the FE solution.

Computations were carried out on an HP Apollo (Series 400) computer. The CPU time required for the optimal FA method was 64 seconds, and 69 seconds were required for the fully implicit FA method. The FEMWATER CPU time is not included here because this code was written for more general applications, and thus may include additional overhead. It is therefore not appropriate to conclude that the optimal FA method is faster than the FE method. However, it is seen that the use of the variable time-weighting factor in the optimal FA method does not increase computational time over the fully implicit FA method.

The next example considered is similar to the first, but it uses the layered system of soil with alternating layers of clay loam and coarse sand shown in Figure 24.1(b). For the coarse sand, the parameters $K_o = 1.20$ cm/min, $\alpha = 0.05$

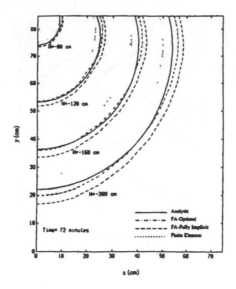

Figure 24.2: Contours of pressure head for simulation of infiltration with constant-flux strip source for clay loam: time = 72 min.

and $A_o = 3.0$ cm/min are used. The clay loam values are the same as in the first problem. A uniform grid spacing $\Delta x = \Delta y = 1$ cm is used for this problem, except fine vertical grid spacing of 0.1 cm is used around the clay loam/coarse sand interfaces. The time interval is $\Delta t = 1$ minute. The pressure head contours computed with the optimal FA method at an elapsed time of 30 minutes are given in Figure 24.3. Note that the pressure head is continuous across the interfaces. The FA solution is very stable, without numerical oscillations occurring along the interfaces.

The final problem treated in this application concerns flow movement in unsaturated porous media from landfill leachate in the field during precipitation. The ground surface has variable slope, as sketched in Figure 24.4(a). A landfill is located in the middle portion of the domain considered. During the rainfall season, the leachate water within the landfill may infiltrate into the unsaturated zone and propagate downward. Finally, the water front may reach the groundwater table and contaminate it.

To predict water flow in the unsaturated zone for a domain with irregular boundaries, a combination of 9-point and 5-point elements in the optimal FA method is used. The properties of the porous media have the same form of $K(H)$ and $\theta(H)$ given in Equations (24.10) and (24.11). The parameters K_o, α and A_o are taken to be 0.06 cm/min, 0.02 and 0.12 cm/min, respectively. During the rainfall season, the leachate water flux from the bottom of the landfill into the unsaturated zone is 1.8 cm/h, while the water flux from the sloping ground surface infiltrating into the unsaturated zone is 0.6 cm/h. The boundary conditions and computational grid are shown in Figure 24.4(b). 5-point FA

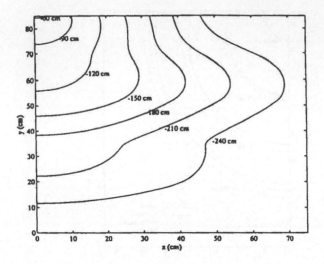

Figure 24.3: Contours of pressure head for simulation of infiltration with constant-flux strip source for a layered system of soil at time = 30 min.

elements are used for the sloping areas (complex geometry), and 9-point FA elements are used for regular boundaries and in the interior of the domain. The computational grid for this problem consists of 22 nodes in the x direction, while nodes in the y direction vary from 10 to 18.

The initial condition for the pressure head H in the unsaturated zone is -200 cm uniformly distributed, representing a very dry initial condition ($\theta = 1 \times 10^{-4}$). The time intervals are $\Delta t = 10$ minutes for the first two hours, and $\Delta t = 1$ hour thereafter. For each time step, the solution is obtained with a line-by-line iteration in the y direction, from the left boundary to the right, until the difference of H between successive iterations is less than 0.001 cm. To improve the accuracy, the advective terms, A and B in Equations (24.5) and (24.6), are renewed after each iterative convergence is achieved. The CPU time required for this simulation was 335 seconds.

The simulated results of pressure head and water content for times corresponding to 3 and 7 days are shown in Figures 24.5 and 24.6. These figures clearly indicate that the numerical solution physically reflects the effects of the ground-surface slopes and boundary conditions on the propagation of water and pressure. This numerical simulation of field infiltration has demonstrated the capability of the FA method to treat complex boundaries, and the potential for simulating the transport of water flow resulting from gravity, imposed water flux and capillary tension in unsaturated porous media. Further, the FA solution is numerically stable for the range of problems investigated. Additional results and more complete detail of the solution procedures can be found in [171].

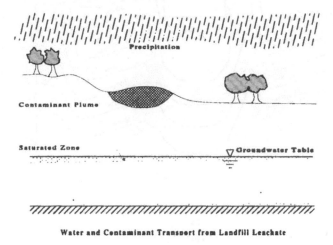

(a) Water and contaminant transport from landfill leachate.

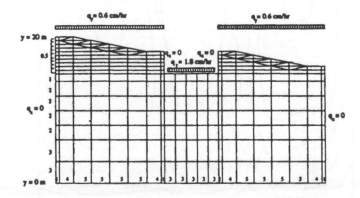

(b) Computational domain and boundary conditions.

Figure 24.4: Landfill leachate and computational domain.

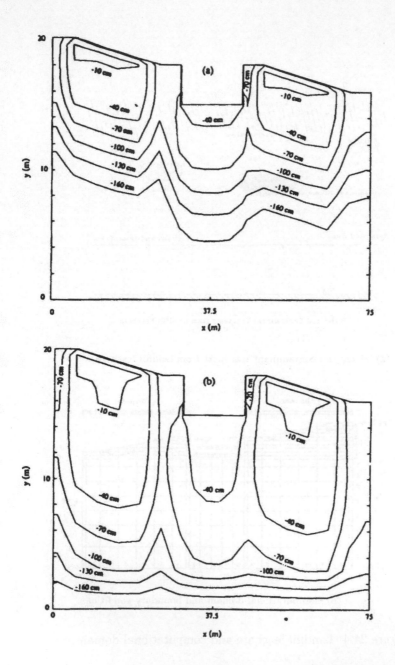

Figure 24.5: Contours of pressure head for simulation of infiltration from landfill leachate and rainfall: (a) time = 3 days; (b) time = 7 days.

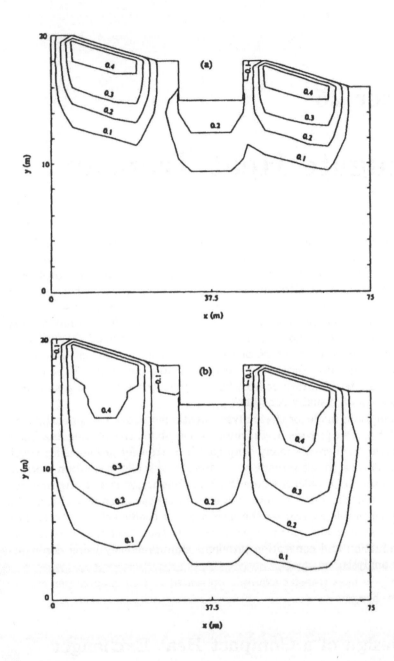

Figure 24.6: Contours of water content for simulation of infiltration from landfill leachate and rainfall: (a) time = 3 days; (b) time = 7 days.

Chapter 25

Conjugate Heat Transfer

Many engineering heat transfer problems involve some combination of solid conduction, fluid convection and radiation. These problems, which involve more than one mode of heat transfer, are known as *conjugate heat transfer* problems. The complexity of heat transfer analysis increases significantly when multiple modes of heat transfer are considered, and thus approximate analyses are frequently performed to circumvent the difficulty. For example, if conduction in a solid is being calculated, convection is usually used as a boundary condition, with either the surface temperature or the heat transfer coefficient h assumed to be known (and usually constant). Likewise, if fluid convection is being calculated, either the temperature or heat flux at the surface is often assumed to be known (and, again, usually constant).

If the assumptions made for the analyses mentioned above are approximately valid, the solution can provide useful information about the phenomena being simulated. However, there are many conjugate heat transfer problems for which these approximations do not provide an accurate and realistic solution, such as conduction-convection problems where the heat conduction is a strong function of the convection in the ambient fluid that encounters the boundary of the solid domain. In such problems, neither the heat transfer coefficient nor the surface temperature/heat flux are known or constant. It then becomes necessary to solve the conduction and convection problems simultaneously using conjugate heat transfer analysis.

The conjugate heat transfer example presented in this chapter pertains to finned heat exchangers.

25.1   Design of a Compact Heat Exchanger

Finned heat exchangers are devices commonly used in the electronics and computer industries to remove heat from a device. For applications with limited space, compact heat exchangers are frequently used. Flow in compact heat exchangers is often designed to be laminar. This is beneficial because laminar flow causes a smaller pressure drop across the heat exchanger compared to turbulent flow, and hence decreases power consumption. In addition, less noise is generated. However, less mixing of the fluid occurs in laminar flow, which decreases the heat transfer to the fluid. A common method for enhancing heat transfer is to add fins which are staggered in the streamwise direction. An example of this is shown in Figure 25.1(a). Adding fins enhances the fluid mixing, enlarges the contact area over which heat transfer can occur and increases the time it takes the fluid to meander through the heat exchanger. All of these changes augment heat transfer. The trade-off in adding fins is an increase in the pressure drop across the heat exchanger. It is important to optimize this trade-off using an accurate conjugate heat transfer analysis of the compact heat exchanger.

Bravo and Chen [21] performed a numerical study on a short heat exchanger such as the one shown in Figure 25.1(a). The overall dimensions of their heat exchanger were $5L \times 2L$, with wall and fin thicknesses of $0.125L$. They considered a heat exchanger without fins, one with one pair of fins and one with two pairs of fins. The fin heights used were $0.5L$, L and $1.5L$. The study was done for Prandtl numbers of 0.7 (air) and 4 (water), and Reynolds numbers 100, 150, 300 and 500. Bravo and Chen found that increasing the Prandtl number or the Reynolds number increased the average Nusselt number (and hence the heat transfer). Additionally, increasing the number of fins or the fin height resulted in an increase in the Nusselt number and an associated increase in the pressure drop across the heat exchanger. However, the increase in the pressure drop was several orders of magnitude larger than the increase in the heat transfer. The same conclusions were reached by Kelkar and Patankar [102], in a similar study of the periodic fully-developed region of a longer staggered fin heat exchanger. Of the different geometries studied, Bravo and Chen [21] concluded that possibly the most efficient design (significant heat transfer augmentation with minimal increase in pressure drop) was the one shown in Figure 25.1(a).

The study of Bravo and Chen [21] assumed that the wall and fin temperatures were constant. In other words, the thermal conductivity of the fins and walls was considered to be infinite. However, examining the fins in Figure 25.1(a), one expects that the fin temperature should decrease near the tip, since the fin conductivity is finite. Carlson et al. [28] investigated the conjugate effects of fin and wall conduction on heat transfer to the fluid flowing through the compact heat exchanger shown in Figure 25.1(a). It was found that the constant wall temperature assumption over-predicts the heat transfer for the compact heat exchanger by 5-8% when Re is between 50 and 500.

The finned heat exchanger increases the bulk fluid temperature at the heat exchanger exit by 70-450% when compared with the same heat exchanger without fins. However, the pressure drop for the finned heat exchanger is considerably larger than for the finless heat exchanger. This may be acceptable if the power required to overcome the additional pressure drop is inexpensive, or if augmentation of heat transfer in a limited amount of space is necessary. If this type of heat exchanger is to be employed in such applications, it would be beneficial to optimize the geometry in order to maximize the heat transfer and minimize the pressure drop. It was found (Carlson et al. [28], Carlson [27]) that

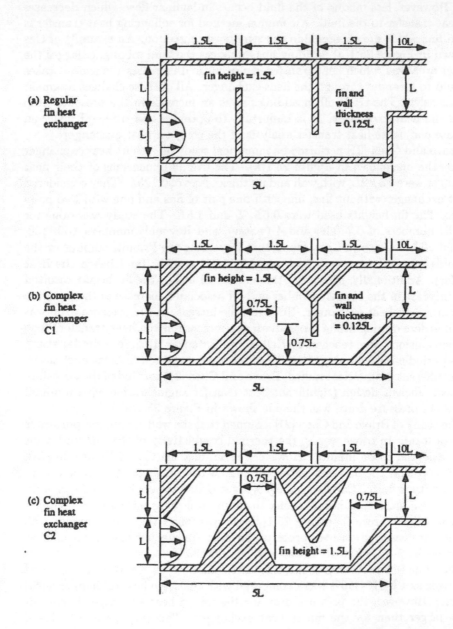

Figure 25.1: Three staggered-fin compact heat exchanger designs.

there is very little heat transfer on the downstream side of the fins and on the wall above the heat exchanger entrance (i.e., in the large recirculation zones). It is also evident that the heat transfer will increase if the fin temperatures can be kept higher near the fin tips. One way to achieve this is to redesign the heat exchanger so that the fins are wider at the base and taper toward the fin tips. This design change will increase the amount of highly conductive material through which the heat can move as it progresses toward the fin tips, and should increase the fin tip temperatures. The wider fin bases should also reduce the size of the recirculation zones, which will bring more rapidly moving fluid into contact with the fins and walls, thus increasing the heat transfer.

Two modified heat exchanger geometries that take these points into consideration are studied here. The first modified geometry is shown in Figure 25.1(b). In this complex fin heat exchanger (which will be referred to as C1), the regular fins are modified so they begin to taper to a wider base half way down the fin. The areas directly above the entrance and below the exit of the heat exchanger are modified in a similar manner. The second complex fin heat exchanger (C2) is shown in Figure 25.1(c). In this heat exchanger, the fins begin to taper at the fin tips instead of in the middle, as in C1. Again, the areas above the entrance and below the exit are modified as well. Conjugate heat transfer and pressure drop will be analyzed for C1 and C2, and compared with results for the regular fin heat exchanger in Figure 25.1(a) and a finless heat exchanger.

25.2 Mathematical Model

Assuming steady laminar flow, the momentum and energy equations were solved using the diagonal Cartesian method described in Chapter 17. The energy equations in the solid and fluid domains were coupled using the conjugate heat transfer methods discussed in Chapter 21. The geometries of the regular fin heat exchanger and heat exchanger C1 can be exactly modeled on a uniform grid, using 9-point FA elements for the vertical and horizontal surfaces in both heat exchangers and 5-point FA elements for the $45°$ diagonal surfaces in C1. The corresponding diagonal surfaces in C2 are not $45°$ angles, and must be approximated. A 196×74 grid was utilized for all the heat exchangers in this study, with a uniform grid inside the heat exchanger and a grid that is uniform in the y direction and stretched in the x direction in the outlet channel. The governing equations were made dimensionless using a characteristic length L, characteristic velocity U_{ref}, characteristic dynamic pressure ρU_{ref}^2 and characteristic temperature difference ΔT. The dimensionless temperature θ is defined as $\theta = (T - T_{in})/(T_w - T_{in})$, where T_{in} is the inlet temperature and T_w is the constant outer wall temperature.

The same boundary conditions were applied to each of the heat exchangers shown in Figure 25.1. For each heat exchanger, the inlet flow has a parabolic velocity profile with a maximum dimensionless velocity $u_{max} = 1.0$ (i.e. the characteristic velocity $U_{ref} = U_{max}$), and a uniform dimensionless temperature of $\theta = 0.0$. The temperature of the outer walls of the heat exchangers was held constant at $\theta = 1.0$. The outlet channels extend a length $10L$ downstream, and the outlet boundary conditions were taken as fully-developed at the end of the channel. The walls of the outlet channel were insulated. The material selected for the heat exchangers was aluminum ($k_s = 273$ W/m-K), and the working fluid was water ($Pr = 4.0$, $k_f = 0.63$ W/m-K), giving a conductivity ratio $K = k_s/k_f = 376$. The Reynolds numbers used in this study are $Re = 50$, 200 and 500, where Re is based on the inlet height L and the maximum inlet velocity U_{max}.

The different heat exchanger designs can be compared in terms of pressure drop and heat transfer. The pressure drop across a heat exchanger is characterized by the pressure coefficient C_{pr}, defined as

$$C_{pr} = \frac{P_{in} - P_{ex}}{\rho U_{ref}^2} = p_{in} - p_{ex} = \Delta p \,, \tag{25.1}$$

where P_{in} and P_{ex} are the average pressures across the heat exchanger inlet and exit, respectively. The bulk temperature of the fluid as it exits the heat exchanger is used to compare heat transfer characteristics, and is given by

$$\theta_{b,ex} = \frac{\int_{A_{ex}} \theta u dA_{ex}}{\int_{A_{ex}} u dA_{ex}} \,,$$

where A_{ex} is the exit area of the heat exchanger.

25.3 Results and Discussion

Figure 25.2 shows the $Re = 200$ streamline plots for (a) the regular fin heat exchanger, (b) heat exchanger C1, and (c) heat exchanger C2. The streamlines in these three plots are similar, but the recirculation zones are smaller in C1 than in the regular fin heat exchanger, and smaller still in C2. This is due to the additional aluminum in the complex fins, above the entrance, and below the exit in C1 and C2. Note that the main flow streamlines in (a) and (b) are strikingly similar. This is because the vertical portions of the fins in C1 force the flow to impinge on the opposing walls rather directly, just as in the regular fin heat exchanger. The angled fins in C2, on the other hand, allow the flow to take a less winding path. The fluctuations in the streamlines near the exit of heat exchanger C2 are the result of numerical integration, not a physical phenomenon.

The pressure coefficients computed for the three different Reynolds number flows through C1 and C2 are compared to the regular fin heat exchanger and a finless heat exchanger in Figure 25.3. The pressure coefficient in C1 is slightly lower than that found in the regular fin heat exchanger for $Re = 50$, but as the Reynolds number increases, the pressure coefficients for the two heat exchangers seem to converge to the same value. The similarity in C_{pr} between C1 and the regular fin heat exchanger is not surprising, considering the similarity of the main flow streams in these two heat exchangers. The pressure coefficient in

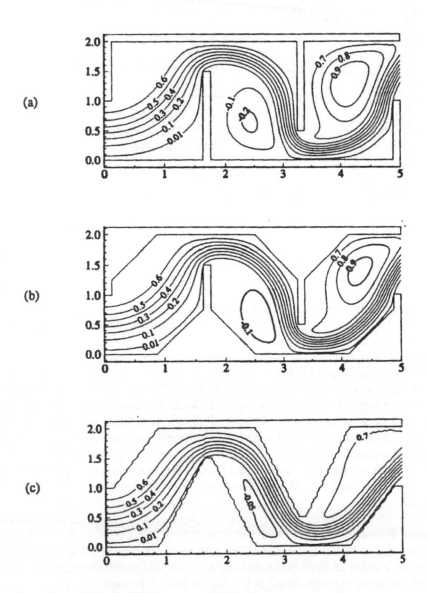

Figure 25.2: Streamlines for $Re = 200$ in (a) the regular fin heat exchanger, (b) heat exchanger C1, and (c) heat exchanger C2.

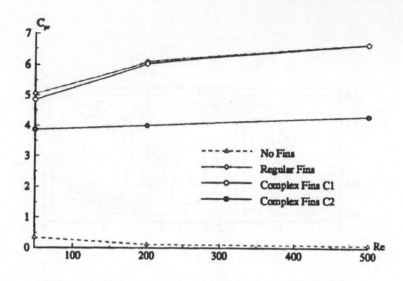

Figure 25.3: Pressure coefficient comparison for different heat exchangers.

C2 is significantly lower for all three Reynolds numbers. It appears that fins tapered all the way to the tip provide a shorter, smoother flow path than regular fins or the fins in C1.

The isotherms for (a) the regular fin heat exchanger, (b) heat exchanger C1, and (c) heat exchanger C2 for $Re = 200$ are presented in Figure 25.4. Notice that the temperatures near the fin tips in both C1 and C2 are indeed higher than for the regular fin heat exchanger. In fact, the fins in C2 are essentially isothermal, because of the large amount of highly conductive aluminum used in these fins. Also, the isotherms near the heat exchanger surfaces in C1 and C2 are packed more closely together, indicating higher temperature gradients than are found with regular fins. Both higher fin tip temperatures and higher thermal gradients in the fluid near the aluminum surfaces will improve heat transfer to the fluid.

Figure 25.5 compares the exit bulk temperatures of the different heat exchangers. It is seen that C1 produces an exit bulk temperature slightly higher than the regular fin heat exchanger. The exit temperatures of C2 are lower than that of the regular fin heat exchanger and C1, but still considerably higher than the exit temperatures generated in the finless heat exchanger.

The results presented in this section indicate that heat exchanger C1 and the regular fin heat exchanger are roughly equivalent, producing about the same exit temperatures and incurring about the same pressure drop. C2, on the other hand, has comparable heat transfer characteristics, but the pressure drop is much lower than in the regular fin heat exchanger or in C1. The choice of heat exchanger depends on the situation. If the cost of the power required to overcome the pressure drop is a concern, then C2 is the best choice. If the

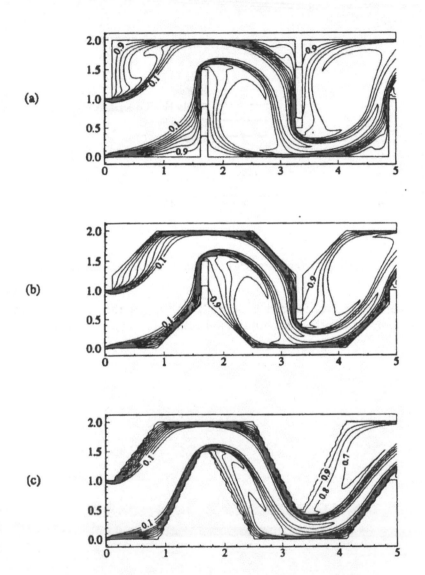

Figure 25.4: Isotherms for $Re = 200$, $Pr = 4$ in (a) the regular fin heat exchanger, (b) heat exchanger C1, and (c) heat exchanger C2. $\Delta\theta = 0.1$.

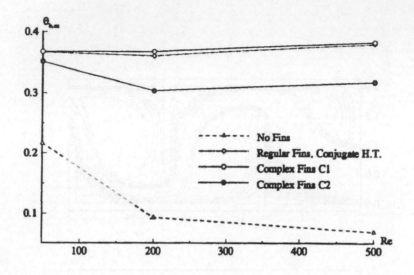

Figure 25.5: Exit bulk temperatures for the heat exchangers studied.

maximum amount of heat transfer is desired, then C1 or the regular fin heat exchanger would be appropriate. Additional results from this analysis can be found in [29].

Appendix A

The One-Dimensional Case

The finite analytic solution for the one-dimensional transport equation,

$$\phi_{xx} = 2A\phi_x + B\phi_t ,$$ (A.1)

is detailed in this appendix. For the one time step formulation, the initial condition ϕ_I and boundary conditions, ϕ_W and ϕ_E are

$$\phi(x,0) = \phi_I(x) = a_S(e^{2Ax} - 1) + b_S x + c_S,$$ (A.2)

$$\phi(-h,t) = \phi_W(t) = a_W + b_W t,$$ (A.3)

and

$$\phi(h,t) = \phi_E(t) = a_E(t) = a_E + b_E t,$$ (A.4)

where

$$a_S = \frac{\phi_{SE} + \phi_{SW} - 2\phi_{SC}}{4 sinh^2 Ah},$$

$$c_S = \phi_{SC},$$

$$b_S = \frac{\phi_{SE} - \phi_{SW} - coth Ah(\phi_{SE} + \phi_{SW} - 2\phi_{SC})}{2h},$$

$$a_W = \phi_{SW},$$

$$b_W = \frac{\phi_{WC} - \phi_{SW}}{\tau},$$

$$a_E = \phi_{SE},$$

and

$$b_E = \frac{\phi_{EC} - \phi_{SE}}{\tau}.$$

With the introduction of a change of variable

$$\phi = we^{Ax - \frac{A^2}{B}t},$$

the convective transport equation (A.1), initial condition (A.2) and boundary conditions (A.3) and (A.4) are transformed to

$$w_{xx} = Bw_t, \tag{A.5}$$

$$w(x,0) = a_S e^{Ax} + b_S x e^{-Ax} + (c_S - a_S)e^{-Ax} = \phi(x), \tag{A.6}$$

$$w(-h,t) = e^{-Ah + \frac{A^2}{B}t}(a_W + b_W t) = \phi_W(t), \tag{A.7}$$

$$w(h,t) = e^{-Ah + \frac{A^2}{B}t}(a_E + b_E t) = \phi_E(t). \tag{A.8}$$

The solution of Equation (A.5) for w can be obtained by superposition of solutions $w = w1 + w2$, where $w1$ solves

$$w1_{xx} = B \cdot w1_t, \tag{A.9}$$
$$w1(x,0) = \phi_I(x), \tag{A.10}$$
$$w1(-h,t) = 0, \tag{A.11}$$
$$w1(h,t) = 0, \tag{A.12}$$

and $w2$ solves

$$w2_{xx} = B \cdot w2_t, \tag{A.13}$$
$$w2(x,0) = 0, \tag{A.14}$$
$$w2(-h,t) = \phi_W(t), \tag{A.15}$$
$$w2(h,t) = \phi_E(t). \tag{A.16}$$

The solution $w1$ is obtained using the method of separation of variables. Assuming $w1 = X(x)T(t)$, and substituting for $w1$ in Equation (A.9), we may separate the variables as

$$\frac{X''}{X} = B\frac{T'}{T} = constant = -\lambda^2. \tag{A.17}$$

Two ordinary differential equations result for Equation (A.17),

$$X'' + \lambda^2 X = 0, \tag{A.18}$$

$$T' + \frac{\lambda^2}{B}T = 0, \tag{A.19}$$

with boundary conditions on X given by

$$X(-h) = X(h) = 0. \tag{A.20}$$

The boundary conditions specified in Equation (A.20) stipulate solution to Equation (A.18) having form

$$X(x) = a_n sin[\lambda_n(x + h)], \tag{A.21}$$

where $\lambda_n = \frac{n\pi}{2h}$ $(n = 1, 2, 3, \dots)$. The solution to Equation (A.19) has the form

$$T(t) = b_n e^{-\frac{\lambda_n^2}{B}t}. \tag{A.22}$$

Using the principle of superposition, the general solution for $w1$ can be written as

$$w1(x, t) = \sum_{n=1}^{\infty} a_n e^{-\frac{\lambda_n^2}{B}t} sin[\lambda_n(x + h)], \tag{A.23}$$

where the coefficients a_n can be determined by applying the initial conditions given in Equation (A.10) as

$$w1(x, 0) = \Phi(x) = \sum_{n=1}^{\infty} a_n sin[\lambda_n(x + h)]. \tag{A.24}$$

Invoking the orthogonality condition of the sine function, this initial condition results in

$$
\begin{aligned}
a_n &= \frac{1}{h} \int_{-h}^{h} \Phi(x) sin[\lambda_n(x + h)] dx \\
&= a_S E_{0n} + b_S h E_{1n} + (c_S - a_S) E_{2n},
\end{aligned}
\tag{A.25}
$$

where

$$
\begin{aligned}
E_{0n} &= \frac{1}{h} \int_{-h}^{h} e^{Ax} sin[\lambda_n(x + h)] dx \\
&= \frac{\lambda_n h}{(Ah)^2 + (\lambda_n h)^2}[e^{-Ah} - (-1)^n e^{Ah}],
\end{aligned}
\tag{A.26}
$$

$$
\begin{aligned}
E_{1n} &= \frac{1}{h^2} \int_{-h}^{h} x e^{-Ax} sin[\lambda_n(x + h)] dx \\
&= \frac{2(Ah)(\lambda_n h}{(Ah)^2 + (\lambda_n h)^2}[e^{Ah} - (-1)^n e^{-Ah}] \\
&\quad - \frac{\lambda_n h}{(Ah)^2 + (\lambda_n h)^2}[e^{Ah} + (-1)^n e^{-Ah}],
\end{aligned}
\tag{A.27}
$$

$$
\begin{aligned}
E_{2n} &= \frac{1}{h} \int_{-h}^{h} e^{-Ax} sin[\lambda_n(x + h)] dx \\
&= \frac{\lambda_n h}{(Ah)^2 + (\lambda_n h)^2}[e^{Ah} - (-1)^n e^{-Ah}].
\end{aligned}
\tag{A.28}
$$

To solve Equation (A.13), note that the boundary conditions specified in Equations (A.15) and (A.16) are functions of time. The solution for this problem can be deduced from the similar constant boundary conditions by the use of Duhamel's theorem. That is, $w2$ solves

$$w2 = \int_0^t \frac{\partial \tilde{w2}}{\partial t}(x, \mu, t - \mu) d\mu, \tag{A.29}$$

where $\tilde{w}2$ satisfies the zero initial condition and constant boundary conditions

$$\tilde{w}2_{xx} = B\tilde{w}2_t, \tag{A.30}$$
$$\tilde{w}2(x,0) = 0, \tag{A.31}$$
$$\tilde{w}2(-h,t) = \Phi_W(\mu), \tag{A.32}$$
$$\tilde{w}2(h,t) = \Phi_E(\mu). \tag{A.33}$$

The solution for $\tilde{w}2$ can be obtained by the superposition of the steady-state solution Φ and the transient solution v, which satisfy homogeneous boundary conditions. That is,

$$\tilde{w}2 = \Phi(x,\mu) + v(x,\mu,t-\mu), \tag{A.34}$$

where

$$\frac{d^2\Phi}{dx^2} = 0, \tag{A.35}$$
$$\Phi(-h,\mu) = \Phi_W(\mu), \tag{A.36}$$
$$\Phi(h,\mu) = \Phi_E(\mu), \tag{A.37}$$

and

$$v_{xx} = B \cdot v_t, \tag{A.38}$$
$$v(x,\mu,-\mu) = -\Phi(x,\mu), \tag{A.39}$$
$$v(\pm h,\mu,t-\mu) = 0. \tag{A.40}$$

The steady-state solution $\Phi(x,\mu)$ for the ordinary differential equation (A.35) is known to be

$$\Phi(x,\mu) = \frac{x}{2h}[\Phi_E(\mu) - \Phi_W(\mu)] + \frac{1}{2}[\Phi_E(\mu) + \Phi_W(\mu)], \tag{A.41}$$

which, in turn, is the initial condition for Equation (A.30). The solution for v is similar to that for $w1$, except that the initial condition $\Phi_I(x)$ is replaced by $-\Phi(x,\mu)$. Therefore

$$v = \sum_{n=1}^{\infty} b_n e^{-\frac{\lambda_n^2}{B}t}\sin[\lambda_n(x+h)], \tag{A.42}$$

where b_n can be obtained from the initial condition given in Equation (A.31). That is,

$$\begin{aligned}
b_n &= \int_{-h}^{h} -\Phi(x,\mu)\sin[\lambda_n(x+h)]dx \\
&= \frac{\Phi_W(\mu) - \Phi_E(\mu)}{2h^2}\int_{-h}^{h} x\sin[\lambda_n(x+h)] \\
&\quad -\frac{\Phi_W(\mu) + \Phi_E(\mu)}{2h}\int_{-h}^{h} \sin[\lambda_n(x+h)] \\
&= \frac{1}{\lambda_n h}[(-1)^n\Phi_E(\mu) - \Phi_W(\mu)].
\end{aligned}$$

Then,

$$\tilde{w}2 = v + \Phi$$

$$= \sum_{n=1}^{\infty} \frac{1}{\lambda_n h}[(-1)^n \Phi_E(\mu) - \Phi_W(\mu)]e^{-\frac{\lambda^2}{B}t}\sin[\lambda_n(x+h)]$$

$$+ \frac{\Phi_E(\mu) - \Phi_W(\mu)}{2h}x + \frac{\Phi_E(\mu) + \Phi_W(\mu)}{2}.$$

Now that $\tilde{w}2$ is known, $w2$ can be determined using Duhamel's Theorem and substituting for $\tilde{w}2$ in Equation (A.29). Thus,

$$w2 = \int_0^t \frac{\partial \tilde{w}2}{\partial t}(x, \mu, t - \mu)d\mu$$

$$= -\sum_{n=1}^{\infty} \frac{\lambda_n}{Bh}\int_0^t [(-1)^n \Phi_E(\mu) - \Phi_W(\mu)]e^{-\frac{\lambda^2}{B}(t-\mu)}\sin[\lambda_n(x+h)]d\mu$$

$$= \sum_{n=1}^{\infty} \frac{\lambda_n}{Bh}e^{-\frac{\lambda^2}{B}t}\sin[\lambda_n(x+h)]\int_0^t e^{\frac{\lambda^2}{B}\mu}[\Phi_W - (-1)^n\Phi_E]d\mu.$$

Now, the analytic solution on the FA element is given by

$$\phi = we^{Az - \frac{A^2}{B}t}$$

$$= (w1 + w2)e^{Az - \frac{A^2}{B}t}$$

$$= \sum_{n=1}^{\infty} e^{Az - \frac{A^2}{B}t}\sin[\lambda_n(x+h)\{a_n$$

$$+ \frac{\lambda_n}{Bh}\int_0^t e^{\frac{\lambda^2}{B}\mu}[\Phi_W(\mu) - (-1)^n\Phi_E(\mu)]d\mu\}. \tag{A.43}$$

When the formula given in Equation (A.43) is evaluated at the node $P(0, \tau)$, a 6-point formula for ϕ_P is obtained.

$$\phi_P = \phi(0, \tau)$$

$$= \sum_{n=1}^{\infty} e^{-\frac{A^2 + \lambda_n^2}{B}\tau}\sin\frac{n\pi}{2}$$

$$* \left[a_n + \frac{\lambda_n}{Bh}\int_0^\tau e^{\frac{\lambda^2}{B}\mu}[\Phi_W(\mu) - (-1)^n\Phi_E(\mu)]d\mu\right] \tag{A.44}$$

Because $\sin\frac{n\pi}{2} = 0$ for even numbers n, Equation (A.44) can be further simplified by letting $n = 2m - 1$, so that

$$\phi_P = \sum_{m=1}^{\infty} (-1)^{m+1}e^{-\frac{A^2 + \lambda_m^2}{B}\tau}\{a_m$$

$$+ \frac{\lambda_m}{Bh}\int_0^\tau e^{\frac{\lambda^2}{B}\mu}[\Phi_W(\mu) + \Phi_E(\mu)]d\mu\}, \tag{A.45}$$

where

$$\int_0^\tau e^{\frac{\lambda_m^2}{B}\mu}[\Phi_W(\mu) + \Phi_E(\mu)]d\mu =$$

$$e^{Ah}a_W + e^{-Ah}a_E)\int_0^\tau e^{F_m\mu}d\mu + (e^{Ah}b_W + e^{-Ah}b_E)\int_0^\tau \mu e^{F_m\mu}d\mu$$

$$= (e^{Ah}a_W + e^{-Ah}a_E)\frac{1}{F_m}(e^{F_m\tau} - 1) + (e^{Ah}b_W + e^{-Ah}b_E)\tau$$

$$*\frac{1}{F_m}\left(e^{F_m\tau} - \frac{e^{F_m\tau}}{F_m\tau} + \frac{1}{F_m\tau}\right),$$

with

$$F_m = \frac{A^2 + \lambda_m^2}{B},$$

and

$$\lambda_m = \frac{m - \frac{\pi}{2}}{h}.$$

Therefore,

$$\phi_P = \sum_{m=1}^\infty (-1)^{m+1}e^{-F_m\tau}\left[a_S E_{0m} + b_S h E_{1m} + (c_S - a_S)E_{2m}\right.$$

$$+\frac{\lambda_m}{BhF_m}\left[(e^{Ah}a_W + e^{-Ah}a_E)(e^{F_m\tau} - 1) + (e^{Ah}b_W\right.$$

$$\left.\left.+ e^{-Ah}b_E)\tau\left(e^{F_m\tau} - \frac{e^{F_m\tau}}{F_m\tau} + \frac{1}{F_m\tau}\right)\right]\right]$$

Define

$$P_i = \sum_{m=1}^\infty \frac{(-1)^{m+1}\lambda_m h e^{-F_m\tau}}{[(Ah)^2 + (\lambda_m h)^2]^i}, \quad i = 1,\ 2\ ,$$

and

$$Q_i = \sum_{m=1}^\infty \frac{(-1)^{m+1}\lambda_m h}{[(Ah)^2 + (\lambda_m h)^2]^i}, \quad i = 1,\ 2\ ,$$

so that

$$\sum_{m=1}^\infty (-1)^{m+1}e^{F_m\tau}E_{0m} = (e^{Ah} + e^{-Ah})P_1\ ,$$

$$\sum_{m=1}^\infty (-1)^{m+1}e^{F_m\tau}E_{1m} = 2Ah(e^{Ah} + e^{-Ah})P_2 - (e^{Ah} - e^{-Ah})P_1\ ,$$

$$\sum_{m=1}^\infty (-1)^{m+1}e^{F_m\tau}E_{2m} = (e^{Ah} + e^{-Ah})P_1\ ,$$

$$\sum_{m=1}^{\infty} (-1)^{m+1} e^{-F_m \tau} \frac{\lambda_m}{BhF_m} (e^{F_m \tau} - 1)$$

$$= \sum_{m=1}^{\infty} \frac{(-1)^{m+1} \lambda_m h}{(Ah)^2 + (\lambda_m h)^2} (1 - e^{-F_m \tau}) = Q_1 - P_1$$

$$\sum_{m=1}^{\infty} (-1)^{m+1} e^{-F_m \tau} \frac{\lambda_m}{BhF_m} (e^{F_m \tau} - \frac{e^{F_m \tau}}{F_m \tau} + \frac{1}{F_m \tau})$$

$$= \sum_{m=1}^{\infty} \frac{(-1)^{m+1} \lambda_m h}{(Ah)^2 + (\lambda_m h)^2} - \frac{Bh^2}{\tau} \sum_{m=1}^{\infty} \frac{(-1)^{m+1} \lambda_m h (1 - e^{-F_m \tau})}{[(Ah)^2 + (\lambda_m h)^2]^2}$$

$$= Q_1 - \frac{Bh^2}{\tau} (Q_2 - P_2).$$

With these relationships, the expression for ϕ_P becomes

$$\phi_P = a_S(e^{Ah} + e^{-Ah})P_1 + b_S h[2Ah(e^{Ah} + e^{-Ah})P_2 - (e^{Ah} - e^{-Ah})P_1]$$
$$+ (c_S - a_S)(e^{Ah} + e^{-Ah})P_1 + (e^{Ah} a_W + e^{-Ah} a_E)(Q_1 - P_1)$$
$$+ (e^{Ah} b_W + e^{-Ah} b_E)\tau \left[Q_1 + \frac{Bh^2}{\tau}(P_2 - Q_2) \right]$$

$$= \frac{1}{2}[\phi_{SE} - \phi_{SW} - \coth(Ah)(\phi_{SE} + \phi_{SW} - 2\phi_{SC})](4Ah\cosh(Ah)P_2$$
$$- 2\sinh(Ah)P_1) + \phi_{SC}(e^{Ah} + e^{-Ah})P_1 + (e^{Ah}\phi_{SW}$$
$$+ e^{-Ah}\phi_{SE})(Q_1 - P_1) + [e^{Ah}(\phi_{WC} - \phi_{SW}) + e^{-Ah}(\phi_{EC}$$
$$- \phi_{SE})] \left[Q_1 + \frac{Bh^2}{\tau}(P_2 - Q_2) \right]$$

$$= (2\cosh(Ah)\phi_{SC} - e^{Ah}\phi_{SW} - e^{-Ah}\phi_{SE})(2Ah\coth(Ah)P_2$$
$$- P_1) + 2\cosh(Ah)P_1\phi_{SC} + (e^{Ah}\phi_{SW} + e^{-Ah}\phi_{SE})(Q_1 - P_1)$$
$$+ [(e^{Ah}\phi_{WC} + e^{-Ah}\phi_{EC}) - (e^{Ah}\phi_{SW} + e^{Ah}\phi_{SE})] \left[Q_1 + \frac{Bh^2}{\tau}(P_2 - Q_2) \right]$$

$$= (e^{Ah}\phi_{SW} + e^{-Ah}\phi_{SE}) \left\{ P_1 - 2Ah\coth(Ah)P_2 + Q_1 - P_1 - Q_1 \right.$$
$$- \frac{Bh^2}{\tau}(P_2 - Q_2) + (e^{Ah}\phi_{WC} + e^{-Ah}\phi_{EC})(Q_1 + \frac{Bh^2}{\tau}(P_2$$
$$\left. - Q_2) \right\} + \phi_{SC}\{2\cosh(Ah)(2Ah\coth(Ah)P_2 - P_1 + P_1)\},$$

or

$$\phi_P = C_{WC}\phi_{WC} + C_{EC}\phi_{EC} + C_{SW}\phi_{SW} + C_{SE}\phi_{WSE} + C_{SC}\phi_{SC}, \qquad (A.46)$$

where

$$C_{WC} = e^{Ah} \left[Q_1 + \frac{Bh^2}{\tau}(P_2 - Q_2) \right], \qquad (A.47)$$

$$C_{EC} = e^{-2Ah}C_{WC}, \tag{A.48}$$

$$C_{SW} = e^{Ah}\left[\frac{Bh^2}{\tau}(Q_2 - P_2) - 2Ah\coth(Ah)P_2\right], \tag{A.49}$$

$$C_{SE} = e^{-2Ah}C_{SW} \quad \text{and} \tag{A.50}$$

$$C_{SC} = 4Ah\cosh(Ah)\coth(Ah)P_2. \tag{A.51}$$

The following closed form expressions for Q_1 and Q_2 are known from investigation

$$Q_1 = \frac{1}{e^{Ah} + e^{-Ah}},$$

$$Q_2 = \frac{e^{Ah} - e^{-Ah}}{2Ah(e^{Ah} + e^{-Ah})^2},$$

so that only one series summation for P_2 is required to calculate the FA coefficients shown in Equations (A.47) - (A.51).

Appendix B

The Two-Dimensional Case

The analytic solution for the linear, homogeneous PDE on a small, uniform FA element is derived in this appendix. The form of the two-dimensional transport equation is shown in Equation (B.1).

$$\tilde{\phi}_{xx} + \tilde{\phi}_{yy} = 2A\tilde{\phi}_x + 2B\tilde{\phi}_y \tag{B.1}$$

Boundary conditions must be specified on the small FA element in order for the problem to be well-posed. Boundary functions are fitted using the three nodal values on a side. Any function that solves Equation (B.1) would be a "natural" choice. However, certain of these functions have been found to given physically unrealistic solutions. The functions used in this solution process will include a constant, linear and exponential term. As an example, for the north boundary we assume

$$\tilde{\phi}_N(x) = a_N(e^{2Ax} - 1) + b_N x + c_N, \tag{B.2}$$

where

$$a_N = \frac{\tilde{\phi}_{NE} + \tilde{\phi}_{NW} - 2\tilde{\phi}_{NC}}{4\sinh^2(Ah)},$$

$$b_N = \frac{\tilde{\phi}_{NE} - \tilde{\phi}_{NW} - \coth(Ah)[\tilde{\phi}_{NE} + \tilde{\phi}_{NW} - 2\tilde{\phi}_{NC}]}{2h},$$

$$c_N = \tilde{\phi}_{NC},$$

and the other three boundary conditions for the south, east and west sides may be similarly approximated

$$\tilde{\phi}_S(x) = a_S(e^{2Ax} - 1) + b_S x + c_S, \tag{B.3}$$

$$\tilde{\phi}_E(x) = a_E(e^{2By} - 1) + b_E x + c_E, \tag{B.4}$$

and

$$\tilde{\phi}_W(x) = a_W(e^{2By} - 1) + b_W x + c_W. \tag{B.5}$$

Introduce the change of variable $\bar{\phi} = we^{Ax+By}$ and Equation (B.1) and boundary the conditions in Equations (B.2)-(B.5) become

$$w_{xx} + w_{yy} = (A^2 + B^2)w, \tag{B.6}$$

$$
\begin{aligned}
w(x, k) &= e^{-Bk}\left[a_N e^{Ax} + b_N x e^{-Ax} + (c_N - a_N)e^{-Ax}\right] \\
&= w_1(x), \\
w(x, -k) &= e^{Bk}\left[a_S e^{Ax} + b_S x e^{-Ax} + (c_S - a_S)e^{-Ax}\right] \\
&= w_2(x), \\
w(h, y) &= e^{-Ah}\left[a_E e^{By} + b_E y e^{-By} + (c_E - a_E)e^{-By}\right] \\
&= w_3(y), \\
w(-h, y) &= e^{-Ah}\left[a_W e^{By} + b_W y e^{-By} + (c_W - a_W)e^{-By}\right] \\
&= w_4(y).
\end{aligned}
\tag{B.7}
$$

The solution procedure is simplified by dividing the problem into four sub-problems, each problem having one non-homogeneous boundary condition and three homogeneous boundary conditions.

Problem I

$$
\left\{
\begin{aligned}
& w^N_{xx} + w^N_{yy} = (A^2 + B^2)w^N , \\
& w^N(x, k) = w_1(x) , \\
& w^N(x, -k) = w^N(h, y) = w^N(-h, y) = 0 .
\end{aligned}
\right.
\tag{B.8}
$$

Problem II

$$
\left\{
\begin{aligned}
& w^S_{xx} + w^S_{yy} = (A^2 + B^2)w^S , \\
& w^S(x, -k) = w_2(x) , \\
& w^S(x, k) = w^S(h, y) = w^S(-h, y) = 0 .
\end{aligned}
\right.
\tag{B.9}
$$

Problem III

$$
\left\{
\begin{aligned}
& w^E_{xx} + w^E_{yy} = (A^2 + B^2)w^E , \\
& w^E(h, y) = w_3(y) , \\
& w^E(-h, y) = w^E(x, k) = w^E(x, -k) = 0 .
\end{aligned}
\right.
\tag{B.10}
$$

Problem IV

$$
\left\{
\begin{aligned}
& w^W_{xx} + w^W_{yy} = (A^2 + B^2)w^W , \\
& w^W(-h, y) = w_4(y) , \\
& w^W(h, y) = w^W(x, k) = w^W(x, -k) = 0 .
\end{aligned}
\right.
\tag{B.11}
$$

The superposition principle for linear PDEs is employed to give

$$w = w^E + w^W + w^N + w^S, \tag{B.12}$$

where w^N, w^S, w^E and w^W solve Problem I through IV, respectively.

The smaller problems can be solved analytically using separation of variables. For example, let $w^N = X(x)Y(y)$ in Problem I, so that the linear PDE is separated into two ordinary differential equations. The first one is

$$X'' + \lambda^2 X = 0, \tag{B.13}$$

with boundary conditions

$$X(-h) = X(h) = 0.$$

The two boundary conditions are used to find the eigenvalues $\lambda_n = \frac{n\pi}{2h}$ ($n = 1, 2, 3, \ldots$) and eigenfunction solutions $\sin(\lambda_n[x + h])$ to Equation (B.13).

The second ordinary differential equation is

$$Y'' - (A^2 + B^2 + \lambda_n^2)Y = 0, \qquad (B.14)$$

with boundary condition

$$Y(-k) = 0.$$

Using the eigenvalues found for Equation (B.13), the eigenfunction solutions to Equation (B.14) are $\sinh(\mu_n[y + k])$ with $\mu_n = \sqrt{A^2 + B^2 + \lambda^2}$ ($n = 1, 2, 3, \ldots$). Combining the eigenfunction solutions gives the formal solution to Equation (B.8)

$$w^N(x, y) = \sum_{n=1}^{\infty} A_n \sinh(\mu_n[y + k]) \sin(\lambda_n[x + h]). \qquad (B.15)$$

The coefficients in Equation (B.15) are easily obtained by applying the non-homogeneous boundary condition in Problem I. That is,

$$w^N(x, k) = w_1(x) = \sum_{n=1}^{\infty} A_n \sinh(2\mu_n k) \sin(\lambda_n[x + h]), \qquad (B.16)$$

where

$$
\begin{aligned}
A_n &= \frac{1}{h \cdot \sinh(2k\mu_n)} \int_{-h}^{h} w_1(x) \sin(\lambda_n[x + h]) dx \\
&= \frac{e^{-Bk}}{\sinh(2k\mu_n} [a_N e_{0n} + b_N h e_{1n} + (c_N - a_N)e_{2n}], \qquad (B.17)
\end{aligned}
$$

with

$$
\begin{aligned}
e_{0n} &= \frac{1}{h} \int_{-h}^{h} e^{Ax} \sin(\lambda_n[x + h]) dx \\
&= \frac{\lambda_n h}{(Ah)^2 + (\lambda_n h)^2} [e^{-Ah} - (-1)^n e^{Ah}], \qquad (B.18)
\end{aligned}
$$

$$
\begin{aligned}
e_{1n} &= \frac{1}{h^2} \int_{-h}^{h} x e^{-Ax} \sin(\lambda_n[x + h]) dx, \\
&= \frac{2(Ah)(\lambda_n h)}{[(Ah)^2 + (\lambda_n h)^2]^2} [e^{Ah} - (-1)^n e^{-Ah}] \\
&\quad - \frac{\lambda_n h}{(Ah)^2 + (\lambda_n h)^2} [e^{Ah} - (-1)^n e^{-Ah}], \qquad (B.19)
\end{aligned}
$$

and

$$
\begin{aligned}
e_{2n} &= \frac{1}{h} \int_{-h}^{h} e^{-Ax} \sin(\lambda_n[x+h]) dx \\
&= \frac{\lambda_n h}{(Ah)^2 + (\lambda_n h)^2} \left[e^{Ah} - (-1)^n e^{-Ah} \right].
\end{aligned} \tag{B.20}
$$

When the solution given by Equation (B.15) in evaluated at the interior node P, whose element coordinates are $(0,0)$, the resulting formula gives the value at node P in terms of the eight surrounding nodes. That is,

$$
w_P^N = w^N(0,0) = \sum_{n=1}^{\infty} A_n \sinh(\mu_n k) \sin(\lambda_n h). \tag{B.21}
$$

Since

$$
\sin(\lambda_n h) = \sin\left(\frac{n\pi}{2}\right) = \left\{ \begin{array}{ll} 0, & n = 2m \\ -(-1)^m, & n = 2m - 1 \end{array} \right. \quad m = 1, 2, 3, , \ldots \tag{B.22}
$$

Equation (B.21) can be further simplified to

$$
\begin{aligned}
w_P^N &= \sum_{m=1}^{\infty} \frac{-(-1)^m e^{-Bk} \sin(h\lambda_m k)}{\sinh(2\mu_m k)} \left[a_N e_{0m} + b_N h e_{1m} + (c_N - a_N) e_{2m} \right] \\
&= e^{-Bk} \sum_{m=1}^{\infty} \frac{-(-1)^m}{2 \cosh(\mu_m k)} \left[a_N e_{0m} + b_N h e_{1m} + (C_N - a_N) e_{2m} \right].
\end{aligned} \tag{B.23}
$$

Define

$$
E_i = \sum_{m=1}^{\infty} \frac{-(-1)^m \lambda_m h}{\left[(Ah)^2 + (\lambda_m h)^2 \right]^i \cosh(\mu_m k)}, \quad i = 1, 2. \tag{B.24}
$$

Then

$$
\begin{aligned}
\sum_{m=1}^{\infty} \frac{-(-1)^m}{\cosh(\mu_m k)} e_{0m} &= \left(e^{Ah} + e^{-Ah} \right) \sum_{m=1}^{\infty} \frac{-(-1)^m \lambda_m h}{\left[(Ah)^2 + (\lambda_m h)^2 \right] \cosh(\mu_m k)} \\
&= 2 \cosh(Ah) \cdot E_1,
\end{aligned} \tag{B.25}
$$

$$
\begin{aligned}
\sum_{m=1}^{\infty} \frac{-(-1)^m}{\cosh(\mu_m k)} e_{1m} &= -\left(e^{Ah} - e^{-Ah} \right) \sum_{m=1}^{\infty} \frac{-(-1)^m \lambda_m h}{\left[(Ah)^2 + (\lambda_m h)^2 \right] \cosh(\mu_m k)} \\
&\quad + 2(Ah) \left(e^{Ah} + e^{-Ah} \right) \sum_{m=1}^{\infty} \frac{-(-1)^m \lambda_m h}{\left[(Ah)^2 + (\lambda_m h)^2 \right]^2 \cosh(\mu_m k)} \\
&= 4Ah \cdot \cosh(Ah) \cdot E_2 - 2 \sinh(Ah) \cdot E_1,
\end{aligned} \tag{B.26}
$$

$$
\begin{aligned}
\sum_{m=1}^{\infty} \frac{-(-1)^m}{\cosh(\mu_m k)} e_{1m} &= \left(e^{Ah} + e^{-Ah} \right) \sum_{m=1}^{\infty} \frac{-(-1)^m \lambda_m h}{\left[(Ah)^2 + (\lambda_m h)^2 \right] \cosh(\mu_m k)} \\
&= 2 \cosh(Ah) \cdot E_1.
\end{aligned} \tag{B.27}
$$

Substituting for a_N, b_N and c_N in Equation (B.23), the solution at node P becomes

$$
\begin{aligned}
\tilde{\phi}_P^N &= w_P^N \\
&= \frac{1}{2}\left\{\frac{1}{2}\left[\tilde{\phi}_{NE} - \tilde{\phi}_{NW} - \coth(Ah)(\tilde{\phi}_{NE} + \tilde{\phi}_{NW} - 2\tilde{\phi}_{NC})\right]\right. \\
&\qquad \left.\left[4Ah\cosh(Ah)E_2 - 2\sinh(Ah)E_1\right) + \tilde{\phi}_{NC}(2\cosh(Ah)E_2)\right]\right\} \\
&= e^{-Bk}\left\{(2\cosh(Ah)\tilde{\phi}_{NC} - e^{Ah}\tilde{\phi}_{NW} - e^{Ah}\tilde{\phi}_{NE})\right. \\
&\qquad \left.\left(Ah\coth(Ah)E_2 - \frac{1}{2}E_1\right) + (\cosh(Ah)E_1)\tilde{\phi}_{NC}\right\} \\
&= e^{-Bk}\left[\left(\frac{1}{2}E_1 - Ah\coth(Ah)E_2\right)\left(e^{-Ah}\tilde{\phi}_{NE} + e^{Ah}\tilde{\phi}_{NW}\right)\right. \\
&\qquad \left. + (2Ah\cosh(Ah)\coth(Ah)E_2)\tilde{\phi}_{NC}\right].
\end{aligned} \tag{B.28}
$$

Using a similar procedure, formulas for $\tilde{\phi}_P^S$, $\tilde{\phi}_P^E$ and $\tilde{\phi}_P^W$ may be derived that incorporate nodal values of $\tilde{\phi}$ on the south, east and west boundaries, respectively.

$$
\begin{aligned}
\tilde{\phi}_P^S &= e^{Bk}\left[\left(\frac{1}{2}E_1 - Ah\coth(Ah)E_2\right)\left(e^{-Ah}\tilde{\phi}_{SE} + e^{Ah}\tilde{\phi}_{SW}\right)\right. \\
&\qquad \left. + (2Ah\cosh(Ah)\coth(Ah)E_2)\tilde{\phi}_{SC}\right], \tag{B.29} \\
\tilde{\phi}_P^E &= e^{-Ah}\left[\left(\frac{1}{2}E_1' - Bk\coth(Bk)E_2'\right)\left(e^{-Bk}\tilde{\phi}_{NE} + e^{Bk}\tilde{\phi}_{SE}\right)\right. \\
&\qquad \left. + (2Bk\cosh(Bk)\coth(Bk)E_2')\tilde{\phi}_{EC}\right], \tag{B.30} \\
\tilde{\phi}_P^W &= e^{Ah}\left[\left(\frac{1}{2}E_1' - Bk\coth(Bk)E_2'\right)\left(e^{-Bk}\tilde{\phi}_{NW} + e^{Bk}\tilde{\phi}_{SW}\right)\right. \\
&\qquad \left. + (2Bk\cosh(Bk)\coth(Bk)E_2')\tilde{\phi}_{WC}\right], \tag{B.31}
\end{aligned}
$$

where

$$
E_i' = \sum_{m=1}^{\infty} \frac{-(-1)^m \lambda_m' h}{[(Ah)^2 + (\lambda_m' h)^2]^i \cosh(\mu_m' k)}, \quad i = 1, 2. \tag{B.32}
$$

and

$$
\lambda_m' = \frac{(2m-1)\pi}{2k}, \quad \mu_m' = \sqrt{A^2 + B^2 + \lambda_m'^2}.
$$

The 9-point FA formula giving the center nodal value $\tilde{\phi}_P$ in terms of the neighboring nodal values is obtained by superimposing the four solutions of Problems I-IV. That is,

$$
\tilde{\phi}_P = \tilde{\phi}_P^N + \tilde{\phi}S_P + \tilde{\phi}_P^E + \tilde{\phi}_P^W
$$

$$= \left(e^{-Ah-Bk}\tilde{\phi}_{NE} + e^{Ah-Bk}\tilde{\phi}_{NW} + e^{-Ah+Bk}\tilde{\phi}_{SE}\right.$$

$$\left. + e^{Ah+Bk}\tilde{\phi}_{SW}\right)\left[\frac{1}{2}(E_1 + E_1') - Ah\coth(Ah)E_2 - Bk\coth(Bk)E_2'\right]$$

$$+ 2Ah\cosh(Ah)\coth(Ah)E_2\left(e^{-Bk}\tilde{\phi}_{NC} + e^{Bk}\tilde{\phi}_{SC}\right)$$

$$+ 2Bk\cosh(Bk)\coth(Bk)E_2'\left(e^{-Ah}\tilde{\phi}_{EC} + e^{Ah}\tilde{\phi}_{WC}\right) \tag{B.33}$$

Since $\tilde{\phi} = 1$ and $\tilde{\phi} = -Bx + Ay$ are two particular solutions of the convective transport equation, and both of them may be represented by the linear and exponential boundary functions given by Equations (B.7), it is useful to use these exact solutions to obtain relationships between the series summations represented by E_1, E_1', E_2 and E_2'.

(a) $\tilde{\phi} = 1$

Since $\tilde{\phi} = 1$ is an analytic solution of Equation (B.1), and may be represented by boundary functions given in Equations (B.7), it should satisfy the FA formula in Equation (B.33) as well. Substituting

$$\tilde{\phi} = \tilde{\phi}_{EC} = \tilde{\phi}_{WC} = \tilde{\phi}_{NC} = \tilde{\phi}_{SC} = \tilde{\phi}_{NE} = \tilde{\phi}_{NW} = \tilde{\phi}_{SE} = \tilde{\phi}_{SW} = 1$$

into Equation (B.33), an analytic relationship between E_1 and E_1' can be obtained.

$$\tilde{\phi}_P = 1$$

$$= (e^{-Ah} + e^{Ah})(e^{-Bk} + e^{Bk})\left[\frac{1}{2}(E_1 + E_1') - Ah\coth(Ah)E_2\right.$$

$$\left. - Bk\coth(Bk)E_2'\right] + 2Ah\cosh(Ah)\coth(Ah)E_2\left(e^{-Bk} + e^{Bk}\right)$$

$$+ 2Bk\cosh(Bk)\coth(Bk)E_2'\left(e^{-Ah} + e^{Ah}\right)$$

$$= 2\cosh(Ah)\cosh(Bk)\left(E_1 + E_1'\right). \tag{B.34}$$

Solving for $E_1 + E_1'$ gives

$$E_1 + E_1' = \frac{1}{2\cosh(Ah)\cosh(Bk)} \tag{B.35}$$

(b) $\tilde{\phi} = -Bx + Ay$

Similarly, $\tilde{\phi} = -Bx + Ay$ satisfies the Equation (B.1), and may be represented in terms of the boundary functions outlined in Equations (B.7). Therefore, it should be represented by the FA formula given in Equation (B.33). Substituting $\tilde{\phi}_P = \tilde{\phi}(0,0) = -B\cdot 0 + A\cdot 0 = 0, \tilde{\phi}_{EC} = -Bh, \tilde{\phi}_{NC} = Ak$, etc. into Equation

(B.33) will yield an analytic relationship between E_2 and E_2'. That is,

$$
\begin{aligned}
\tilde{\phi} &= 0 \\
&= \left[Ak\left(e^{-Ah-Bk} + e^{Ah-Bk} - e^{-Ah+Bk} - e^{Ah+Bk}\right) + Bh\left(e^{Ah-Bk} + \right.\right. \\
&\quad \left.\left. e^{Ah+Bk} - e^{-Ah-Bk} - e^{-Ah+Bk}\right)\right]\left[\frac{1}{2}\left(E_1 + E_1'\right) - Ah\coth(Ah)E_2\right. \\
&\quad \left. - Bk\coth(Bk)E_2'\right] + 2Ah\cosh(Ah)\coth(Ah)E_2 Ak\left(e^{-Bk}\right. \\
&\quad \left. - e^{Bk}\right) + 2Bk\cosh(Bk)\coth(Bk)E_2'Bk\left(e^{Ah} - e^{-Ah}\right) \\
&= \frac{1}{\cosh(Ah)\cosh(Bk)}(Bh\sinh(Ah)\cosh(Bk) - Ak\cosh(Ah)\sinh(Bk) \\
&\quad + 4(Ak)(Bk)\cosh(Ah)\sinh(Bk)\coth(Bk)E_2' \\
&\quad - 4(Ah)(Bh)\cosh(Bk)\sinh(Ah)\coth(Ah)E_2,
\end{aligned}
$$

or

$$
E_2' = \left(\frac{h}{k}\right)^2 E_2 + \frac{Ak\tanh(Bk) - Bh\tanh(Ah)}{4AkBk\cosh(Ah)\cosh(Bk)} \tag{B.36}
$$

Now, define

$$
E = \frac{1}{2}\left(E_1 + E_1'\right) - Ah\coth(Ah)E_2 - Bk\coth(Bk)E_2', \tag{B.37}
$$

$$
EA = 2Ah\cosh(Ah)\cdot\coth(Ah)E_2, \tag{B.38}
$$

$$
EB = 2Bk\cosh(Bk)\cdot\coth(Bk)E_2'. \tag{B.39}
$$

Then, the 9-point FA formula may be summarized as

$$
\begin{aligned}
\tilde{\phi}_P &= C_{NE}\tilde{\phi}_{NE} + C_{NW}\tilde{\phi}_{NW} + C_{SE}\tilde{\phi}_{SE} + C_{SW}\tilde{\phi}_{SW} + C_{NC}\tilde{\phi}_{NC} \\
&\quad + C_{EC}\tilde{\phi}_{EC} + C_{SC}\tilde{\phi}_{SC} + C_{WC}\tilde{\phi}_{WC}, \tag{B.40}
\end{aligned}
$$

where

$$
\begin{array}{ll}
C_{EC} = EBe^{-Ah}, & C_{NE} = Ee^{-Ah-Bk}, \\
C_{WC} = EBe^{Ah}, & C_{NW} = Ee^{Ah-Bk}, \\
C_{SC} = EAe^{Bk}, & C_{SE} = Ee^{-Ah+Bk}, \\
C_{NC} = EAe^{-Bk}, & C_{SW} = Ee^{Ah+Bk}. \tag{B.41}
\end{array}
$$

After applying the analytic relationships in Equations (B.35) and (B.39), E_2 is the only series summation that needs to be approximated.

The FA solution for the unsteady, non-homogeneous convective transport equation on the FA element is obtained from Equation (B.40) by substituting for $\tilde{\phi}$ in terms of ϕ. The result is

$$
\begin{aligned}
\phi_P &= C_{NE}\phi_{NE} + C_{NW}\phi_{NW} + C_{SE}\phi_{SE} + C_{SW}\phi_{SW} + C_{NC}\phi_{NC} \\
&\quad + C_{EC}\phi_{EC} + C_{SC}\phi_{SC} + C_{WC}\phi_{WC} + C_P g, \tag{B.42}
\end{aligned}
$$

where

$$C_P = \frac{1}{2(A^2 + B^2)} [Ah (C_{NW} + C_{SW} + C_{WC} - C_{NE} - C_{SE} - C_{EC})$$
$$+ Bk (C_{SE} + C_{SW} + C_{SC} - C_{NE} - C_{NW} - C_{NC})] \qquad \text{(B.43)}$$

Recall that $g = f_P - R\frac{(\phi_P^m - \phi^{m-1})}{\tau}$, so that substituting for g will yield a 10-point (counting ϕ_P^{m-1}) FA formula for the unsteady, nonhomogeneous convective transport equation.

$$\phi_P = \frac{1}{1 + \frac{R}{\tau}C_P} [C_{NE}\phi_{NE} + C_{NW}\phi_{NW} + C_{SE}\phi_{SE} + C_{SW}\phi_{SW}$$

$$+ C_{NC}\phi_{NC} + C_{EC}\phi_{EC} + C_{SC}\phi_{SC} + C_{WC}\phi_{WC} + \frac{R}{\tau}C_P\phi_P^{m-1}$$

$$- - C_P f_P], \qquad \text{(B.44)}$$

where

$$f_P = f^{m-1}(x, y, \phi_j)\big|_P,$$

and the nodal values without a superscript denote those values on the $t = m$ time plane, while the superscript $m - 1$ means the value is taken on the $t = m - 1$ time plane.

Bibliography

[1] First International Symposium on Finite Element Methods in Flow Problems. Huntsville, 1974. University of Alabama in Huntsville Press.

[2] S. Abdallah. Numerical Solutions for the Incompressible Navier-Stokes Equations in Primitive Variables Using a Non-staggered Grid, II. *Journal of Computational Physics*, 70:193–202, 1987.

[3] S. Abdallah. Numerical Solutions for the Pressure Poisson Equation with Neumann Boundary Conditions Using a Non-staggered Grid, I. *Journal of Computational Physics*, 70:182–192, 1987.

[4] M. J. Aftosmis, J. E. Melton, and M. J. Berger. Adaptation and Surface Modeling for Cartesian Mesh Methods. *AIAA*, 95:1725, 1996.

[5] R. K. Agarwal. A Third-Order-Accurate Upwind Scheme for Navier-Stokes Solutions at High Reynolds Numbers . *AIAA No. 81-0012*, 1981.

[6] H. Aksoy and C. J. Chen. Finite Analytic Numerical Solution of Navier Stokes Equations Using Non-staggered Grids. *Numerical Methods in Thermal Problems*, 6:1633–1643, 1989.

[7] H. Aksoy and C.J. Chen. Numerical Solution of Navier-Stokes Equations with Nonstaggered Grids Using Finite Analytic Method. *Numerical Heat Transfer, Part B*, 21:287–306, 1992.

[8] D.A. Anderson, J.C. Tannehill, and R.H. Pletcher. *Computational Fluid Mechanics and Heat Transfer*. Hemisphere Publishing Co., 1984.

[9] S.P. Arya. *Introduction to Micrometeorology*. Academic Press, New York, 1988.

[10] J. F. Baker. *Finite Element Computational Fluid Mechanics*. McGraw-Hill Co., New York, 1983.

[11] T. J. Baker. Prospects and Expectations for Unstructured Methods. In *The Surface Modeling, Grid Generation and Related Issues in Computational Fluid Dynamics Workshop*, pages 273–287, NASA Lewis Research Center, Cleveland, OH, May 1995. NASA Conference Publication 3291.

[12] R. Balu and C. Unnikrishnan. An Algebraic Grid Generation Technique for Multi Body Launch Vehicle Configurations. In B.K. Soni, J.F. Thompson, Hauser J., and Eiseman P., editors, *5th International Conference on Numerical Grid Generation in Computational Field Simulations*, pages 901–910, Mississippi, USA, April 1-April 5 1996.

[13] W. D. Barfield. An optimal mesh generator for langrangian hydrodynamic calculations in two space dimensions. *Journal of Computational Physics*, 6:417–429, 1970.

[14] E. Bauer. Dispersion of tracers in the atmosphere and ocean: Survey and comparison of experimental data. *J. Geophys. Rev.*, 79:789–795, 1974.

[15] J. Bear. *Hydraulics of Groundwater Water*. McGraw-Hill Co., New York, 1979.

[16] D. M. Belk. Automated Assembly of Structured Grids for Moving Body Problems . In *AIAA-95-1680-CP*, pages 381–390, 1995.

[17] D. M. Belk. The Role of Overset Grids in the Development of the General Purpose CFD Code. In *The Surface Modeling, Grid Generation and Related Issues in Computational Fluid Dynamics Workshop*, page p. 193, NASA Lewis Research Center, Cleveland, OH, May 1995. NASA Conference Publication 3291.

[18] R. A. Bernatz. *Development of the Finite Analytic Method for Turbulent Forced and Free Convection*. PhD thesis, The University of Iowa, 1991.

[19] R. A. Bernatz, C. J. Chen, and H. M. Cekirge. Finite Analytical Solution of a Two-Dimensional Sea Breeze on a Regular Grid. *Mathematical Computational Modeling*, 19:71–87, 1994.

[20] R. D. Blevins. *Flow Induced Vibration*. Van Norstand Reinhold Company, 1977.

[21] R. H. Bravo and C. J. Chen. Heat Flow Characteristics of a Finned Heat Exchanger. In *Proceedings of ASME Heat Transfer in Convective Flows.*, pages 185–190, Aug. 6-9 1989.

[22] R. H. Bravo, A. Sanchez, C. J. Chen, and T. F. Smith. Convection and Radiation Heat Transfer Analysis in Three-Dimensional Arrays of Electronics Chips. In *3rd Intersociety Conference on Thermal Phenomena in Electronic Systems (ITHERM)*, Austin, Texas, Feb. 5-7 1992.

[23] William L. Briggs. *A Multigrid Tutorial*. SIAM, Philadelphia, 1987.

[24] James Ward Brown and Ruel V. Churchill. *Fourier Series and Boundary Value Problems*. McGraw-Hill, Inc., New York, 5 edition, 1993.

[25] A. H. Bruneau and P. Fabrie. Effective Downstream Boundary Conditions for Incompressible Navier-Stokes Equations. *International Journal for Numerical Method in Fluids*, pages 693–705, 1994.

[26] K. D. Carlson. Finite Analytic Numerical Simulation of Two-Dimensional Flow Problems with Irregular Boundaries in Cartesian Coordinates. Master's thesis, University of Iowa, 1993.

[27] K. D. Carlson. *Numerical Simulation of Fluid Flow and Conjugate Heat Transfer for Complex Geometries*. PhD thesis, Department of Mechanical Engineering, The Florida A & M University - Florida State University College of Engineering, 1997.

[28] K. D. Carlson, W. L. Lin, and C. J. Chen. Pressure Boundary Conditions of Incompressible Fluid Flows with Conjugate Heat Transfer on Non-staggered Grids, Part II: Applications. *Journal of Numerical Heat Transfer*, 1997. In Press.

[29] K. D. Carlson, W. L. Lin, and C. J. Chen. Numerical Modeling of Conjugate Heat Transfer on Complex Geometries with Diagonal Cartesian Method, Part II: Applications. *Journal of Heat Transfer*, 121(2):261–267, 1999.

[30] José E. Castillo, editor. *Mathematical Aspects of Grid Generation*. SIAM, 1991.

[31] J. B. Cazalbou, M. Braza, and H. H. Mihn. A Numerical Method for Computing Three-Dimensional Navier-Stokes Equations Applied to Cubic Cavity Flows with Heat Transfer. In C. Taylor, J. A. Johnson, and W. R. Smith, editors, *Numerical Methods in Laminar and Turbulent Flows*, pages 786–797, Swansea, U. K., 1983. Pineridge Press.

[32] M. A. Celia, L. R. Ahuja, and G. F. Pinder. Orthogonal Collocation and Alternating-Direction Procedures for Unsaturated Flow Problems. *Advances in Water Resources*, 10:178–187, 1987.

[33] J. C. Chai, H. S. Lee, and S. V. Patankar. Treatment of Irregular Geometries Using A Cartesian Coordinates Finite-Volume Radiation Heat Transfer Procedure. *Numerical Heat Transfer,Part B*, 26:179–197, 1994.

[34] C. J. Chen and R. H. Bravo. Heat Transfer Study of Staggered Thin Rectangular Blocks in a Channel Flow. *ASME Transaction, Journal of Electronic Packaging*, 113:294–300, 1991.

[35] C. J. Chen, R. H. Bravo, H. C. Chen, and Z. Xu. Accurate Discretization of Incompressible Three-Dimensional Navier-Stokes Equations. *J. of Numerical Heat Transfer*, 27:4:371–392, 1995.

[36] C. J. Chen and L. S. Chang. Prediction of Turbulent Flows in Rectangular Cavity with $k - \epsilon$–A and $k - \epsilon$–E Models. In C. J. Chen, L. D. Chen, and F. H. Holly, editors, *2nd International Symposium on Refined Flow Modeling and Turbulent Measurements*, pages 611–620, Iowa City, Iowa, 1985. Hemisphere Pub. Corp.

[37] C. J. Chen and H. C. Chen. Development of Finite Analytic Numerical Method for Unsteady Two-Dimensional Navier-Stokes Equations. *J. of Computational Physics*, 53:2:209–226, 1984.

[38] C. J. Chen and Bravo R. H. Heat Transfer Study of Staggered Thin Rectangular Blocks in a Channel Flow . In *AICHE Symposium Series*, volume 269:85, pages 442–447, 1989.

[39] C. J. Chen, Y. S. Haik, and R. H. Bravo. Numerical Visualization of Heat Transfer Characteristics of a Heated Rectangular Block in Channel Flow. In *National Heat Transfer Conference on Heat Transfer, Measurement, Analysis and Flow Visualization*, volume 112, pages 149–154, Philadelphia, PA, August 6-9 1989.

[40] C. J. Chen, H. Naseri-Neshat, and K. S. Ho. Finite Analytic Numerical Solution of Heat Transfer in Two-Dimensional Cavity Flow. *Numerical Heat Transfer*, 4:179–197, 1981.

[41] C. J. Chen and K. Obasih. Finite Analytic Numerical Solution of Heat Transfer and Flow Past a Square Channel Cavity. In *7th Int. Heat Transfer Conf., paper no. F65 82-IHTC-43*, pages 1–6, Munich, Germany, 1982.

[42] C. J. Chen, M. Z. Sheikholeslami, and R. B. Bhiladvala. Finite Analytic Numerical Method for Linear and Nonlinear Ordinary Differential Equations. In *8th International Conference on Computing Method in Applied Science and Engineering*, Versailles, France, 1987.

[43] C. J. Chen and K. Singh. Prediction of buoyant free shear flows by k-ϵ model based on two turbulence scale concept. In *Proceedings of the International Symposium on Buoyant Flows*, pages 26–36, Athens, Greece, Sept. 1986.

[44] C. J. Chen and K. Singh. Development of a two-scale turbulence model and prediction of buoyant shear flows. In *AIAA/ASME Thermophysics and Heat Transfer Conference*, 1990.

[45] C. J. Chen and V. Talaie. Finite Analytic Numerical Solutions of Laminar Natural Convection in Two-Dimensional Inclined Rectangular Enclosures. In *1985 National Heat Transfer Conf., ASME paper 85-HT-10*, Denver, CO, 1985.

[46] C. J. Chen and T. S. Wung. Finite Analytic Solutions of Convective Heat Transfer for Tube Arrays in Crossflow:part II–Heat Transfer Analysis. In

25th National Heat Transfer Conference, ASME paper, Houston, Texas, 1988.

[47] C. J. Chen and Y. H. Yoon. Prediction of Turbulent Heat Transfer in Flow Past a Cylindrical Cavity. In B. F. Armaly and L. S. Yao, editors, *Symposium on Mixed Convective Heat Transfer, 1985 Winter Annual Meeting, ASME 85-HTD*, volume 53, pages 1–8, Miami, FL, 1985.

[48] C. J. Chen, C. H. Yu, and K. B. Chandran. Finite Analytic Numerical Method and Fluid Dynamics of Disc Type Valves. In R. L. Spilker and B. R. Simon, editors, *Computational Methods in Bioengineering, ASME Winter Annual Meeting*, volume 9, pages 347–358, 1988.

[49] C. J. Chen, C. H. Yu, and K. B. Chandran. Steady Turbulent Flow Through a Disc-Type Valve. I: Finite Analytic Solution. *Journal of Engineering Mechanics*, 114(5):777–796, 1988.

[50] C. J. Chen, C. H. Yu, and K. B. Chandran. Finite Analytic Numerical Solution of Unsteady Laminar Flow Past Disc-Values. *J. of Engineering Mechanics*, 8:113:1147–1162, Aug., 1987.

[51] Ching Jen Chen and Sheng Yuh. Jaw. *Fundamentals of Turbulence Modeling*. Taylor & Francis, 1998.

[52] H. C. Chen and W. S. Cheng. Finite Analytic Prediction of Turbulent Flow Past an Inclined Cylinder. In *Third Symposium Numerical and Physical Aspects Aerodynamics Flows*, pages 3–27, Long Beach, California State University, 1985.

[53] H. C. Chen and V. C. Patel. Near-wall turbulence models for complex flows including separation. *AIAA Journal*, 26(6):641–648, June 1988.

[54] W. C. Chen and C. J. Chen. Prediction of Supersonic Oblique Shock Wave in Arbitrary Internal Passage by Method of Characteristics. In *Proceedings of the 5th International Conference on Numerical Methods in Laminar and Turbulent Flow*, pages 1009–1020, Montreal, Canada, July 6-10 1987.

[55] S. K. Choi and C. J. Chen. Finite Analytic Numerical Solution of Turbulence Flow Past Axisymmetric Bodies by Zonal Modeling Approach. In *ASME Winter Annual Meeting*, Chicago, Illinois, November 28- December 2 1988.

[56] S. K. Choi and C. J. Chen. A navier-stokes numerical analysis of a complete turbulent flow past finite axisymmetric bodies. *AIAA Journal*, 29(6):998–1001, 1991.

[57] B. A. Cipra. A Rapid-Deployment Force For CFD: Cartesian Grids. *SIAM(Society for Industrial and Applied Mathematics) News*, 28:1–4, 1995.

[58] Gianni Comini, Stefano Del Giudice, and Carlo Nonino. *Finite Element Analysis in Heat Transfer: Basic Formulation and Linear Problems*. Taylor & Francis, 1994.

[59] I. G. Currie. *Fundamental Mechanics of Fluids*. McGraw-Hill, 1974.

[60] W. Dai and C.J. Chen. Finite analytic scheme for viscous bergers' equation. In *ASCE Engineering Mechanics Specialty Conference*, Columbus, Ohio, May 19-22 1991.

[61] A. W. Date. Solution of Navier-Stokes Equations on Non-staggered Grid. *Int. J. Heat Mass Transfer*, 7:1913–1922, 1993.

[62] M. K. Denham and M. A. Patrick. Laminar flow over a downstream facing step in a two-dimensional flow channel. *Tran. Instit. Chem. Engrs.*, 52:361–367, 1974.

[63] J. P. Doormaal and G. D. Raithby. Enhancements of the SIMPLE Method for Predicting Incompressible Fluid Flows. *Numerical Heat Transfer*, 7:147–163, 1984.

[64] A. J. Dyer. A review of flux-profile relationships. *Bound.-Layer Meteor.*, 7:363–372, 1974.

[65] M.A. Estoque. The sea breeze as a function of the prevailing synoptic situation. *J.Atmos. Sci.*, 19:244–250, 1962.

[66] B.A. Finlayson. *The Method of Weighted Residuals and Variational Techniques*. Academic Press, New York, 1972.

[67] E.L. Fisher. An observational study of the sea breeze. *J. of Meteor.*, 17:645–660, 1960.

[68] Francis J. Flanigan. *Complex Variables: Harmonic and Analytic Functions*. Dover Publications, Inc., New York, 1983.

[69] U. Frisch, P. L. Sulem, and M. Nelkim. A simple dynamical model of intermittent fully developed turbulence. *Journal of Fluid Mechanics*, 87:719–736, 1978.

[70] J.A. Frizzola and E.L. Fisher. A series of sea breeze observations in the new york city area. *J. of Climate Appl. Meteor.*, 2:722–739, 1963.

[71] P. R. Garabedian. *Partial Differential Equations*. John Wiley & Sons, Inc., 1967.

[72] J. Girard and R. Curlet. Etude des courants de recirculation dans une cavite. Technical report, Institute De Mecanique De Grenoble, Grenoble, France, 1975.

[73] K.. Goda. A Multistep Technique with Implicit Difference Schemes for Calculating Two- or Three-Dimensional Cavity Flow. *Journal of Computational Physics*, 30:76–95, 1979.

[74] D. Grand. *Contribution a L'Etude des courants recirculation*. PhD thesis, Universite de Grenoble, 17 April 1975.

[75] P. M. Gresho. Incompressible Fluid Dynamics: Some Fundamental Formulation Issues. *Ann. Rev. Fluid Mech.*, 23:413–453, 1991.

[76] P. M. Gresho. Some Current CFD Issues Relevant to the Incompressible Navier-Stokes Equations. *Computer Methods in Applied Mechanics and Engineering*, 87:201–252, 1991.

[77] P. M. Gresho and R. L. Sani. On Pressure Boundary Conditions for the Incompressible Navier-Stokes Equations. *International Journal for Numerical Methods in Fluids*, 7:1111–1145, 1987.

[78] S.E. Gryning. The oresund experiment - a nordic mesoscale dispersion experiment over a land-water area. *Bull. Am. Meteorol. Soc.*, 66:403–407, 1985.

[79] K. Gustafson and K. Halasi. Vortex Dynamics of Cavity Flows. *Journal of Computational Physics*, 64:279–319, 1986.

[80] K. Hanjalic and B. E. Launder. A reynolds stress model of turbulence and its application to thin shear. *Journal of Fluid Mechanics*, 52(4):609–638, 1972.

[81] F.H. Harlow and J. E. Welch. Numerical Calculation of Time-Dependent Viscous Incompressible Flow of Fluid with Free Surface. *The Physics of Fluids*, 8:2182–2189, 1965.

[82] F. Hildebrand. *Introduction to Numerical Analysis*. McGraw-Hill Pub. Co., New York, 1956.

[83] K. S. Ho and C. J. Chen. Finite analytic numerical solution of two-dimensional channel flow over backward-facing step. In *Proceedings of the 4th International Conference on Applied Numerical Modeling*, Tainan, Republic of China, 1984.

[84] S. Hsu. Coastal Air Circulation System: Observations and Empirical Model. *Monthly Weather Review*, 98:487–509, 1970.

[85] D. P. Hwang and H. T. Huynh. A Finite Difference Scheme for Three-Dimensional Steady Laminar Incompressible Flow. 1987.

[86] J. C. Hwang, Z. M. Chen, C. J. Sheikholeslami, and B. K. Panigraphi. Finite Analytic Numerical Solution for Two Dimensional Groundwater Solute Transport. *Water Resource Research*, 9:21:1354–1360, September, 1985.

[87] R. I. Issa. Solution of the Implicitly Discretized Fluid Flow Equations by Operator-Splitting. *Journal of Computational Physics*, 62:40–65, 1985.

[88] R. I. Issa, A. D. Gosman, and A. P. Watkins. The Computation of Compressible and Incompressible Recirculating Flows by a Non-Iterative Implicit Scheme. *Journal of Computational Physics*, 62:66–82, 1986.

[89] R. Iwatsu, J. M. Hyun, and K. Kuwahara. Analysis of Three-Dimensional Flow Calculations in a Driven Cavity. *Computer Physics Communications*, 53:329–336, 1989.

[90] R. Iwatsu, K. Ishii, T. Kawamura, K. Kuwahara, and J.M. Hyun. Numerical Simulation of Three-Dimensional Flow Structure in a Driven-Cavity. *Fluid Dynamics Research*, 5:173–189, 1989.

[91] Jr. J. Douglas. In F. L. Alt., editor, *Advances in Computers*, page 1, New York, 1961. Academic Press.

[92] D. S. Jang, R. Jetli, and S. Acharya. Comparison of the PISO, SIMPLER, and SIMPLEC Algorithms for the Treatment of the Pressure-Velocity Coupling in Steady Flow Problems. *Numerical Heat Transfer*, 10:209–228, 1986.

[93] J. Y. Jang, W. J. Chang, and M. S. Lin. A Numerical Analysis of Three-Dimensional Turbulent Fluid Flow and Heat Transfer in Plate-Fin and Tube Heat Exchangers. In B.K. Soni, J.F. Thompson, Hauser J., and Eiseman P., editors, *5th International Conference on Numerical Grid Generation in Computational Field Simulations*, pages 963–972, Mississippi, USA, April 1-April 5 1996.

[94] S. Y. Jaw. *Development of an Anisotropic Turbulence Model for Prediction of Complex Flow*. PhD thesis, University of Iowa, Iowa City, December 1991.

[95] S. Y. Jaw and C. J. Chen. Development of turbulence model including fractal and kolmogorov scale. In *Symposium on Advances and Applications in Computational Fluid Dynamics, Winter Annual Meeting of ASME*, Dallas, Texas, November 25-30 1990.

[96] S. Y. Jaw and C. J. Chen. On the Determination of Turbulence Model Coefficients. In *Proceedings of the ASCE Engineering Mechanics Specialty Conference, ASCE*, Reston, Virginia, 1991.

[97] S. Y. Jaw and C. J. Chen. Present Status of Second-Order Closure Turbulence Models. I: Overview. *Journal of Engineering Mechanics*, 124(5):485–501, 1998.

[98] S. Y. Jaw and C. J. Chen. Present Status of Second-Order Closure Turbulence Models. II: Applications. *Journal of Engineering Mechanics*, 124(5):502–512, 1998.

[99] D.A. Mansfield J.E. Simpson and J.R. Milford. Inland penetration of the sea breeze fronts. *Quart. J. Royal Meteor. Soc.*, 103:47–76, 1977.

[100] W. P. Jones and B. E. Launder. The Calculation of Low Reynolds Number Phenomena with a Two Equation Model of Turbulence. *International Journal of Heat and Mass Transfer*, 16:119–130, 1973.

[101] W. M. Kays and M. E. Crawford. *Convective Heat and Mass Transfer*. McGraw-Hill Books, Inc., 1980.

[102] K. M. Kelkar and S. V. Patankar. Numerical Prediction of Flow and Heat Transfer in a Parallel Plate Channel With Staggered Fins. *Journal of Heat Transfer*, 109:25–30, 1987.

[103] S. H. Kim, K. B. Chandran, and C. J. Chen. Numerical Simulation of Steady Flow in a Two-Dimensional Total Artificial Heart Model. *Journal of Biomechanical Engineering*, 1992.

[104] G. H. L. Kimble. Tropical land and sea breeze (with special reference to the east indies). *Bull. Amer. Meteor. Soc.*, 27:99–113, 1946.

[105] T. Kitada. Turbulent structure of sea breeze front and its implication in air polution transport-application of the $k - \epsilon$ turbulence model. *Bound.-Layer Meteor.*, 41:217–239, 1987.

[106] Patrick Knupp and Stanley Steinberg. *Fundamentals of Grid Generation*. CRC Press, 1994.

[107] A. N. Kolmogorov. Equations of turbulent motion of an incompressible fluid. *IZV Akad. Nauk. USSR, Ser. Phys.*, 6:56–58, 1942.

[108] H. C. Ku, R. S. Hirsh, and T. D. Taylor. A Pseudospectoral Method for Solution of the Three-Dimensional Incompressible Navier-Stokes Equations. *Journal of Computational Physics*, 70:439–462, 1987.

[109] Lapidus L. and G. F. Pinder. *Numerical Solution of Partial Differential Equations in Science and Engineering*. Springer-Verlag Inc., New York, 1982.

[110] N.M. Larrentier. *Some Improperly Posed Problems*. Springer-Verlag, New York, 1967.

[111] B. E. Launder and D. B. Spalding. The numerical computation of turbulent flows. *Computer Methods in Applied Mechanics and Engineering*, 3:269–289, 1974.

[112] David C. Lay. *Linear Algebra and its Applications*. Addison-Wesley, 2 edition, 1996.

[113] B. P. Leonard. A Stable and Accurate Convective Modelling Procedure Based on Quadratic Upstream Interpolation . *Comp. Methods Appl. Mech., Eng.*, 12:59–98, 1979.

[114] P. Li and C. J. Chen. The Finite Differential Method-A Numerical Solution to Differential Equations. In *Proceedings, 7th Canadian Congress of Applied Mechanics*, Sherbrooke, Canada, 1979.

[115] S. G. Li, F. Ruan, and D. McLaughlin. A Space-Time Accurate Method for Solving Solute Transport Problems. *Water Resources Research*, 28(9):2297–2306, 1992.

[116] W. L. Lin, K. D. Carlson, R. H. Bravo, and C. J. Chen. Finite Analytic Method in Computational Fluid Dynamics (Invited Paper). In *Sixth National Conference on Computational Heat Transfer*, Xian, China, 1995.

[117] W. L. Lin, K. D. Carlson, and C. J. Chen. Pressure Boundary Conditions of Incompressible Fluid Flows with Conjugate Heat Transfer on Non-staggered Grids, Part I: Methods. *Journal of Numerical Heat Transfer*, April, 1997. In Press.

[118] W. L. Lin and C. J. Chen. Numerical Simulation of Complex Geometries. In *Proceeding of 1996 ASME Computer in Engineering Conference*, Irvine, CA, August 18-22 1996.

[119] W. L. Lin and C. J. Chen. Numerical Simulation of Complex Geometries in Fluid Flows. In C.J. Chen, C. Shih, J. Lienau, and Kung, editors, *Flow Modeling and Turbulence Measurements VI*, pages 847–854. 1996 Balkema, Rotterdam. ISBN 9054108266, September 8-10 1996.

[120] W. L. Lin and C. J. Chen. Treatments of Pressure Boundary Conditions for Incompressible Fluid Flows over Complex Boundaries. *Int. Journal of Numerical Methods in Fluids*, 1997. under review.

[121] W. L. Lin and C. J. Chen. Automatic Grid Generation of Complex Geometries in Cartesian Coordinates. *Int. Journal of Numerical Methods in Fluids*, 1997. under review.

[122] W.L. Lin, K. Carlson, and C.J. Chen. Diagonal Cartesian Method for Modeling of Incompressible Flows over Complex Boundaries. *Journal of Numerical Heat Transfer*, 1996. Accepted for Publication.

[123] Peter Linz. *Theoretical Numerical Analysis : An Introduction to Advanced Techniques*. John Wiley and Sons, 1978.

[124] M. L.Mansour and A. Hamed. Implicit solution of the unsteady incompressible navier-stokes equations in primitive variables. In *Proc. 5th Int. Conf. in Numerical Methods in Laminar and Turbulent Flow*, volume 5, pages 300–311, 1987.

[125] W. A. Lyons. The Climatology and Prediction of Chicago Lake Breeze . *Journal of Climate and Applied Meteorology*, 11:1259–1270, 1972.

[126] N. N. Mansour, J. Kim, and P. Moin. Reynolds-stress and dissipation rate budgets in a turbulent channel flow. *J. of Fluid Mechanics*, 194:15–44, September 1988.

[127] C.W. Mastin and J.F. Thompson. Transformation of three-dimensional regions onto rectangular regions by elliptic systems. *Numer. Math.*, 29:397–407, 1978a.

[128] D. J. Mavriplis. Unstructured Grid Techniques. *Annu. Rev. Fluid Mech.*, 29:473–514, 1997.

[129] Stephen F. McCormick. *Multilevel Adaptive Methods for Partial Differential Equations*. SIAM, Philadelphia, 1989.

[130] J. E. Melton, M. J. Berger, M. J. Aftosmis, and M. D. Wong. 3D Applications of A Cartesian Grid Euler Method. *AIAA*, 95-0853, January 1995.

[131] T.F Miller and F.W. Schmidt. Use of pressure weighted interpolation method for the solution of the incompressible navier-stokes equations on a nonstaggered grid system. *Numerical Heat Transfer*, 14:213–233, 1988.

[132] R. D. Mills. On the closed motion of a fluid in a cavity. *J. of Royal Aeronautical Society*, 69:116–120, February 1965.

[133] P. Moin and J. Kim. On the Numerical Solution of Time-dependent Viscous Incompressible Fluid Flows Involving Solid Boundaries. *Journal of Computational Physics*, 3:381–392, 1980.

[134] P. Moin and J. Kim. Numerical investigation of turbulent channel flow. *Journal of Fluid Mechanics*, 118:341–377, 1982.

[135] W. J. Moroz. A Lake Breeze on the Easter Shore of Lake Michigan: Observations and Model . *Journal of Computational Physics*, 3:381–392, 1980.

[136] T.J. Mueller, J.R. Lloyd, J.L. Lower, W.T. Strouble, and F.N. Underwood. On the Sperate Flow Produced by Fully Open Disc-Type Prosthetic Heart Valve. In *Biomech. symp. ASME AMD*, volume 2, pages 97–98, 1973.

[137] S. Nakamura. Marching grid generation using parabolic partial differential equations. In J.F. Thompson, editor, *Numerical Grid Generation*, pages 775–786. North-Holland, New York, 1982.

[138] K. Nakatsuji, T. Sueyoshi, and K. Muraoka. Numerical Experiments of Residual Circulation and Its Formulation Mechanism in Tidal Estuary. In *Proceedings of the 5th International Symposium on Refined Flow Modelling and Turbulence Measurements*, pages 695–702, Paris, France, September 7-10 1993.

[139] J Neumann and Y. Mahrer. A theoretical study of the land and sea breeze circulation. *J. Atmos. Sci.*, 28:532–542, 1971.

[140] N. H. Ng and D. B. Spalding. Some applications of a model of turbulence to boundary layers near walls. *Physics of Fluids*, 15:20, 1972.

[141] M. Normandin. *Etude experimentale de l'ecoulement Turbulent dans une cavite profonde*. PhD thesis, L'Universite Scientifique Et Medicale L'institut National Polytechnique De Grenoble, 1978.

[142] S. Orszag and M. Israeli. Numerical Simulation of Viscous Incompressible Flows. *Journal of Atmospheric Science*, 24:337–355, 1972.

[143] M. Necati Özişik. *Heat Conduction*. John Wiley and Sons, New York, 1980.

[144] J. G. Papageorgiou. A 3-d sea breeze model of the PBL including pollutant dispersion. *Bound.-Layer Meteor.*, 45:9–29, 1988.

[145] S. V. Patankar. *Numerical Heat Transfer and Fluid Flow*. Taylor & Francis, 1980.

[146] S. V. Patankar. A calculation Procedure for Two-Dimensional Elliptic Situations. *Numerical Heat Transfer*, 14:409–425, 1985.

[147] S. V. Patankar and D. B. Spalding. A Calculation Procedure for Heat, Mass and Momentum Transfer in Three-Dimensional Parabolic Flows. *Int. J. Heat Mass Transfer*, 15:1787–1806, 1972.

[148] E. Paterson and F. Stern. Computation of Unsteady Viscous Flow with Application to the MIT Flapping-Foil Experiment. In *Proceedings of the 6th International Conference on Numerical Ship Hydrodynamics*, Iowa City, Iowa, August 1993.

[149] D.W. Peaceman and H.H. Rachford. The numerical solution of parabolic and elliptic differential equations. *J. Soc. Indust. Appl. Math.*, 3:28–41, 1955.

[150] M. Peric. *A Finite Volume Method for the Prediction of Three-Dimensional Flow in Complex Ducts*. PhD thesis, University of London, 1985.

[151] W. L. Physick. Numerical experiments on the inland penetration of the sea breeze. *Quart. J. Royal Meteor. Soc.*, 106:735–746, 1980.

[152] R.A. Pielke. A three-dimensional numerical model of the sea breezes over south florida. *Mon. Wea. Rev.*, 102:115–139, 1974.

[153] G. D. Raithby and G. E. Schneider. Numerical Solution of Problems in Incompressible Fluid Flow: Treatment of the Velocity-Pressure Coupling. *Numerical Heat Transfer*, 2:417–440, 1979.

[154] O. Reynolds. On the dynamical theory of incompressible fluids and the determination of the criterion. *Philosophical Transactions of the Royal Society*, 186, 1894.

[155] C. M. Rhie and W. L. Chow. Numerical Study of the Turbulent Flow Past an Airfoil with Trailing Edge Separation. *AIAA J.*, 21:1525–1532, 1983.

[156] W. Rodi. *The prediction of free turbulent boundary layers by use of a two-equation model of turbulence*. PhD thesis, Imperial College, London, 1972. Department of Mechanical Engineering.

[157] W. Rodi. Examples of turbulence models for incompressible flows. *AIAA Journal*, 20(7):872–879, 1981.

[158] H. Schlichting. *Boundary-Layer Theory*. McGraw-Hill Book Co., 1979. Translated by J. Kestin.

[159] W.M. Sha and T.K. Kawamura. A numerical study on sea/land breezes as a gravity current: Kelvin-helmholtz billows and inland penetration of the sea breeze front. *J. Atmos. Sci.*, 48:1649–1665, 1991.

[160] F. Sotiropoulos and S. Abdallah. The Discrete Continuity Equation in Primitive Variable Solutions of Incompressible Flow. *Journal of Computational Physics*, 95:212–227, 1991.

[161] D. B. Spalding. A Novel Finite Difference Formulation for Differential Expressions Involving Both First and Second Derivatives. *International Journal of Numerical Methods in Engineering*, 4:551–559, 1972.

[162] D. B. Spalding. The vorticity-fluctuations (kw) model of turbulence. Technical Report CFDW Report CFD 82/17, Imperial College, 1982.

[163] J. L. Steger and D.S. Chausee. Generation of body-fitted coordinates using hyperbolic partial differential equations. *SIAM J. Sci. Stat. Comput.*, pages 431–437, 1980.

[164] S. Takanashi and M. Takemoto. An Automatic Grid Generation Procedure for Complex Aircraft Configurations . *Computers and Fluids*, 24:393–400, 1995.

[165] V. Talaie and C. J. Chen. Finite Analytic Solutions of Steady and Transient Natural Convection in Two-Dimensional Rectangular Enclosures. In *1985 ASME Winter Annual Meeting, ASME paper 85-WA/HT-68*, Miami, FL, 1985.

[166] F. C. Thompson, J. F.and Thames and Mastin C. W. Automatic Numerical Generation of Body-fitted Curvilinear Coordinate System for Field Containing Any Number of Arbitrary Two-Dimensional Bodies. *J. of Comp. Physics*, 15:299–319, 1974.

[167] J. F. Thompson. A Reflection of Grid Generation in the 90s: Trends, Needs, and Influences. In B.K. Soni, J.F. Thompson, Hauser J., and Eiseman P., editors, *5th International Conference on Numerical Grid Generation in Computational Field Simulations*, pages 1029–1110, Mississippi, USA, April 1-April 5 1996.

[168] J.F. Thompson, Z.U.A. Warsi, and C.W. Mastin. *Numerical Grid Generation: Foundations and Applications*. North-Holland, Amsterdam, 1985.

[169] H. C. Tien. *Finite Analytic Method for Two-Dimensional Flow with Irregular Boundaries*. PhD thesis, University of Iowa, 1993.

[170] W. F. Tsai and C. J. Chen. Finite Analytic Numerical Solution for Contaminant Transport in 2D Groundwater Flow. In *Proceedings of 25th IAHR (International Assoc. Hydraulic Res.) Congress*, Tokyo, Japan, Aug. 30-September 3 1993.

[171] W. F. Tsai, C. J. Chen, and H. C. Tien. Finite Analytic Numerical Solutions for Unsteady Flow with Irregular Boundaries. *Journal of Hydraulic Engineering*, 119:1274–1298, 1993.

[172] A. W. Warrick and D. O. Lomen. Time-Dependent Linearized Infiltration: III. Strip and Disc Sources. *Soil Sci. Soc. Am. J.*, 40:639–643, 1976.

[173] A. *et al.* Weill. A mesoscale shear convective cell observed during the c.o.a.s.t. experiment: Acoustic sounder measurements. *Bound.-Layer Meteor.*, 44:359–371, 1988.

[174] R. Wexler. Theory and observation of land and sea breezes. *Bull. amer. Meteor. Soc.*, 27:272–287, 1946.

[175] A. M. White. *Viscous Fluid Flow*. McGraw-Hill Co., New York, 1991.

[176] Whittaker and Watson. *A Course of Modern Analysis*. Cambridge University Press, 1927.

[177] M. Wolfshtein. The velocity and temperature distribution in one-dimensional flow with turbulence augmentation and pressure gradient. *International Journal of Heat and Mass Transfer*, 12:301–318, 1969.

[178] G. T. Yeh. Femwater: A finite-element model of water flow through saturated-unsaturated porous media. Technical Report Report ORNL-5567/R1, 1st Revision, Oak Ridge National Laboratory, Oak Ridge, Tennessee, 1987.

[179] G. T. Yeh and D. S. Ward. Femwater: A finite-element model of water flow through saturated-unsaturated porous media. Technical Report Report ORNL-5567, Oak Ridge National Laboratory, Oak Ridge, Tennessee, 1980.

[180] H Yoshikado and K. Hiroaki. Inland penetration of the sea breeze over suburban area of tokyo. *Bound.-Layer Meteor.*, 48:389–407, 1989.

[181] C. H. Yu, C. J. Chen, and K. B. Chandran. Steady Turbulent Flow Through a Disc-Type Valve. II: Study on Disc Size and Position. *Journal of Engineering Mechanics*, 114(5):797–811, 1988.

[182] Y. Zang, R. L. Street, and J. R. Koseff. A Non-staggered Grid, Fractional Step Method for Time-Dependent Incompressible Navier-Stokes Equations in Curvilinear Coordinates. *Journal of Computational Physics*, 114:18–33, 1994.

[183] X. J. Zeng and W. Li. The Stability and Convergence of Finite Analytic Method for Unsteady Two-Dimensional Convective Transport Equations. In C. J. Chen, L. D. Chen, and E. M. Holly, editors, *Turbulence Measurements and Flow Modelling*, pages 427–433. Hemisphere Publishing Corporation, 1987.

[184] Zienkiewics. *The Finite Element Method in Engineering Science.* McGraw-Hill, 1971.

[185] A. A. Zukauskas, R.V. Ulinskas, and E. S. Bubelis. Average heat transfer and pressure drop in cross flow of viscous fluid over a tube bundle at low re. *Heat Transfer-Soviet Research*, 10(6):90–101, 1978.

Index

T - #0172 - 101024 - C0 - 229/152/19 [21] - CB - 9781560328988 - Gloss Lamination